# Springer Texts in Business and Economics

More information about this series at http://www.springer.com/series/10099

Clifford S. Ang

# Analyzing Financial Data and Implementing Financial Models Using R

Second Edition

 Springer

Clifford S. Ang
Compass Lexecon
Oakland, CA, USA

ISSN 2192-4333          ISSN 2192-4341   (electronic)
Springer Texts in Business and Economics
ISBN 978-3-030-64154-2          ISBN 978-3-030-64155-9   (eBook)
https://doi.org/10.1007/978-3-030-64155-9

This Springer imprint is published by the registered company Springer Nature Switzerland AG
The registered company address is: Gewerbestrasse 11, 6330 Cham, Switzerland

*To my family and friends.*

*With special gratitude to my wife and children without whom nothing in this world matters.*

*And, to my parents and brother, all of whom have been extremely supportive of all my endeavors.*

# Preface

The first edition of this book was written in 2015. To say many things have changed since then is an understatement. As I am finalizing the second edition, we are still in the midst of the COVID-19 pandemic and are experiencing what some believe will be the new normal. This event is a reminder that the only constant thing is change and we have to learn to adapt.

This second edition is one of adaptation as well. I took into consideration comments I received from those who used my book to hopefully make the second edition more helpful. The biggest visual change is that charts are now in color. Chapters 1–7 and 9–10 are chapters that were also in the first edition, but that is where most of the similarities end. The contents in these chapters have been almost completely re-written. The examples are all new. I now use 5 years instead of 3 years of data for many examples. Some of the techniques have also been revised. I have added some new topics as well. For example, I have added examples of two-axis charting in Chap. 1, winsorization and truncation in Chap. 2, time- and money-weighted rates of return in Chap. 3, testing for heteroskedasticity, non-normality, and autocorrelation as well as selecting best regressors in factor models in Chap. 5, short rate models in Chap. 9, and American option pricing in Chap. 10.

There are also three new chapters in the second edition. Chapter 8 discusses equities. In that chapter, I show how to scrape financial statement data from SEC filings, analyze projections, perform relative valuation using regression analysis, and identify shocks in stock returns. Chapter 11 is on simulation, where I demonstrate how to simulate stock prices and how to use simulations in Value-at-Risk, Monte Carlo options pricing, and exotic option valuation. I then explore trading strategies in Chap. 12 and show the basics of building a trading strategy and machine learning techniques, like k-nearest neighbor, regression with k-fold cross validation, and artificial neural networks.

Not all things have changed though, as I am staying true to the promise of the first edition. I still handhold you through every step of the process. We start with obtaining the raw data, importing the data, getting the data into shape, and then performing our analysis. I still show intermediate output to make sure you do not get lost along the way. My intended audience is still anyone who wants to learn financial modeling and data analysis in R. Aside from your investment in this book, R is free and I still use data that is freely accessible. Prior background in finance is helpful, but not necessary. I also try to explain in layperson's terms most of the math used in this book.

What does this all mean? For students, this means you can learn how to use the tools and techniques you learn in the classroom using real world data with real world challenges. For professors, this means you can concentrate on teaching the theory and concepts and let this book handhold your students through the implementation of the various models. For practitioners, this means you do not need to go to a formal class to learn how to perform data analysis and implement financial models in an organized fashion.

I hope you find this book helpful and insightful. To contact me or find supplemental materials for this second edition, please visit www.cliffordang.com.

Oakland, CA, USA                                                                          Cliff Ang
September 2020

# Acknowledgments

The second edition of this book would not have been possible without the assistance and support of many people.

I am thankful to my wife and kids for once again putting up with me throughout the entire process.

I would like to thank the wonderful team at Springer Nature. First, my editor, Nitza Jones-Sepulveda, for giving me the opportunity to write this second edition and to work with her again. I also owe a debt of gratitude to her predecessor, Nicholas Philipson, who was my editor for the first edition. I would also like to thank Faith Su, Kirthika Selvaraju, and others who provided assistance throughout the process.

I am grateful to the following for helping improve this book by providing their input, comments, feedback, and corrections to drafts of the second edition and to the first edition: Dexter Agcaoili, Dan Benton, Elijah Brewer, Jian Cai, Michael Crain, Adam Fleck, Marcus Hennig, Karl-Werner Hansmann, Wei Hu, Chien-Ling Lo, Merritt Lyon, Andria van der Merwe, Vince Warther, and Marie Winters.

This book also would not have been possible if I were not given permission to use third-party data. I would especially like to thank Nassrin Berns at CSI, who assisted me in obtaining permission for both editions of my book. Data on Yahoo Finance is from CSI, and I use that data extensively in this book. I would also like to thank Professor Ken French at Dartmouth, the Chicago Board of Options Exchange, S&P Dow Jones, and Moody's Analytics for granting me permission to use their data.

I am sure I have forgotten to mention many other people that in some way have contributed to this book. For that, my sincerest apologies.

# Contents

# Chapter 1
# Prices

Market prices are relied upon by market participants in many ways. For example, incentive-based compensation could be tied to the performance of the company's stock price. When some investors buy and sell a company's stock, they could compare the stock's price relative to their own valuation of the stock. Other investors could compare the company's stock price with comparable companies to determine whether the company's stock is overvalued or undervalued relative to its peers. Some investors view the traded price of an asset as an indication of the asset's value. Therefore, understanding how to analyze prices is critical for investments and asset pricing.

In this chapter, we will focus on working with prices of stocks and exchange traded funds (ETFs). One of our goals in this chapter is to gain familiarity with Yahoo Finance price data. Doing so will help us gain expertise in the data, which we could then leverage in later applications. Moreover, although the techniques in this chapter are applied to price data, the same techniques can be applied to other types of data as well.

The rest of the chapter is organized as follows. We start with a discussion of the distinction between the price and the value of a security in Sect. 1.1. In Sect. 1.2, we show how to obtain stock price data from Yahoo Finance. We then discuss how to check raw security price data in Sect. 1.3 and how we can manipulate such raw data to fit some basic analyses using prices in Sect. 1.4.

Next, we discuss several applications of price data. In Sect. 1.5, we show several ways of presenting normalized price charts. These types of charts allow us to compare the performance of different securities over time. Next, we discuss how to calculate moving averages in Sect. 1.6. We then show how to calculate volume-weighted average prices in Sect. 1.7. In Sect. 1.8, we show an example of how to create a candlestick chart, which requires the open, high, low, and close prices. Finally, we end the chapter by showing how to plot a stock's price and volume in a two-axis chart in Sect. 1.9.

© The Author(s), under exclusive license to Springer Nature Switzerland AG 2021
C.S. Ang, *Analyzing Financial Data and Implementing Financial Models Using R*,
Springer Texts in Business and Economics,
https://doi.org/10.1007/978-3-030-64155-9_1

## 1.1  Price Versus Value

Price is not the same as value. These are two different concepts that are often used interchangeably. **Price** is the amount we pay to buy an asset or receive when we sell an asset. **Value** is how much the asset is worth.

Value is driven by fundamentals, like expected cash flows, growth, and risk. The value of an asset is equal to the present value of expected cash flows generated by the asset discounted at the appropriate risk-adjusted discount rate. On the other hand, prices are partly driven by fundamentals, but prices could also be affected by investors' mood and momentum.

Securities can be overvalued or undervalued depending on its price. This happened during the dot-com bubble of the early 2000s. At that time, prices of technology firms were much higher than what their fundamentals implied. Eventually prices caught up and the dot-com bubble burst in 2001, which brought the sky high stock prices down to more reasonable levels relative to the value of those firms' stocks.

In fact, we have a more recent example. As I was finalizing this book in September 2020, it has been six months since the US stock market began feeling the effects of the COVID-19 pandemic. As shown in Table 1.1, from March 9 to 16, 2020, the S&P 500 Index gyrated from one day to the next. Over that period, the market declined by 20%. Such a large percentage change over a very short window is unlikely 100% attributable to a change in the fundamental values of the underlying stocks. Instead, at least a portion of the price change is likely caused by the changing risk premiums being factored into stock prices. We can confirm this by looking at the level of the CBOE Volatility Index (VIX), which is sometimes called the "investors' fear gauge." Using volatility implied by options on the S&P 500 Index, the VIX tells us the amount of risk that investors are embedding in market prices. The VIX peaked at an all-time high of 82.7 or an annualized volatility of 82.7% on March 16, 2020. For context, the last time the VIX exceeded 80 was during the 2008–2009 financial crisis.

Some argue that price is the best estimate of value because the market price is determined by thousands of investors putting their money on the line. There is intuitive appeal to that argument, but, when we look at the types of investors that trade, we can see that many investors trade without regard for how close price is to value.

As a threshold matter, because of transactions costs, it is unlikely that price would equal value for buyers and sellers. You would only buy a stock if you think the price is less than your valuation of the stock. On the other hand, you would only sell a stock if you think the price is more than your valuation of the stock. Even in special

**Table 1.1**  Percentage changes in the S&P 500 Index, March 9, 2020–March 16, 2020

|          | Mar 9 | Mar 10 | Mar 11 | Mar 12 | Mar 13 | Mar 16 |
|----------|-------|--------|--------|--------|--------|--------|
| % Change | −7.6  | 4.9    | −4.9   | −9.5   | 9.3    | −12.0  |

circumstances when the above does not hold (e.g., needing to sell stock to cover an upcoming loan payment or tax liability), the purchase or sale decision is not primarily driven by the relationship between price and value. Ignoring these special circumstances, the more typical case suggests that price may be *close* to value if transactions costs are low. Thus, if all traders can fit into this simple characterization, the intuitive appeal of the "thousands of investors putting their money on the line" argument certainly could hold for actively traded stocks.

However, the argument becomes less clear when we think about the different types of investors that trade in the market. One group are **value investors**. The most famous example of a value investor is Warren Buffett, who is arguably one of the most successful value investors of all time. Value investors care about price and value, but they care because they believe that price and value can diverge and they seek to identify stocks that are currently undervalued. The value investor hopes to profit when the stock's price converges to their estimate of the stock's value at some point in the future, which may not happen for years (or it may never happen at all).

Second, we also have relative price investors that are less concerned with value and focused primarily or solely on price. You could think of these as the typical "trader" in the market. Traders look at many factors, like momentum indicators, moving average prices, and changes in mood, but are unlikely to look at the value of the underlying stock. This group trades because the price of a security is misaligned relative to the prices of comparable securities. These trades could occur even if the prices for all securities are overvalued or undervalued. The key for these types of traders is that there is an indicator that identifies some sort of mispricing and that the trader expects to profit from exploiting that mispricing. I would also put high-frequency traders (HFTs) in this group, which are traders that seek to profit off mispricings that appear over very short time intervals. For example, HFTs can compare prices of futures and ETFs that track the S&P 500 Index and create an algorithm that trades when there is some level of mispricing between the prices of both securities. Because HFTs trade in infinitesimal time windows, the price of the securities is the primary or sole factor that drives the decision to buy or sell.

Third, you have investors in the efficient market camp. I am going to make an important point here that is often confused by people. Market efficiency *does not mean* prices cannot diverge from value. What market efficiency implies is that it is very difficult for investors to employ a trading strategy that consistently profits from such mispricings in the market. Thus, in the limit, an implication for those that believe in efficient markets would be to invest in index funds because they do not believe they can do better in a consistent fashion.

Having made the distinction between price and value above, for purposes of this book, we will use the terms price and value interchangeably. This is mostly because the finance literature mixes the two together and it is difficult and more confusing to change commonly used terminology. For example, Black-Scholes and Binomial methods are often labeled options *pricing* models, not options valuation models.

Moreover, prices may still be a good way to sanity check your valuations. For example, if you want to buy a stock, you could come up with your valuation and you would only buy a stock if you think your valuation of the stock is higher than the stock's price. However, if your valuation is too far away from the price, you may want to find a justification as to the source of the differential. Oftentimes, one justification would be that there is material private information, but we and all other public investors would not have access to that information. Thus, if you are basing your valuation on publicly available information alone that other investors also have access to, you would have to justify to yourself why others' interpretation of the same information is incorrect or why that information is not properly factored into the price even though the information has been publicly disseminated.

## 1.2   Importing Price Data from Yahoo Finance

Before we can analyze price data, we have to first learn how to obtain price data. Throughout this book, we only use publicly available sources of data that you can obtain for free. In the first few chapters, we will focus on price data for stocks and exchange traded funds (ETF) from Yahoo Finance. The goal is to get comfortable with a single public data source of price data. This will allow you to begin familiarizing yourself with data that is not generated for a specific analysis, which will generally be the type of data that you will encounter outside of the classroom setting.

In this chapter, we will go through how to obtain Yahoo Finance data and perform basic data cleaning and manipulation. This is a seemingly boring and uninteresting task to learn. However, in practice, this is one of the most important skills to have and, in reality, getting the data into proper shape is where most people spend their time. According to a May 23, 2016 article in *Forbes*, data scientists spend 19% of their time collecting data and 60% of their time on cleaning and organizing data. This means that, on average, approximately 80% of your time would be spent collecting and preparing the data so they can be used in whatever analysis you would like to do. The reason for doing so is we want to avoid a GIGO (garbage in, garbage out) situation.

Yahoo Finance stock and ETF data are provided by Commodity Systems, Inc. (CSI), a commercial provider of market data and trading software. Therefore, because Yahoo Finance data is free to us does not mean it is any less reliable than any other subscription-based databases. Being the freely accessible public version does come with some disadvantages. For example, when firms get acquired, the historical data for those firms are no longer available on Yahoo Finance. This is a problem for analyses that need to incorporate the data of companies that no longer trade to be correct, such as when we backtest trading strategies. Regardless, for our purposes and for many other practical purposes, using data from Yahoo Finance will be sufficient.

One important aspect of data analysis is **domain expertise**. This means that you have knowledge about the subject of the data. In terms of stock prices, we would want to gain familiarity with the data in terms of what it contains, how it is calculated, what are its limitations, etc. Most of the time, we learn this through experience working with the data or reading about the data. Thus, working with the Yahoo Finance data and other data in this book will move you in that direction. Ultimately, the goal is to acquire the skills that allow us to extend our experience in learning about Yahoo Finance data to new data sources we encounter going forward.

Surprisingly, the lack of domain expertise or lack of even an attempt at understanding the data being used is a common problem observed in practice. Many people simply make assumptions about the data without really understanding what the data contains, so they do not know the relevance of the assumptions made to construct the data in their analysis. For example, let us say you were standing on September 1, 2020 and you wanted to calculate the market capitalization of Apple stock on August 28, 2020, which is the trading day before Apple's stock was split 4:1 (i.e., one share of Apple stock became four shares) on August 31, 2020. You then go to Apple's SEC Form 10-Q for 2Q 2020 filed July 31, 2020 (a likely publicly available source of the latest shares outstanding data) and see that Apple had 4.275 billion shares outstanding as of July 17, 2020. You then pull Yahoo Finance close price data for August 28, 2020 and you get $124.81. This implies that Apple's market capitalization was $533 billion. The problem with this calculation is Yahoo Finance's close price is split-adjusted. This means after the stock split, historical close prices are adjusted to be consistent with the post-split prices. This means that the price that is consistent with the 4.275 billion shares is $499.24, which is four times the $124.81 reported on Yahoo Finance. Using these figures yields the correct market capitalization for Apple of $2.1 trillion.

Hopefully, by using the same data source for many of the applications in this book, we would gain familiarity with the data and so we can focus on learning the techniques down the road. Much of our analysis will be using common stock price data for Amazon.com. The ticker for Amazon is AMZN. I sometimes refer to Amazon by its name or by its ticker. There is no particular reason why I selected AMZN price data. Likely, this is because of inertia. I used AMZN in the first edition, so it made sense to use AMZN for the second edition as well. For some applications, we will use AMZN data for a five-year period from December 31, 2014 to December 31, 2019, so we might as well obtain all that data at the beginning. Below, we show the different steps we take to obtain Yahoo Finance data for AMZN over the above date range.

**Step 1: Download CSV File from Yahoo Finance**   To obtain the data, one would need to know either the company name or the security's ticker symbol. In our case, we are downloading data for Amazon with ticker symbol AMZN. In order to replicate the results we report in this book on future dates, we are going to download historical stock price data for Amazon for the five-year period from December 31, 2014 to December 31, 2019.

The following general steps lay out how to retrieve a CSV file of the historical data for AMZN. Depending on the browser we use, the steps or terms that appear on the screen may be slightly different.

1. On your web browser, enter the web address: https://finance.yahoo.com.
2. Enter the ticker symbol "AMZN" in the Quote Lookup box. This will take us to the page of the latest AMZN price quote.
3. On the page with the latest AMZN price quote, click on the Historical Prices link.
4. Look for "Time Period" and change the Start Date and End Date to December 31, 2014 and December 31, 2019, respectively. After entering the date range, click the Done button.
5. Click the Apply button, which will update the price data to the date range we selected.
6. Click on Download Data and save the CSV file to the R working directory. To identify the R working directory, type `getwd()` at the command prompt in the R console. We should also place our R code in this working directory, so R can find the CSV files of the data it needs to import.
7. Label the saved file **AMZN Yahoo.csv**.

---

**Tip to Download Yahoo Finance Data for Other Symbols**

This would be a good time to download the Yahoo Finance data for the other securities we will use in this book. Since all our examples are between December 31, 2014 and December 31, 2019, we can download the rest of the data by repeating some steps after Step 7 above. To start, on the upper right side of the Historical Prices screen, we will find a box labeled Quote Lookup with a magnifying glass in a blue box. In that box, we enter the symbol of the security whose data we want to download. Doing this keeps us on the same web page and allows us to download the new security's data using the same date range. More importantly, this saves us from repeating all the steps outlined above and should save us a decent amount of time.

The list of other securities we will use is as follows: Apple (AAPL), Alphabet (GOOG), iShares Long-Term Corporate Bond ETF (IGLB), SPDR S&P 500 ETF (SPY), SPDR S&P 500 Value ETF (SPYV), SPDR S&P 500 Growth ETF (SPYG), SPDR MSCI ACWI ex-US ETF (CWI), SPDR Bloomberg Barclays 1-3 Month T-Bill ETF (BIL), and Tesla (TSLA).

For consistency with the format we use in this book, we should save the files using the format [Ticker] and then add "Yahoo.csv." For example, in the case of the AAPL data, we save the data under the filename **AAPL Yahoo.csv**.

---

When we make use of the Yahoo Finance data, we will import all the data in the CSV file. The reason is that we want to preserve the actual or raw data we downloaded. We then track all the changes from the raw data in the R script/program (see gray box below for using the R Editor). For example, suppose we want to delete all data prior to 2018. One way is to delete those observations in Excel and save that to another CSV file. Alternatively, we can tell R to import the raw data and then delete the data prior to 2018 after we import the data. The advantage of the latter approach is we have a record of what we did with the raw data. Several months or years down the road, we may want to look back and may not know what we did just by looking at the CSV file we created, but the R script would tell us exactly what we did. This helps in ensuring we understand any edits or modifications to the data that we have done. In practice, the issue of why we would like to preserve raw data may not be as simple as what we discussed above. The important takeaway is having a place where you document what you did with the raw data could be helpful in avoiding memory tests down the road.

---

**Using the R Editor for Writing Programs**

For convenience, we should use the R Editor to write and execute the codes in this book. The R Editor can be opened using the CTRL+N (Windows) or Command+N (OS X). We can type the lines of code that appear in this text in the R Editor, so the codes can be saved, easily be modified, and re-run at any point in time. More details about the R Editor can be found in Appendix A.

In this book, we show intermediate output as shown in the R console. For example, in the output below, lines that are preceded by a command prompt (>) or plus sign (+) are lines of code. We type those into the R editor without the command prompt and plus sign. The command prompt denotes the start of a line of code. Multiple lines of code that belong together but broken up into separate lines are identified by the lines that begin with a plus sign. An example of this is in Step 5 below in which we separate our code to rename variables into multiple lines of code. Separating into multiples lines make it easier to read and understand the code, but is not necessary to run the code properly.

To execute the code, we can highlight the portion of code we want to run and hit CTRL+R (Windows) or CTRL+ENTER (OS X). The output in the R console when we run specific lines of code in the different steps should be identical to the output we report in this text. Lines that do not begin with a command prompt or plus sign will be output of the code we run.

---

**Step 2: Import Data into R**  We use `read.csv()` to import the data from the CSV file that contains the AMZN price data. The argument `header = TRUE` tells R to treat the first row as the column headers. In Appendix B, we create the `head.tail()` function, which outputs the first three and last three observations of each object. There is nothing special about this function, except that it helps us be more efficient throughout this book when we want to output a snapshot of the contents of datasets. The `head.tail()` function is equivalent to running `head(x, 3)` and `tail(x, 3)` in one line of code, where $x$ is the dataset whose contents we want to view.

```
1   > data.amzn <- read.csv("AMZN Yahoo.csv", header = TRUE)
```

```
 2  > head.tail(data.amzn)
 3            Date   Open   High    Low  Close Adj.Close  Volume
 4  1 2014-12-31 311.55 312.98 310.01 310.35    310.35 2048000
 5  2 2015-01-02 312.58 314.75 306.96 308.52    308.52 2783200
 6  3 2015-01-05 307.01 308.38 300.85 302.19    302.19 2774200
 7            Date   Open   High    Low  Close Adj.Close  Volume
 8  1257 2019-12-27 1882.92 1901.40 1866.01 1869.80   1869.80 6186600
 9  1258 2019-12-30 1874.00 1884.00 1840.62 1846.89   1846.89 3674700
10  1259 2019-12-31 1842.00 1853.26 1832.23 1847.84   1847.84 2506500
```

**Step 3: Change the Date Variable** We use `str()` to show the structure of the data. Note that the data in the Date column is considered a Factor, which means R is not reading this in as a date variable. To do this, we first create a separate *Date* variable. We do this in Line 11. As you can see, this *Date* variable is considered a date by R. We now combine this *Date* variable with the other six columns of *data.amzn* (i.e., dropping the first column, which was the date variable in the original dataset by putting a $-1$ to the right of the comma inside the square brackets) in Line 14. When we run `str()` again on *data.amzn*, we can see in Line 26 that the date is now being recognized by R as a date variable.

```
 1  > str(data.amzn)
 2  'data.frame': 1259 obs. of  7 variables:
 3   $ Date     : Factor w/ 1259 levels "2014-12-31","2015-01-02",..: 1 2 3 4 ...
 4   $ Open     : num  312 313 307 302 298 ...
 5   $ High     : num  313 315 308 303 301 ...
 6   $ Low      : num  310 307 301 292 295 ...
 7   $ Close    : num  310 309 302 295 298 ...
 8   $ Adj.Close: num  310 309 302 295 298 ...
 9   $ Volume   : int  2048000 2783200 2774200 3519000 2640300 3088400 ...
10  >
11  > Date <- as.Date(data.amzn$Date, format = "%Y-%m-%d")
12  > str(Date)
13   Date[1:1259], format: "2014-12-31" "2015-01-02" "2015-01-05" ...
14  > data.amzn <- cbind(Date, data.amzn[, -1])
15  > head.tail(data.amzn)
16            Date   Open   High    Low  Close Adj.Close  Volume
17  1 2014-12-31 311.55 312.98 310.01 310.35    310.35 2048000
18  2 2015-01-02 312.58 314.75 306.96 308.52    308.52 2783200
19  3 2015-01-05 307.01 308.38 300.85 302.19    302.19 2774200
20            Date   Open   High    Low  Close Adj.Close  Volume
21  1257 2019-12-27 1882.92 1901.40 1866.01 1869.80   1869.80 6186600
22  1258 2019-12-30 1874.00 1884.00 1840.62 1846.89   1846.89 3674700
23  1259 2019-12-31 1842.00 1853.26 1832.23 1847.84   1847.84 2506500
24  > str(data.amzn)
25  'data.frame': 1259 obs. of  7 variables:
26   $ Date     : Date, format: "2014-12-31" "2015-01-02" ...
27   $ Open     : num  312 313 307 302 298 ...
28   $ High     : num  313 315 308 303 301 ...
29   $ Low      : num  310 307 301 292 295 ...
30   $ Close    : num  310 309 302 295 298 ...
31   $ Adj.Close: num  310 309 302 295 298 ...
32   $ Volume   : int  2048000 2783200 2774200 3519000 2640300 3088400 ...
```

**Step 4: Convert into an xts Object** We use the xts package to convert the data into an extensible time series, which is a common class of objects in R that we use when dealing with financial data. We convert the object to xts in order to make our data identical to the output of using getSymbols() to obtain stock prices. This is so that those that would rather use getSymbols() can follow along what we do in this book as well. Line 2 applies xts() to the six data columns (Columns 2 to 7) in *data.amzn* and tells R to order it by the date column (Column 1). This moves the date into the index (row names) portion and the dataset now only has six columns, which contain the open, high, low, close, adjusted close, and volume data for Amazon.

```
1  > library(xts)
2  > data.amzn <- xts(data.amzn[, 2:7], order.by = data.amzn[, 1])
3  > head.tail(data.amzn)
4               Open   High    Low  Close Adj.Close  Volume
5   2014-12-31 311.55 312.98 310.01 310.35    310.35 2048000
6   2015-01-02 312.58 314.75 306.96 308.52    308.52 2783200
7   2015-01-05 307.01 308.38 300.85 302.19    302.19 2774200
8               Open   High    Low  Close Adj.Close  Volume
9   2019-12-27 1882.92 1901.40 1866.01 1869.80  1869.80 6186600
10  2019-12-30 1874.00 1884.00 1840.62 1846.89  1846.89 3674700
11  2019-12-31 1842.00 1853.26 1832.23 1847.84  1847.84 2506500
12  > class(data.amzn)
13  [1] "xts" "zoo"
```

**Step 5: Rename the Variables** To be compatible with getSymbols(), we rename the headers with the ticker first and then the variable type (e.g., AMZN.Open, AMZN.High, and so on).

```
1  > names(data.amzn) <-
2  +   c("AMZN.Open", "AMZN.High", "AMZN.Low",
3  +     "AMZN.Close", "AMZN.Adjusted", "AMZN.Volume")
4  > head.tail(data.amzn)
5             AMZN.Open AMZN.High AMZN.Low AMZN.Close AMZN.Adjusted AMZN.Volume
6   2014-12-31    311.55    312.98   310.01     310.35        310.35     2048000
7   2015-01-02    312.58    314.75   306.96     308.52        308.52     2783200
8   2015-01-05    307.01    308.38   300.85     302.19        302.19     2774200
9             AMZN.Open AMZN.High AMZN.Low AMZN.Close AMZN.Adjusted AMZN.Volume
10  2019-12-27   1882.92   1901.40  1866.01    1869.80       1869.80     6186600
11  2019-12-30   1874.00   1884.00  1840.62    1846.89       1846.89     3674700
12  2019-12-31   1842.00   1853.26  1832.23    1847.84       1847.84     2506500
```

**Step 6: Exchange the Fifth and Sixth Columns** getSymbols() has the adjusted close price in Column 6 and the volume data in Column 5. To make our data consistent with getSymbols(), we switch the position of these two variables.

```
1  > data.amzn <- cbind(data.amzn[, 1:4], data.amzn[, 6], data.amzn[, 5])
2  > head.tail(data.amzn)
3             AMZN.Open AMZN.High AMZN.Low AMZN.Close AMZN.Volume AMZN.Adjusted
4   2014-12-31    311.55    312.98   310.01     310.35     2048000        310.35
5   2015-01-02    312.58    314.75   306.96     308.52     2783200        308.52
6   2015-01-05    307.01    308.38   300.85     302.19     2774200        302.19
```

```
7                AMZN.Open AMZN.High AMZN.Low AMZN.Close AMZN.Volume AMZN.Adjusted
8   2019-12-27    1882.92   1901.40  1866.01   1869.80     6186600       1869.80
9   2019-12-30    1874.00   1884.00  1840.62   1846.89     3674700       1846.89
10  2019-12-31    1842.00   1853.26  1832.23   1847.84     2506500       1847.84
```

In Appendix B, we create a function called load.data() that we can pre-load to the R environment every time we run R. The load.data() function will perform the above steps to load CSV files obtained from Yahoo Finance.

Now, we can load our data in **AMZN Yahoo.csv** using load.data() and compare our results to what we get after Step 6 above. As we can see, the outputs of both approaches are identical.

```
1   > amzn.load.data <- load.data("AMZN Yahoo.csv", "AMZN")
2   > head.tail(amzn.load.data)
3                AMZN.Open AMZN.High AMZN.Low AMZN.Close AMZN.Volume AMZN.Adjusted
4   2014-12-31    311.55    312.98   310.01    310.35      2048000        310.35
5   2015-01-02    312.58    314.75   306.96    308.52      2783200        308.52
6   2015-01-05    307.01    308.38   300.85    302.19      2774200        302.19
7                AMZN.Open AMZN.High AMZN.Low AMZN.Close AMZN.Volume AMZN.Adjusted
8   2019-12-27    1882.92   1901.40  1866.01   1869.80     6186600       1869.80
9   2019-12-30    1874.00   1884.00  1840.62   1846.89     3674700       1846.89
10  2019-12-31    1842.00   1853.26  1832.23   1847.84     2506500       1847.84
```

As we can see, the above data is identical to the data below that we obtain when using getSymbols(). Note that Yahoo Finance has some recent quirk in which the end date sometimes needs to be one day after the end date that you want. For example, normally, we would use December 31, 2019 in the to field (Line 2). However, when you try that, we somehow only download data through December 30, 2019 (i.e., one day before the end date that we want). This may make sense if December 31, 2019 was not a trading day, but we know from the above that December 31, 2019 is a trading day and, therefore, should have data. Thus, we have to be careful when we use getSymbols() to make sure we have all the data that we need.

```
1   > gSymb <- getSymbols("AMZN", from = "2014-12-31",
2   +      to = "2020-01-01", auto.assign = FALSE)
3   > head.tail(gSymb)
4                AMZN.Open AMZN.High AMZN.Low AMZN.Close AMZN.Volume AMZN.Adjusted
5   2014-12-31    311.55    312.98   310.01    310.35      2048000        310.35
6   2015-01-02    312.58    314.75   306.96    308.52      2783200        308.52
7   2015-01-05    307.01    308.38   300.85    302.19      2774200        302.19
8                AMZN.Open AMZN.High AMZN.Low AMZN.Close AMZN.Volume AMZN.Adjusted
9   2019-12-27    1882.92   1901.40  1866.01   1869.80     6186600       1869.80
10  2019-12-30    1874.00   1884.00  1840.62   1846.89     3674700       1846.89
11  2019-12-31    1842.00   1853.26  1832.23   1847.84     2506500       1847.84
```

## 1.3   Checking the Data

Since we are obtaining new real world data, we have to perform at least some tests to make sure we have correct and complete data. It is worth emphasizing again that it is important to start being vigilant about the reliability of the data for our intended purpose. Even fee-based data services are not immune to data issues. Thus, we should not assume that because we pay for something that it is free from error (i.e., paying a fee does not necessarily make something better). Also, even a data source we use regularly needs to be checked as things may have changed or that there are special cases in which issues arise. An example is the missing one trading day issue when pulling Yahoo Finance data using getSymbols().

Although not comprehensive, this section discusses some checks we can implement to make sure that the data is not wrong on its face. Depending on how comfortable we are with the results, we do not need to do all these tests. The important thing is that we do tests that fit whatever analysis we are doing. If our analysis only uses month-end prices, then we may not care as much if there are issues with a data point in the middle of the month if it does not affect month-end prices.

### 1.3.1 Check the Start and End Data

One check to ensure that we have all the data we would need is to look at the first few observations in the beginning and the last few observations in the end. In R, head() outputs the first six observations and tail() outputs the last six observations.

```
1  > head(data.amzn)
2            AMZN.Open AMZN.High AMZN.Low AMZN.Close AMZN.Volume AMZN.Adjusted
3  2014-12-31   311.55    312.98   310.01    310.35     2048000       310.35
4  2015-01-02   312.58    314.75   306.96    308.52     2783200       308.52
5  2015-01-05   307.01    308.38   300.85    302.19     2774200       302.19
6  2015-01-06   302.24    303.00   292.38    295.29     3519000       295.29
7  2015-01-07   297.50    301.28   295.33    298.42     2640300       298.42
8  2015-01-08   300.32    303.14   296.11    300.46     3088400       300.46
9  > tail(data.amzn)
10            AMZN.Open AMZN.High AMZN.Low AMZN.Close AMZN.Volume AMZN.Adjusted
11 2019-12-23  1788.26   1793.00  1784.51   1793.00     2136400      1793.00
12 2019-12-24  1793.81   1795.57  1787.58   1789.21      881300      1789.21
13 2019-12-26  1801.01   1870.46  1799.50   1868.77     6005400      1868.77
14 2019-12-27  1882.92   1901.40  1866.01   1869.80     6186600      1869.80
15 2019-12-30  1874.00   1884.00  1840.62   1846.89     3674700      1846.89
16 2019-12-31  1842.00   1853.26  1832.23   1847.84     2506500      1847.84
```

We can also change the number of observations R outputs. We do this by adding a second argument to the head() and tail() functions. For example, to output the first three observations and the last three observations we can use the code in Lines 1 and 6 below, respectively. This is essentially the code we use in the head.tail() function we created earlier.

```
1  > head(data.amzn, 3)
2            AMZN.Open AMZN.High AMZN.Low AMZN.Close AMZN.Volume AMZN.Adjusted
3  2014-12-31   311.55    312.98   310.01    310.35     2048000       310.35
4  2015-01-02   312.58    314.75   306.96    308.52     2783200       308.52
5  2015-01-05   307.01    308.38   300.85    302.19     2774200       302.19
6  > tail(data.amzn, 3)
7            AMZN.Open AMZN.High AMZN.Low AMZN.Close AMZN.Volume AMZN.Adjusted
8  2019-12-27  1882.92   1901.40  1866.01   1869.80     6186600      1869.80
9  2019-12-30  1874.00   1884.00  1840.62   1846.89     3674700      1846.89
10 2019-12-31  1842.00   1853.26  1832.23   1847.84     2506500      1847.84
```

### 1.3.2   Plotting the Data

Another simple test we can perform is to visually look at the data. Looking at a chart of the data allows us to get a sense of whether the data we are using is complete. It also allows us to see if there are odd things with the data. Thus, knowing the patterns in the data could help explain the reasonableness of the results of our analysis. To chart the data, we use plot(). To plot the AMZN close price, we want to plot the variable labeled AMZN.Close.

```
1   > plot(data.amzn$AMZN.Close)
```

Without any additional arguments, the above line of code plots a simple line chart of AMZN close prices. Figure 1.1 shows the output of the above code. The chart that is generated is that of an xts object and is the default chart if we do not add other features to it. Looking at the chart, the data does not show anything obviously suspicious with the AMZN close price data. This is what we would expect from a heavily traded stock like AMZN. The chart shows that AMZN trades everyday as the stock price bounces up and down with no visible flat spots. Note that the Yahoo Finance data excludes trading holidays and weekends, but when we are looking at a chart of daily prices over a five-year period, we cannot visibly see two days of no data. If there are missing data, R connects the last two data points by a line. Hence, if we are missing, say 200 trading days of data in the middle, there would be a long straight line in the chart. We can illustrate this by arbitrarily deleting observations 400 to 600 from the AMZN data. Since we are using a range, we can use a colon (:) to represent the "to" in the range. Also, to delete rows we place a negative sign in front of the observation numbers. Finally, to delete rows, we need to place all these commands to the left of the comma inside the square brackets. We put all of this together in Line 1 of the output below.

```
1   > data.missing <- data.amzn[-400:-600, ]
2   > plot(data.missing$AMZN.Close)
```

Line 2 of the code generates a basic plot of the AMZN close prices using the data in the *data.missing* dataset. As shown in Fig. 1.2, there is a long straight line in the middle of the chart, which would have been very suspicious for a highly traded stock like AMZN. Had this outcome been observed from the raw data pull, this would have been a sign that we should investigate the raw data to determine whether the source files itself did not contain the observations or we should look for the causes as to why the data was missing (e.g., was the data inadvertently deleted).

Another point worth considering is, when we are the only ones that are looking at the data, we do not need to spend time making the charts look pretty. That is why in the above we used plot() without any additional arguments. We are not distributing the output, so there is no need to think about what color the line should be, what title we should use, etc.

**data.amzn$AMZN.Close**                    2014-12-31 / 2019-12-31

Dec 31 2014  Dec 01 2015  Dec 01 2016  Dec 01 2017  Dec 03 2018  Nov 29 2019

**Fig. 1.1** AMZN Stock Price, December 31, 2014–December 31, 2019. Data source: Price data reproduced with permission of CSI ©2020. www.csidata.com

### 1.3.3  Checking the Dimension

We can learn a lot about the completeness of the data also by looking at the number of columns and rows or the **dimension** of the data object. For Yahoo Finance data, the number of columns tells us how many variables are in the data. We know by looking at the data above this number is six. If the number of variables is not equal to six, then we should investigate what the additional or missing variables are and how they got added or deleted.

In a time series, the number of rows represents the dates of the observation. Most stock price data will show only trading days. A good rule of thumb is there are approximately 252 trading days in a calendar year. We can use this number to estimate how many rows we expect to observe over our entire date range. Since we are using five years of data, we expect to see approximately 1,260 [= 252 trading days * 5 years] rows.

In R, we use dim() to output the dimension of the dataset. The output below shows we have 1,259 rows (the number on the left), which is one less than our 1,260

**Fig. 1.2** AMZN Stock Price after removing 200 observations in the middle of the period. Data source: Price data reproduced with permission of CSI ©2020. www.csidata.com

row estimate. Having one less observation than what our estimate over a five-year period would not in and of itself raise a red flag. As for the number of columns (the number on the right), the output shows we have six columns in the data object, which is the exact number we expected.

```
1  > 252*5
2  [1] 1260
3  > dim(data.amzn)
4  [1] 1259    6
```

### 1.3.4 Outputting Summary Statistics

Another useful check is to look at the **summary statistics** of the data. In R, we do this by using summary(). R will then output, for each variable we select, the variable's minimum and maximum values, the interquartile range (i.e., the values in the 25th percentile and 75th percentile), mean, and median. These metrics give a sense of whether there is anything abnormal with the data. For example, a negative price or volume or an absurdly high price or volume may be a sign that there is something for us to investigate. In the example below, we output the summary statistics for the AMZN close price. Note that in an xts object like *data.amzn*, even if we choose to only output summary statistics for the column AMZN.Close, R also outputs the summary statistic for the dates, which is labeled Index below because the dates are the names of the rows in an xts object. Looking at the date ranges also helps us know whether there are issues with the data. For example, from the output, we know that the data in *data.amzn* is from December 31, 2014 to December 31, 2019, which is the date range that we would expect to see. We can also see that the median and mean are the dates that we would expect to be in the middle of our date range.

```
1    > summary(data.amzn$AMZN.Close)
2          Index                 AMZN.Close
3    Min.    :2014-12-31    Min.    : 286.9
4    1st Qu.:2016-04-02     1st Qu.: 664.6
5    Median :2017-06-30     Median : 967.8
6    Mean    :2017-07-01    Mean    :1114.4
7    3rd Qu.:2018-09-29     3rd Qu.:1676.2
8    Max.    :2019-12-31    Max.    :2039.5
```

### 1.3.5   Checking the Ticker Symbol

**Ticker symbols** or **tickers** are a common identifier for various securities. However, we have to be careful when using tickers for several reasons. First, tickers are not unique identifiers because firms change ticker symbols for various reasons and old ticker symbols get recycled from one company to another.

Second, tickers may not be the most obvious variation of the company name. One of the most interesting and recent examples of this is the confusion investors had between Zoom Video Communications with ticker ZM and Zoom Technologies with ticker ZOOM. ZM is the video conferencing service that saw its usage increase dramatically when shelter-in-place orders were instituted during COVID. ZM saw its shares double from the start of January 2020 to March 23, 2020, which is an impressive return. On the other hand, ZOOM, which develops games and electronic components for mobile phones, saw their shares rise from $3 to over $20 by mid-March 2020, which is an even more impressive 7x increase, because investors confused ZOOM for ZM. Consequently, the SEC had to step in and suspend trading in ZOOM on March 26, 2020.

Third, ticker symbols may not be the same across data sources. This is less of a problem for stocks, because the stock tickers are used in major exchanges for trading. For non-traded instruments like indexes, this can be a bigger problem. For example, the S&P 500 Index has the ticker SPX on Bloomberg and its ticker on

Yahoo Finance is ˆGSPC. To add to the confusion, the ticker SPX used to be for SPX Corporation, but SPX Corporation has since changed its ticker symbol.

Using the wrong ticker symbol in the analysis will surely result in incorrect results. Therefore, it is good practice to check to make sure we are obtaining data for the correct security. If available, looking at other descriptors could be helpful. If we have some sense of the level of the price of the security at some point in time, it may also be helpful to visually check the data. We can also check the data we are downloading from a trusted source. For example, many publicly traded firms have an Investor Relations section on their website in which stock prices are sometimes reported or obtainable. We can then validate at least some of the data from the new source with data reported on the company's website. Alternatively, we can pull the data from another source or look at the firm's SEC filings, which may give us some data on the performance of the firm's stock price over some period of time.

## 1.4  Basic Data Manipulation Techniques

In many classroom settings, we are given data that has already been cleaned and organized in a way that makes running a specific task easy. For example, to run a regression of the returns of AMZN stock on the returns of a market index over the period 2019, we may be provided two time series of returns. One will be returns data for AMZN and the other returns data for the S&P 500 Index and the data will only be for 2019. All we have to do is to tell whatever program we are using what the dependent variable is and what the independent variable is. The regression result will then be outputted.

In practice, importing raw data often means reading in more data than you need. This could be in the form of too many observations or too many variables. In our example of AMZN stock data obtained from Yahoo Finance, *data.amzn* contains 1,259 observations and six variables. If we only need one year of data, we can still use *data.amzn* but have to subset the data to only contain the data in the year we are interested in. There is no need to go back to Yahoo Finance and download a new dataset that only contains that one year of data. Also, by having the CSV file locally on your computer, you can work offline like on an airplane or where an internet connection is spotty. As another example, below we show how to calculate a volume-weighted average price (VWAP). A VWAP calculation requires only the close price and volume data. Thus, we only need two of the six variables in *data.amzn*. Downloading a CSV file of the data again from Yahoo Finance is not going to help because we will always get six variables when we do that. Thus, we have to learn how to delete or keep variables from the original dataset.

In what follows, we show how to perform a number of data manipulation techniques.

### 1.4.1  Keeping and Deleting One Row

Subsetting data in R requires the addition of square brackets after the data object name (i.e., data.amzn[, ]). The terms inside the square bracket are separated by a comma. To the left of the comma are arguments we use for the rows and to the right of the comma are argument for the columns.

To keep or delete one row, we place the row number to the left of the comma inside the square brackets. For example, if we want to keep only the first row, we type [1, ].

```
1  > data.amzn[1, ]
2            AMZN.Open AMZN.High AMZN.Low AMZN.Close AMZN.Volume AMZN.Adjusted
3  2014-12-31   311.55    312.98   310.01     310.35     2048000        310.35
```

A negative sign in front of the row number will tell R to delete that row. For example, if we want to delete the first row, we type [−1, ]. Using head(), we output the first six observations of the data. From the above, we know *data.amzn* started with December 31, 2014. Below, we see that the data started with January 2, 2015, which is the second observation because January 1, 2015 (New Year's Day), is a holiday.

```
1  > head(data.amzn[-1, ])
2            AMZN.Open AMZN.High AMZN.Low AMZN.Close AMZN.Volume AMZN.Adjusted
3  2015-01-02   312.58    314.75   306.96     308.52     2783200        308.52
4  2015-01-05   307.01    308.38   300.85     302.19     2774200        302.19
5  2015-01-06   302.24    303.00   292.38     295.29     3519000        295.29
6  2015-01-07   297.50    301.28   295.33     298.42     2640300        298.42
7  2015-01-08   300.32    303.14   296.11     300.46     3088400        300.46
8  2015-01-09   301.48    302.87   296.68     296.93     2592400        296.93
```

### 1.4.2  Keeping First and Last Rows

In some instances, we may want to know what the first and last observation of the data object are. For example, if we want to see the change in the AMZN price in the five-year period from December 31, 2014 to December 31, 2019, we do not need to look at all 1,259 observations. We only need to look at the first and last observation.

To combine a number of different elements, we use c(). For example, c(1, 2) creates a vector with two elements: 1 and 2.

```
1  > c(1, 2)
2  [1] 1 2
```

We can put as many row numbers inside c() as long as we separate those row numbers by commas. For example, we can include the numbers 1 through 5 as five elements in a vector.

```
1  > c(1, 2, 3, 4, 5)
2  [1] 1 2 3 4 5
```

Given the above, if we wanted to only output the first and last rows of the data, we use c(1, 1259).

```
1  > data.amzn[c(1, 1259), ]
2              AMZN.Open AMZN.High AMZN.Low AMZN.Close AMZN.Volume AMZN.Adjusted
3  2014–12–31    311.55    312.98   310.01     310.35     2048000        310.35
4  2019–12–31   1842.00   1853.26  1832.23    1847.84     2506500       1847.84
```

However, we only know that the last row in *data.amzn* is 1,259 from our earlier discussion. If we did not know the exact number or if that number may change, we can use `nrow()` to identify the number of rows in a dataset. We can then use that to replace 1,259 above inside the `c()`.

```
1  > nrow(data.amzn)
2  [1] 1259
3  > data.amzn[c(1, nrow(data.amzn)), ]
4              AMZN.Open AMZN.High AMZN.Low AMZN.Close AMZN.Volume AMZN.Adjusted
5  2014–12–31    311.55    312.98   310.01     310.35     2048000        310.35
6  2019–12–31   1842.00   1853.26  1832.23    1847.84     2506500       1847.84
```

### 1.4.3  Keeping Contiguous Rows

We can use `c()` to keep rows that are not contiguous (i.e., side-by-side). If the rows are contiguous, we use a colon (:) to simplify the process. Suppose we want to output data for the first week of January and we know that those are rows 2 through 6 in *data.amzn*, we would add `[2:6, ]`.

```
1  > data.amzn[2:6, ]
2              AMZN.Open AMZN.High AMZN.Low AMZN.Close AMZN.Volume AMZN.Adjusted
3  2015–01–02    312.58    314.75   306.96     308.52     2783200        308.52
4  2015–01–05    307.01    308.38   300.85     302.19     2774200        302.19
5  2015–01–06    302.24    303.00   292.38     295.29     3519000        295.29
6  2015–01–07    297.50    301.28   295.33     298.42     2640300        298.42
7  2015–01–08    300.32    303.14   296.11     300.46     3088400        300.46
```

We can also use numeric variables when subsetting data. For example, if we wanted to calculate the 30-day VWAP, we would need the last 30 close prices and last 30 volume data. From the above, we know that we have to use a colon (:), as these are contiguous rows. However, given that we are using the last 30 trading days, the answer is not easily obtainable without checking the data for how many non-trading days there were and those change depending on the time period we were looking at. One approach is for us to create a variable *last30*, which equals `(nrow(data.AMZN) - 30 + 1)`. Using `nrow()` returns the number of rows in the data. We then subtract 30 from that number, which get us the 31st to the last trading day of 2019. From there, we add 1 so we get the 30th trading day from the end of 2019. We then use *last30* when we subset *data.amzn*. Inside the square brackets, we put *last30* to the left of the colon and `nrow(data.AMZN)` to the right of the colon. We can check by counting the rows of data that we indeed have the last 30 trading days of 2019.

```
1  > (last30 <- nrow(data.amzn) − 30 + 1)
2  [1] 1230
3  > data.amzn[last30:nrow(data.amzn), ]
```

| 4 | | AMZN.Open | AMZN.High | AMZN.Low | AMZN.Close | AMZN.Volume | AMZN.Adjusted |
|---|---|---|---|---|---|---|---|
| 5 | 2019-11-18 | 1738.30 | 1753.70 | 1722.71 | 1752.53 | 2839500 | 1752.53 |
| 6 | 2019-11-19 | 1756.99 | 1760.68 | 1743.03 | 1752.79 | 2270800 | 1752.79 |
| 7 | 2019-11-20 | 1749.14 | 1762.52 | 1734.12 | 1745.53 | 2790000 | 1745.53 |
| 8 | 2019-11-21 | 1743.00 | 1746.87 | 1730.36 | 1734.71 | 2662900 | 1734.71 |
| 9 | 2019-11-22 | 1739.02 | 1746.43 | 1731.00 | 1745.72 | 2479100 | 1745.72 |
| 10 | 2019-11-25 | 1753.25 | 1777.42 | 1753.24 | 1773.84 | 3486200 | 1773.84 |
| 11 | 2019-11-26 | 1779.92 | 1797.03 | 1778.35 | 1796.94 | 3181200 | 1796.94 |
| 12 | 2019-11-27 | 1801.00 | 1824.50 | 1797.31 | 1818.51 | 3025600 | 1818.51 |
| 13 | 2019-11-29 | 1817.78 | 1824.69 | 1800.79 | 1800.80 | 1923400 | 1800.80 |
| 14 | 2019-12-02 | 1804.40 | 1805.55 | 1762.68 | 1781.60 | 3925600 | 1781.60 |
| 15 | 2019-12-03 | 1760.00 | 1772.87 | 1747.23 | 1769.96 | 3380900 | 1769.96 |
| 16 | 2019-12-04 | 1774.01 | 1789.09 | 1760.22 | 1760.69 | 2670100 | 1760.69 |
| 17 | 2019-12-05 | 1763.50 | 1763.50 | 1740.00 | 1740.48 | 2823800 | 1740.48 |
| 18 | 2019-12-06 | 1751.20 | 1754.40 | 1740.13 | 1751.60 | 3117400 | 1751.60 |
| 19 | 2019-12-09 | 1750.66 | 1766.89 | 1745.61 | 1749.51 | 2442800 | 1749.51 |
| 20 | 2019-12-10 | 1747.40 | 1750.67 | 1735.00 | 1739.21 | 2514300 | 1739.21 |
| 21 | 2019-12-11 | 1741.67 | 1750.00 | 1735.71 | 1748.72 | 2097600 | 1748.72 |
| 22 | 2019-12-12 | 1750.00 | 1764.00 | 1745.44 | 1760.33 | 3095900 | 1760.33 |
| 23 | 2019-12-13 | 1765.00 | 1768.99 | 1755.00 | 1760.94 | 2745700 | 1760.94 |
| 24 | 2019-12-16 | 1767.00 | 1769.50 | 1757.05 | 1769.21 | 3145200 | 1769.21 |
| 25 | 2019-12-17 | 1778.01 | 1792.00 | 1777.39 | 1790.66 | 3644400 | 1790.66 |
| 26 | 2019-12-18 | 1795.02 | 1798.20 | 1782.36 | 1784.03 | 3351400 | 1784.03 |
| 27 | 2019-12-19 | 1780.50 | 1792.99 | 1774.06 | 1792.28 | 2652800 | 1792.28 |
| 28 | 2019-12-20 | 1799.62 | 1802.97 | 1782.45 | 1786.50 | 5150800 | 1786.50 |
| 29 | 2019-12-23 | 1788.26 | 1793.00 | 1784.51 | 1793.00 | 2136400 | 1793.00 |
| 30 | 2019-12-24 | 1793.81 | 1795.57 | 1787.58 | 1789.21 | 881300 | 1789.21 |
| 31 | 2019-12-26 | 1801.01 | 1870.46 | 1799.50 | 1868.77 | 6005400 | 1868.77 |
| 32 | 2019-12-27 | 1882.92 | 1901.40 | 1866.01 | 1869.80 | 6186600 | 1869.80 |
| 33 | 2019-12-30 | 1874.00 | 1884.00 | 1840.62 | 1846.89 | 3674700 | 1846.89 |
| 34 | 2019-12-31 | 1842.00 | 1853.26 | 1832.23 | 1847.84 | 2506500 | 1847.84 |

Note that the code for *last30* in Line 1 above is wrapped with parentheses. This is not necessary to run the code. What this does is output whatever code we write inside the parenthesis. If we did not wrap the code in Line 1 in parenthesis, the number will simply be stored in *last30* and we would not know what the number is.

```
1  > last30 <- nrow(data.amzn) - 30 + 1
```

We would then have to enter `last30` as a second step if we want to see the output. Thus, we cut the process from two steps to one step by wrapping the code with parentheses.

   In practice, we may sometimes want to output results, while other times we may not want to. Similarly, in this book, we output more intermediate output at the beginning of the book compared to later in the book. In many times, we would repeat the same lines of code to setup the data. Since we would have seen that data earlier (possibly multiple times also), there would be less need to show intermediate output of such a step. For example, in the next chapter, we will be calculating returns data for AMZN and certain number of securities often. To calculate a return, we typically would need to import the data, calculate the percentage change, delete the first observation, and rename the variable. It is more critical to see the output of each step the first time we do this. However, when we have already seen the steps and have a sense of what the data looks like at these steps, showing the intermediate output only after new or major steps makes going through the implementation less cumbersome.

## 1.4.4 Keeping One Column

Now we turn to showing examples of subsetting the columns of our data. For our purposes, columns represent variable names. Using names() command, we can see what the names of the variables are. Notice there is a [1] and [5] on the left side of the output below. These represent the variable number of the first observation on each line. As we know, the *data.amzn* dataset has six columns. The first line shows the names of columns 1 through 4, while the last line shows the names for columns 5 and 6. How many column names show up on each line depends on the size of your monitor and how wide your R console is. The wider your monitor and/or R console, the more column names that could be fit into one line before R begins wrapping the line. In other words, if you have a large enough monitor or wide enough R console, the entire output of names() may all be on one line. Accordingly, we should not be surprised to see a different layout of the output on the screen. However, the actual column names and the number of column names should be the same.

```
1  > names(data.amzn)
2  [1] "AMZN.Open"    "AMZN.High"     "AMZN.Low"     "AMZN.Close"
3  [5] "AMZN.Volume"  "AMZN.Adjusted"
```

As an example of keeping one column, suppose we want to plot only the AMZN close price, which is Column 4 in *data.amzn*. We can do so by typing in data.amzn[, 4]. Since we are subsetting columns, we enter 4 to the right of the comma. We use head() to report the first six observations.

```
1  > head(data.amzn[, 4])
2              AMZN.Close
3  2014-12-31    310.35
4  2015-01-02    308.52
5  2015-01-05    302.19
6  2015-01-06    295.29
7  2015-01-07    298.42
8  2015-01-08    300.46
```

An alternative way is to specify the variable name preceded by a dollar sign ($). For example, the close price is under the column labeled AMZN.Close, so we enter data.amzn$AMZN.Close. As we can see, the outputs of both approaches are identical.

```
1  > head(data.amzn$AMZN.Close)
2              AMZN.Close
3  2014-12-31    310.35
4  2015-01-02    308.52
5  2015-01-05    302.19
6  2015-01-06    295.29
```

```
7   2015–01–07    298.42
8   2015–01–08    300.46
```

### 1.4.5  Deleting One Column

We can now look at an example of when we want to delete one column. Later in this chapter, we will create an open, high, low, and close (OHLC) price chart with volume. The sixth column in our data is the adjusted close price, which we will not need for such an application. To delete a column we put a negative sign in front of the column number to delete that column.

```
1   > head(data.amzn[, -6])
2               AMZN.Open AMZN.High AMZN.Low AMZN.Close AMZN.Volume
3   2014–12–31     311.55    312.98   310.01     310.35     2048000
4   2015–01–02     312.58    314.75   306.96     308.52     2783200
5   2015–01–05     307.01    308.38   300.85     302.19     2774200
6   2015–01–06     302.24    303.00   292.38     295.29     3519000
7   2015–01–07     297.50    301.28   295.33     298.42     2640300
8   2015–01–08     300.32    303.14   296.11     300.46     3088400
```

### 1.4.6  Keeping Non-Contiguous Columns

In most applications, we likely will need more than one variable. Suppose we want to compare the daily open and close prices for AMZN. We would then need to keep the open price (Column 1) and close price (Column 4). For non-contiguous columns, we use c() to combine the column numbers that we would like to keep. In this case, we enter c(1, 4) to the right of the comma.

```
1   > head(data.amzn[, c(1, 4)])
2               AMZN.Open AMZN.Close
3   2014–12–31     311.55     310.35
4   2015–01–02     312.58     308.52
5   2015–01–05     307.01     302.19
6   2015–01–06     302.24     295.29
7   2015–01–07     297.50     298.42
8   2015–01–08     300.32     300.46
```

### 1.4.7  Keeping Contiguous Columns

Sometimes we may want to keep variables that are next to each other. For example, suppose we want to analyze the AMZN daily close price (Column 4) and volume (Column 5). We can use the colon (:) to denote that these columns are contiguous (i.e., data.amzn[, 4:5]).

```
1  > head(data.amzn[, 4:5])
2              AMZN.Close AMZN.Volume
3  2014-12-31    310.35     2048000
4  2015-01-02    308.52     2783200
5  2015-01-05    302.19     2774200
6  2015-01-06    295.29     3519000
7  2015-01-07    298.42     2640300
8  2015-01-08    300.46     3088400
```

## 1.4.8  Keeping Contiguous and Non-Contiguous Columns

In some instances, we may end up wanting to keep several columns that are not all contiguous but some are. To do this, we combine the techniques we learned for keeping contiguous and non-contiguous data. For example, if we want to keep the AMZN open price, close price, and volume data, we enter c(1, 4:5) to the right of the comma.

```
1  > head(data.amzn[, c(1, 4:5)])
2              AMZN.Open AMZN.Close AMZN.Volume
3  2014-12-31   311.55     310.35     2048000
4  2015-01-02   312.58     308.52     2783200
5  2015-01-05   307.01     302.19     2774200
6  2015-01-06   302.24     295.29     3519000
7  2015-01-07   297.50     298.42     2640300
8  2015-01-08   300.32     300.46     3088400
```

Depending on which one is easier, we can get to the same result as keeping Columns 1, 4, and 5 by deleting Columns 2, 3, and 6. To do so, we put a negative sign either in front of each column number (Line 1) or we put the negative sign in front of the c() (Line 9).

```
1  > head(data.amzn[, c(-2:-3, -6)])
2              AMZN.Open AMZN.Close AMZN.Volume
3  2014-12-31   311.55     310.35     2048000
4  2015-01-02   312.58     308.52     2783200
5  2015-01-05   307.01     302.19     2774200
6  2015-01-06   302.24     295.29     3519000
7  2015-01-07   297.50     298.42     2640300
8  2015-01-08   300.32     300.46     3088400
9  > head(data.amzn[, -c(2:3, 6)])
10             AMZN.Open AMZN.Close AMZN.Volume
11 2014-12-31   311.55     310.35     2048000
12 2015-01-02   312.58     308.52     2783200
13 2015-01-05   307.01     302.19     2774200
14 2015-01-06   302.24     295.29     3519000
15 2015-01-07   297.50     298.42     2640300
16 2015-01-08   300.32     300.46     3088400
```

The choice of whether to keep rows that we want or to delete rows we do not want may boil down to preference and convenience. In this case, we are keeping three out of the six columns. We could either keep three or delete three without any loss in efficiency, so the choice becomes a matter of preference. However, if we have 30 columns and we want to keep three, we would choose to specify the three columns we wanted to keep unless the columns we are deleting are contiguous, then it may become a matter of preference again. Given that we are not working with gigantic datasets in this book, we will not sweat over how we get to the ultimate dataset that we need for our analysis.

### 1.4.9   Keeping Rows and Columns

So far we have subset rows only or columns only. It is likely that in many applications we would have to keep some rows and some columns at the same time. For example, suppose we want to calculate the VWAP over the last 30 trading days of 2019. What we would need in this analysis is the last 30 days of data in 2019 and AMZN's close price (Column 4) and volume (Column 5).

```
1  > (last30 <- nrow(data.amzn) - 30 + 1)
2  [1] 1230
3  > vwap <- data.amzn[last30:nrow(data.amzn), c(4, 5)]
4  > head.tail(vwap)
5              AMZN.Close AMZN.Volume
6  2019-11-18    1752.53     2839500
7  2019-11-19    1752.79     2270800
8  2019-11-20    1745.53     2790000
9              AMZN.Close AMZN.Volume
10 2019-12-27    1869.80     6186600
11 2019-12-30    1846.89     3674700
12 2019-12-31    1847.84     2506500
13 > dim(vwap)
14 [1] 30   2
```

We use a function here labeled `head.tail()`, which is a function we created in Appendix B. In this book, we will be reporting the intermediate output frequently. We could have used `head(, 3)` and `tail(, 3)` separately to output the first three and last three observations. However, to save time, we are using `head.tail()` instead. This allows us to get a sense of where the dataset starts and ends. In this case, we can see that the last 30 trading days of 2019 start on November 18, 2019 and ends on December 31, 2019. Using `dim()` (Line 13) confirms the *vwap* dataset has 30 rows (i.e., trading days) and 2 columns.

### 1.4.10   Subsetting Time Series Data Using Dates

In many financial applications, we will deal with **time series** data. This is also partly the reason why we are using xts objects. We sometimes convert the objects into other classes (e.g., data frame), because certain functions work better in non-xts objects.

We use the term time series in the more general sense as data that can be indexed by some time interval, such as daily, weekly, monthly, quarterly, or annual. When dealing with time series data, it is often easier to subset data when the object we are subsetting has some tangible meaning. For example, it is easier to use dates to subset data in 2018 rather than figuring out which row numbers in *data.amzn* represent the 2018 data. In this book, we sometimes refer to this as "xts-style" date subsetting to differentiate it from the alternative subsetting approach we discuss later.

Since *data.amzn* is an xts object, we can subset the data by putting a date range inside square brackets. The start and end dates inside the square brackets are separated by a slash (/) and the date range is placed inside quotes.

```
 1  > amzn.2018a <- data.amzn["2018-01-01/2018-12-31"]
 2  > head.tail(amzn.2018a)
 3            AMZN.Open AMZN.High AMZN.Low AMZN.Close AMZN.Volume AMZN.Adjusted
 4  2018-01-02    1172.0   1190.00  1170.51    1189.01     2694500       1189.01
 5  2018-01-03    1188.3   1205.49  1188.30    1204.20     3108800       1204.20
 6  2018-01-04    1205.0   1215.87  1204.66    1209.59     3022100       1209.59
 7            AMZN.Open AMZN.High AMZN.Low AMZN.Close AMZN.Volume AMZN.Adjusted
 8  2018-12-27   1454.20   1469.00  1390.31    1461.64     9722000       1461.64
 9  2018-12-28   1473.35   1513.47  1449.00    1478.02     8829000       1478.02
10  2018-12-31   1510.80   1520.76  1487.00    1501.97     6954500       1501.97
```

Alternatively, whether the data is an xts object or not, we can use `subset()` to, as the name indicates, subset the data. Because *data.amzn* is an xts object, the date is called using `index()`. We have two conditions when we subset the data in this case because the data we want is in the middle of the dataset. We combine different conditions using an ampersand (&) when we want both conditions to hold. That is, the date must be greater than or equal to January 1, 2018 and must be less than or equal to December 31, 2018. As we can see, the output below is identical to the output above.

```
 1  > amzn.2018b <- subset(data.amzn,
 2  +     index(data.amzn) >= "2018-01-01" &
 3  +     index(data.amzn) <= "2018-12-31")
 4  > head.tail(amzn.2018b)
 5            AMZN.Open AMZN.High AMZN.Low AMZN.Close AMZN.Volume AMZN.Adjusted
 6  2018-01-02    1172.0   1190.00  1170.51    1189.01     2694500       1189.01
 7  2018-01-03    1188.3   1205.49  1188.30    1204.20     3108800       1204.20
 8  2018-01-04    1205.0   1215.87  1204.66    1209.59     3022100       1209.59
 9            AMZN.Open AMZN.High AMZN.Low AMZN.Close AMZN.Volume AMZN.Adjusted
10  2018-12-27   1454.20   1469.00  1390.31    1461.64     9722000       1461.64
11  2018-12-28   1473.35   1513.47  1449.00    1478.02     8829000       1478.02
12  2018-12-31   1510.80   1520.76  1487.00    1501.97     6954500       1501.97
```

We can also subset using other column names. Suppose we want to know on which dates AMZN closed higher than $2,000. The output shows there are 12 instances during the 5-year period from 2015 to 2019 on which AMZN closed above $2,000. Note that because we already called *data.amzn* as the first argument in `subset()` and because AMZN.Close is in *data.amzn*, it is superfluous to prefix AMZN.Close with *data.amzn* (i.e., `data.amzn$AMZN.Close`).

```
 1  > (amzn.high.prc <- subset(data.amzn, AMZN.Close > 2000))
 2            AMZN.Open AMZN.High AMZN.Low AMZN.Close AMZN.Volume AMZN.Adjusted
 3  2018-08-30   1997.42   2025.57  1986.90    2002.38     7277300       2002.38
```

| 4  | 2018–08–31 | 2007.00 | 2022.38 | 2004.74 | 2012.71 | 4204400 | 2012.71 |
|----|------------|---------|---------|---------|---------|---------|---------|
| 5  | 2018–09–04 | 2026.50 | 2050.50 | 2013.00 | 2039.51 | 5721100 | 2039.51 |
| 6  | 2018–09–27 | 1993.24 | 2016.16 | 1988.58 | 2012.98 | 4329400 | 2012.98 |
| 7  | 2018–09–28 | 2004.41 | 2026.52 | 1996.46 | 2003.00 | 4085100 | 2003.00 |
| 8  | 2018–10–01 | 2021.99 | 2033.19 | 2003.60 | 2004.36 | 3460500 | 2004.36 |
| 9  | 2019–07–10 | 1996.51 | 2024.94 | 1995.40 | 2017.41 | 4931900 | 2017.41 |
| 10 | 2019–07–11 | 2025.62 | 2035.80 | 1995.30 | 2001.07 | 4317800 | 2001.07 |
| 11 | 2019–07–12 | 2008.27 | 2017.00 | 2003.87 | 2011.00 | 2509300 | 2011.00 |
| 12 | 2019–07–15 | 2021.40 | 2022.90 | 2001.55 | 2020.99 | 2981300 | 2020.99 |
| 13 | 2019–07–16 | 2010.58 | 2026.32 | 2001.22 | 2009.90 | 2618200 | 2009.90 |
| 14 | 2019–07–24 | 1969.30 | 2001.30 | 1965.87 | 2000.81 | 2631300 | 2000.81 |

The above condition when subsetting dates assumes both conditions must hold. We can also subset the data when either of the two conditions hold. Suppose we want to subset only to data when the AMZN close price was over $2,000 or if its volume was over 20 million. Here, we use a vertical bar (|) to tell R to output whatever observations fit both conditions.

```
1  > subset(data.amzn,
2  +    AMZN.Close > 2000 |
3  +    AMZN.Volume > 20000000)
4           AMZN.Open AMZN.High AMZN.Low AMZN.Close AMZN.Volume AMZN.Adjusted
5  2015–01–30   346.32   359.50   340.74    354.53    23856100       354.53
6  2015–07–24   578.99   580.57   529.35    529.42    21909400       529.42
7  2018–08–30  1997.42  2025.57  1986.90   2002.38     7277300      2002.38
8  2018–08–31  2007.00  2022.38  2004.74   2012.71     4204400      2012.71
9  2018–09–04  2026.50  2050.50  2013.00   2039.51     5721100      2039.51
10 2018–09–27  1993.24  2016.16  1988.58   2012.98     4329400      2012.98
11 2018–09–28  2004.41  2026.52  1996.46   2003.00     4085100      2003.00
12 2018–10–01  2021.99  2033.19  2003.60   2004.36     3460500      2004.36
13 2019–07–10  1996.51  2024.94  1995.40   2017.41     4931900      2017.41
14 2019–07–11  2025.62  2035.80  1995.30   2001.07     4317800      2001.07
15 2019–07–12  2008.27  2017.00  2003.87   2011.00     2509300      2011.00
16 2019–07–15  2021.40  2022.90  2001.55   2020.99     2981300      2020.99
17 2019–07–16  2010.58  2026.32  2001.22   2009.90     2618200      2009.90
18 2019–07–24  1969.30  2001.30  1965.87   2000.81     2631300      2000.81
```

We can also subset data that combines different types of conditions. Suppose we want to look at data when AMZN close price was above $2,000 or AMZN trading volume was over 20 million, but only data in 2019. In this case, we want to satisfy the first two conditions first, so we put those inside parentheses and then use an ampersand (&) to join the third condition that is applicable after running the first set of conditions. Thus, the output below will be the last six rows of the above output because those are the only observations that meet the first two conditions in 2019.

```
1  > subset(data.amzn,
2  +    (AMZN.Close > 2000 |
3  +     AMZN.Volume > 20000000) &
4  +     index(data.amzn) >= "2019–01–01")
5           AMZN.Open AMZN.High AMZN.Low AMZN.Close AMZN.Volume AMZN.Adjusted
6  2019–07–10  1996.51  2024.94  1995.40   2017.41     4931900      2017.41
7  2019–07–11  2025.62  2035.80  1995.30   2001.07     4317800      2001.07
8  2019–07–12  2008.27  2017.00  2003.87   2011.00     2509300      2011.00
9  2019–07–15  2021.40  2022.90  2001.55   2020.99     2981300      2020.99
10 2019–07–16  2010.58  2026.32  2001.22   2009.90     2618200      2009.90
11 2019–07–24  1969.30  2001.30  1965.87   2000.81     2631300      2000.81
```

### 1.4.11   Converting to Weekly Prices

We sometimes want to convert the daily prices we obtained from Yahoo Finance to other frequencies. For example, to calculate weekly returns, we would need weekly prices. To convert daily price data into weekly price data in R, we use `to.weekly()`. We first create a new dataset labeled *wk*. After applying `to.weekly()`, the output below shows that the prefix of the column names becomes *wk*. Note that December 31, 2019 is a Tuesday, so the values reported for that "week" are not a full week of data.

```
1   > wk <- data.amzn
2   > head.tail(to.weekly(wk))
3               wk.Open wk.High wk.Low wk.Close wk.Volume wk.Adjusted
4    2015-01-02  311.55  314.75 306.96   308.52   4831200      308.52
5    2015-01-09  307.01  308.38 292.38   296.93  14614300      296.93
6    2015-01-16  297.56  301.50 285.25   290.74  20993900      290.74
7               wk.Open wk.High  wk.Low wk.Close wk.Volume wk.Adjusted
8    2019-12-20 1767.00 1802.97 1757.05 1786.50  17944600     1786.50
9    2019-12-27 1788.26 1901.40 1784.51 1869.80  15209700     1869.80
10   2019-12-31 1874.00 1884.00 1832.23 1847.84   6181200     1847.84
```

In practice, creating a new dataset is unnecessary but it may be worth considering just so the output does not get out of hand. The reason we use a shorter dataset name like *wk* is because `to.weekly()` uses the dataset name as the prefix when it renames the column header. The output below shows the first three rows after we apply `to.weekly()` to the *data.amzn* dataset. The data is the same but the output looks much messier than the above.

```
1   > head(to.weekly(data.amzn), 3)
2               data.amzn.Open data.amzn.High data.amzn.Low data.amzn.Close
3    2015-01-02        311.55         314.75        306.96          308.52
4    2015-01-09        307.01         308.38        292.38          296.93
5    2015-01-16        297.56         301.50        285.25          290.74
6               data.amzn.Volume data.amzn.Adjusted
7    2015-01-02         4831200             308.52
8    2015-01-09        14614300             296.93
9    2015-01-16        20993900             290.74
```

To understand how `to.weekly()` works, let us take a closer look at the first full week of January 2015 (row 2). From the daily data, these are rows 3 through 7 of *data.amzn*. The weekly open price is the first open price that week (January 5, 2015) of $307.01. The weekly high price is the highest of the high prices during that week (January 5, 2015) of $308.38. The weekly low price is the lowest of the low prices during that week (January 6, 2015) of $292.38. The weekly close price and weekly adjusted close price are the corresponding values for the last day during that week (January 9, 2015) of $296.93. The weekly volume is equal to the *sum* of the daily volumes for the week. We confirmed that using the code in Line 11 and the output in Line 12 (i.e., the sum of the daily trading volume is 14,614,300).

```
 1   > to.weekly(wk)[2]
 2            wk.Open wk.High wk.Low wk.Close wk.Volume wk.Adjusted
 3   2015-01-09  307.01  308.38 292.38   296.93 14614300       296.93
 4   > data.amzn[3:7, ]
 5            AMZN.Open AMZN.High AMZN.Low AMZN.Close AMZN.Volume AMZN.Adjusted
 6   2015-01-05    307.01    308.38   300.85     302.19     2774200        302.19
 7   2015-01-06    302.24    303.00   292.38     295.29     3519000        295.29
 8   2015-01-07    297.50    301.28   295.33     298.42     2640300        298.42
 9   2015-01-08    300.32    303.14   296.11     300.46     3088400        300.46
10   2015-01-09    301.48    302.87   296.68     296.93     2592400        296.93
11   > sum(data.amzn[3:7, 5])
12   [1] 14614300
```

### 1.4.12   Converting to Monthly Prices

To convert data into monthly data, we use to.monthly(). Similar to *wk* above, the choice of creating the *mo* dataset is to make the data easier to read on the R console. How we get to the value of each of the monthly variables is essentially analogous to the weekly case but instead of the data for one week we use data for one month.

```
 1   > mo <- data.amzn
 2   > head.tail(to.monthly(mo))
 3            mo.Open mo.High mo.Low mo.Close mo.Volume mo.Adjusted
 4   Dec 2014  311.55  312.98 310.01   310.35   2048000      310.35
 5   Jan 2015  312.58  359.50 285.25   354.53 103057100      354.53
 6   Feb 2015  350.05  389.37 350.01   380.16  70846200      380.16
 7            mo.Open mo.High  mo.Low mo.Close mo.Volume mo.Adjusted
 8   Oct 2019 1746.00 1798.85 1685.06  1776.66  70360500     1776.66
 9   Nov 2019 1788.01 1824.69 1722.71  1800.80  52076200     1800.80
10   Dec 2019 1804.40 1901.40 1735.00  1847.84  68149600     1847.84
```

## 1.5   Comparing Capital Gains Between Securities

Before we begin with our first application of the price data, we may want to consider clearing the R environment. We can think of the R environment as R's memory. This is where all functions and objects created are stored during our R session. Clearing the slate is sometimes necessary to make sure that no residual objects are in the R environment that could somehow influence the analysis we are doing. We can use ls() to show what objects are in the R environment. Below is what I have in my R environment. You may have a different set of objects, depending if you created fewer objects or more objects.

```
 1   > ls()
 2   [1] "amzn.2018a"     "amzn.2018b"    "amzn.high.prc"  "amzn.load.data"
 3   [5] "amzn.ohlc"      "before.day"    "data.amzn"      "data.missing"
 4   [9] "Date"           "gSymb"         "head.tail"      "last30"
```

```
5  [13] "load.data"        "mo"              "ohlc"            "one.day"
6  [17] "vwap"             "wk"
```

We can delete a single object using rm() and putting the object name inside the parenthesis. For example, if we wanted to remove *data.missing* we would use rm(data.missing). As the output below shows, we now only have 17 objects instead of 18 and *data.missing* is no longer in the R environment. This is handy if we accidentally made a dataset and decided not to use it or we want to remove an object whose name we are going to recycle.

```
1  > rm(data.missing)
2  > ls()
3   [1] "amzn.2018a"       "amzn.2018b"       "amzn.high.prc"  "amzn.load.data"
4   [5] "amzn.ohlc"        "before.day"       "data.amzn"      "Date"
5   [9] "gSymb"            "head.tail"        "last30"         "load.data"
6  [13] "mo"               "ohlc"             "one.day"        "vwap"
7  [17] "wk"
```

We can also delete all objects in the R environment using rm(list = ls()). However, we have to be careful when including this command in the code, especially if we send the code around to other people. Unknowingly running this code could be disastrous as this completely wipes out all functions, datasets, and any other objects in the R environment. In other words, everything gets wiped clean. This means that we have to pre-load the codes in Appendix B again. We can do this by copying and pasting the code that we have previously written.

```
1  > rm(list = ls())
2  > ls()
3  character(0)
```

Sometimes in this book I suggest to clear the R environment. Even with such a suggestion, we have to be careful to make sure we only clear the R environment when we are sure we no longer need anything in the R environment. Now, we are ready to begin our analysis of comparing the capital gains of multiple securities over time.

A common analysis using prices is to compare the performance of our various investments over time. This means comparing the price on the date of investment $P_0$ to the price at some later date $P_t$. The difference in price from the purchase date to a later date ($P_t - P_0$) is called **capital gains**. Thus, to track the value of our investment in a particular security against our cost of acquiring the security, we can create a **normalized price chart**. The beauty of a normalized price chart is that we can compare the performance of multiple securities simultaneously because by "normalizing" we put all the securities on the same starting point, which is $1 in our case. Because how to present the data is usually a matter of preference, we show below several examples of how to create a normalized price chart.

In our example, let us suppose we made an investment in Amazon (AMZN), Alphabet (GOOG), Apple (AAPL), and S&P 500 ETF (SPY) on December 31, 2014. How would we know how these investments performed through the end of 2019 and which of these investments performed better over that period?

**Step 1: Import Price Data for Each Security**   We use `load.data()` to import the price data for each of the four securities. `load.data()` is a function we created using code in Appendix B to simplify how we import data we obtained from Yahoo Finance. In addition, `load.data()` generates output that is essentially identical to what we would get had we instead used `getSymbols()` from the `quantmod` package.

```
 1   > data.amzn <- load.data("AMZN Yahoo.csv", "AMZN")
 2   > head.tail(data.amzn)
 3            AMZN.Open AMZN.High AMZN.Low AMZN.Close AMZN.Volume AMZN.Adjusted
 4   2014-12-31    311.55    312.98   310.01     310.35     2048000        310.35
 5   2015-01-02    312.58    314.75   306.96     308.52     2783200        308.52
 6   2015-01-05    307.01    308.38   300.85     302.19     2774200        302.19
 7            AMZN.Open AMZN.High AMZN.Low AMZN.Close AMZN.Volume AMZN.Adjusted
 8   2019-12-27   1882.92   1901.40  1866.01    1869.80     6186600       1869.80
 9   2019-12-30   1874.00   1884.00  1840.62    1846.89     3674700       1846.89
10   2019-12-31   1842.00   1853.26  1832.23    1847.84     2506500       1847.84
11   > data.goog <- load.data("GOOG Yahoo.csv", "GOOG")
12   > head.tail(data.goog)
13            GOOG.Open GOOG.High GOOG.Low GOOG.Close GOOG.Volume GOOG.Adjusted
14   2014-12-31  529.7955  531.1417 524.3604   524.9587     1368200      524.9587
15   2015-01-02  527.5616  529.8154 522.6650   523.3731     1447500      523.3731
16   2015-01-05  521.8273  522.8944 511.6552   512.4630     2059800      512.4630
17            GOOG.Open GOOG.High GOOG.Low GOOG.Close GOOG.Volume GOOG.Adjusted
18   2019-12-27   1362.99   1364.53 1349.310    1351.89     1038400       1351.89
19   2019-12-30   1350.00   1353.00 1334.020    1336.14     1050900       1336.14
20   2019-12-31   1330.11   1338.00 1329.085    1337.02      962468       1337.02
21   > data.aapl <- load.data("AAPL Yahoo.csv", "AAPL")
22   > head.tail(data.aapl)
23            AAPL.Open AAPL.High AAPL.Low AAPL.Close AAPL.Volume AAPL.Adjusted
24   2014-12-31    112.82    113.13   110.21     110.38    41403400     101.41906
25   2015-01-02    111.39    111.44   107.35     109.33    53204600     100.45430
26   2015-01-05    108.29    108.65   105.41     106.25    64285500      97.62434
27            AAPL.Open AAPL.High AAPL.Low AAPL.Close AAPL.Volume AAPL.Adjusted
28   2019-12-27    291.12    293.97   288.12     289.80    36566500        289.80
29   2019-12-30    289.46    292.69   285.22     291.52    36028600        291.52
30   2019-12-31    289.93    293.68   289.52     293.65    25201400        293.65
31   > data.spy <- load.data("SPY Yahoo.csv", "SPY")
32   > head.tail(data.spy)
33            SPY.Open SPY.High SPY.Low SPY.Close SPY.Volume SPY.Adjusted
34   2014-12-31   207.99   208.19  205.39    205.54  130333800     186.2590
35   2015-01-02   206.38   206.88  204.18    205.43  121465900     186.1593
36   2015-01-05   204.17   204.37  201.35    201.72  169632600     182.7974
37            SPY.Open SPY.High SPY.Low SPY.Close SPY.Volume SPY.Adjusted
38   2019-12-27   323.74   323.80  322.28    322.86   42528800       322.86
39   2019-12-30   322.95   323.10  320.55    321.08   49729100       321.08
40   2019-12-31   320.53   322.13  320.15    321.86   57077300       321.86
```

**Step 2: Combine Close Prices of All Securities into One Dataset**   For our analysis, we only need the close prices for each of the four securities. We can identify which column we want to keep by using the variable name (Lines 1–2). We can then combine each security's close price using `cbind()`. We use `names()` in Line 3 to rename the variables with the security's ticker for simplicity.

```
1  > close.prc <- cbind(data.amzn$AMZN.Close, data.goog$GOOG.Close,
2  +    data.aapl$AAPL.Close, data.spy$SPY.Close)
3  > names(close.prc) <- c("AMZN", "GOOG", "AAPL", "SPY")
4  > head.tail(close.prc)
5               AMZN      GOOG    AAPL     SPY
6  2014-12-31 310.35  524.9587 110.38 205.54
7  2015-01-02 308.52  523.3731 109.33 205.43
8  2015-01-05 302.19  512.4630 106.25 201.72
9               AMZN      GOOG    AAPL     SPY
10 2019-12-27 1869.80 1351.89 289.80 322.86
11 2019-12-30 1846.89 1336.14 291.52 321.08
12 2019-12-31 1847.84 1337.02 293.65 321.86
```

Since these are actively traded securities, each security has a price on each trading day. Thus, using `cbind()` here works seamlessly as these securities have the same number of observations.

**Step 3: Normalize Prices**  The starting prices of the different securities are not the same. For example, AAPL starts at \$110.38, while GOOG starts at \$524.96. This makes it hard to compare how well the different securities perform. For example, a \$5 increase in price is a larger gain for a stock we purchased for \$10 than one we purchased for \$100. Thus, to put things on an apples-to-apples basis, we **normalize** the prices to each security's December 31, 2014 price. Doing so assumes that we made an investment in each security of \$1 on December 31, 2014. The resulting normalized prices then show how that \$1 changed over time. By the end of 2019, we can see that a \$1 investment in AMZN would have increased to almost \$6, while that same \$1 invested in GOOG or AAPL would have increased to approximately \$2.60. To put how well these firms have done in context, a \$1 investment in SPY would have only made us \$1.57.

```
1  > norm.prc <- close.prc
2  > norm.prc$AMZN <- close.prc$AMZN / as.numeric(normalize[1, 1])
3  > norm.prc$GOOG <- close.prc$GOOG / as.numeric(normalize[1, 2])
4  > norm.prc$AAPL <- close.prc$AAPL / as.numeric(normalize[1, 3])
5  > norm.prc$SPY <- close.prc$SPY / as.numeric(normalize[1, 4])
6  > head.tail(norm.prc)
7                AMZN      GOOG      AAPL       SPY
8  2014-12-31 1.0000000 1.0000000 1.0000000 1.0000000
9  2015-01-02 0.9941034 0.9969795 0.9904875 0.9994648
10 2015-01-05 0.9737071 0.9761967 0.9625838 0.9814148
11                AMZN      GOOG      AAPL       SPY
12 2019-12-27 6.024811 2.575231 2.625476 1.570789
13 2019-12-30 5.950991 2.545229 2.641058 1.562129
14 2019-12-31 5.954052 2.546905 2.660355 1.565924
```

A normalized price chart allows us to easily convert the price of the security to any investment amount. For example, if you invested \$100 in SPY on December 31, 2014, you would have \$157 at the end of 2019. That is, you multiply the normalized ending price by the amount of your investment to see how much that investment would have made at the end of the period.

**Reducing the Number of Decimals R Outputs**

The output above has many decimal places. We can't easily control the number of decimal places to output in R is. What we can control though are the number of digits that is reported. For example, if we want to show at least three digits for each observation displayed, we can enter `options(digits = 3)`. This option remains in effect until we change it. So to go back to R's default digits of seven, we add the code in Line 7.

```
1  > options(digits = 3)
2  > head(norm.prc, 3)
3              AMZN  GOOG  AAPL   SPY
4  2014-12-31 1.000 1.000 1.000 1.000
5  2015-01-02 0.994 0.997 0.990 0.999
6  2015-01-05 0.974 0.976 0.963 0.981
7  > options(digits = 7)
```

**Step 5: Plot the Capital Appreciation of Each Security** We create a line chart using `plot()` with `type = "l"`. Before charting, we first determine the range of normalized prices in the dataset. The reason is that `plot()` takes the range of the chart from the first security that we plot. If we picked a security that has too small of a range relative to the other securities, plots of securities with values that fall outside that range would appear truncated. To avoid this problem, we create a variable called *y.range*. That variable uses `range()` to get the minimum and maximum prices in *norm.prc*. Note that we applied `range()` to the entire dataset *norm.prc* because all the data in that dataset are normalized prices that we are plotting. This gives us a normalized price range of \$0.82 to \$6.57. If there is a mix of normalized prices and other variables, say, dates, we would have to specify the columns that we want the range to be calculated over. For the x-axis, we need the date values. We thus create the variable *dt*. Using `index()` calls the row labels of *norm.prc*, which are dates. Line 11 changes the thickness of the AMZN line by using `lwd = 2`. The higher the number, the thicker the line.

```
1  > (y.range <- range(norm.prc))
2  [1] 0.8184454 6.5716448
3  > dt <- index(norm.prc)
4  >
5  > plot(x = dt,
6  +     y = norm.prc$AMZN,
7  +     xlab = "Date",
8  +     ylab = "Value of Investment",
9  +     ylim = y.range,
10 +     type = "l",
11 +     lwd = 2,
12 +     main = "Value of $1 Investment in AMZN, GOOG, AAPL, and SPY")
```

**Step 6: Add Lines for the Other Securities and a Reference Line at $1**  We use
lines() to add the normalized prices of the other three securities to the chart.
We have to execute these after we execute plot() above. The default color is
black. However, to easily differentiate the other securities, we can change the color
of the lines using col. For our purposes, we use "blue," "red," and "darkgreen."
R allows you to choose a wide array of colors. For example, the following PDF
file (http://www.stat.columbia.edu/~tzheng/files/Rcolor.pdf) provides you with the
R names for many colors.

```
1  > lines(x = dt, y = norm.prc$GOOG, col = "blue")
2  > lines(x = dt, y = norm.prc$AAPL, col = "red")
3  > lines(x = dt, y = norm.prc$SPY, col = "darkgreen")
4  > abline(h = 1, lty = 2)
```

We can add a reference line to easily see when the investment has made money
(above the horizontal line) or lost money (below the horizontal line). To do so, we
use abline(). The h = 1 tells R to draw a horizontal line with a value of $1.
The lty = 2 tells R to draw a dashed line.

**Step 7: Add a Legend**  A legend is useful to inform the reader looking at the chart
which line references which company's stock. The first thing we would have to do
is figure out where to position the chart. We do this by looking at the chart and
see which area of the chart has space to fit in a legend. The chart we have should
have a bit of white space on the top-left portion of the chart, which makes that area
a good candidate for where the legend should be placed. To add a legend, we use
legend(). We can customize how we define the elements of the legend. Line 2
tells R what the labels in the legend should be. We can order the securities, however,
which way we want. Line 3 tells R how thick each of the lines are and Line 4 is what
color each security's line is. Note that in Lines 3 and 4, the order of the elements has
to be the same as the order the securities appear in Line 2. For example, anything
related to AMZN will be in the first element, GOOG will be in the second element,
etc. If we flip the order in Line 2, we also have to flip the order in Lines 3 and 4.

```
1  > legend("topleft",
2  +       c("AMZN", "GOOG","AAPL", "SPY"),
3  +       lwd = c(2, 1, 1, 1),
4  +       col = c("black", "blue", "red", "darkgreen"))
```

Just like when we are adding the lines, we have to run the code for the legend
immediately after Step 6. The chart output from the above code can be seen in
Fig. 1.3.
    There is no single way to present charts. It is often a matter of preference. We
present below three alternative ways to show a normalized price chart of these four
securities.

**Value of $1 Investment in AMZN, GOOG, AAPL, and SPY**

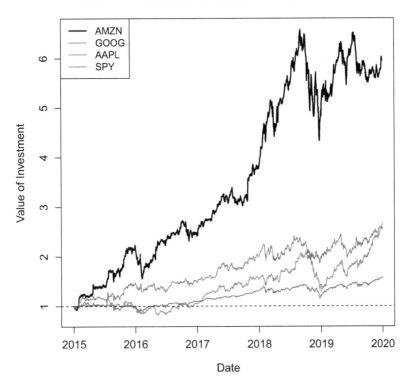

**Fig. 1.3** Capital appreciation of AMZN, GOOG, AAPL, and SPY, December 31, 2014 to December 31, 2019. This chart assumes a $1 investment in each of the four securities by dividing each stock's close price with that stock's price on December 31, 2014. The above shows that the AMZN price increased substantially compared to the prices of GOOG and AAPL. All three, however, did better than the overall market (SPY). Data source: Price data reproduced with permission of CSI ©2020. www.csidata.com

## 1.5.1  Alternative 1—Using xts-Style Chart

The chart in Fig. 1.4 is simpler to create. As an xts object, we can use plot(norm.prc), which will create a chart that similar to Fig. 1.4 that has four lines in different colors. However, to replicate the elements and theme of the chart in Fig. 1.3, we can use this xts plot but add a few more lines of code. Just like the above code, we can draw each of the four lines separately. For consistency with what we did previously, we can also use the same colors as our initial chart.

```
1  > plot(norm.prc$AMZN,
2  +    ylim = y.range,
3  +    lwd = 2,
4  +    main = "Value of $1 Investment in\n AMZN, GOOG, AAPL, and SPY")
5  > lines(norm.prc$GOOG, col = "blue")
```

**Fig. 1.4** Alternative 1—Normalized Price Chart. This chart is an alternative presentation of the normalized price chart using plot() on an xts object. The format looks similar in many respects, but has gridlines, no outer border, and two identical y-axes. Data source: Price data reproduced with permission of CSI ©2020. www.csidata.com

```
 6  > lines(norm.prc$AAPL, col = "red")
 7  > lines(norm.prc$SPY, col = "darkgreen")
 8  > addLegend("topleft",
 9  +     legend.names = c("AMZN", "GOOG", "AAPL", "SPY"),
10  +     lwd = c(2, 1, 1, 1),
11  +     col = c("black", "blue", "red", "darkgreen"),
12  +     bg = "white",
13  +     bty = "o")
```

Note the last two lines of code in the legend. Because an xts-style chart will have gridlines, we can use bg = "white" to fill the background of the legend with a white color. The code bty = "o" puts a border around the legend.

**Value of $1 Investment in AMZN, GOOG, AAPL and SPY**

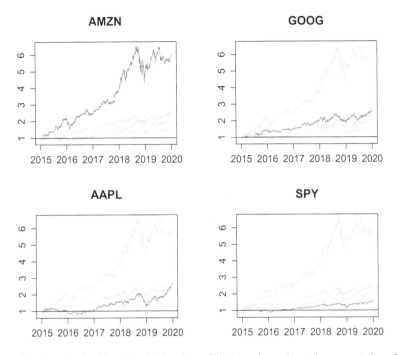

**Fig. 1.5** Alternative 2—Normalized Price Chart. This chart is an alternative presentation of the normalized price chart by creating four separate mini-charts with each chart focusing on the relative price of one security at a time. In each mini-chart, the subject security is in blue (e.g., AMZN in the top-left chart, GOOG in the top-right chart, etc.) while the remaining three securities are in gray. Data source: Price data reproduced with permission of CSI ©2020. www.csidata.com

## 1.5.2  Alternative 2—Plotting Four Mini-Charts

The chart in Fig. 1.5 allows us to focus on the security represented by the blue line in each panel. Each of the other securities gets its turn to be the focus by making them blue in the other panels. By graying out the other securities in each mini-chart, we can de-emphasize the identity of the other stocks but still present their results so we can make a comparison. For example, in Fig. 1.5, we can see that AMZN did better than the other three securities, while SPY did worse than the other three.

To create the chart in Fig. 1.5, we create four mini-charts as follows:

**Step 1: Tell R How We Want the Charts To Be Laid Out** We use oma (Line 1) so that we can add a title that covers all four charts. The third argument where we put the 3 is for the top margin and tells R to leave us some space on top. There is no magic as to how we arrived at 3. We used trial-and-error until we arrive at a satisfactory output. The mfrow tells R to lay the charts out in a 2×2 format.

```
1  > par(oma = c(0, 0, 3, 0))
2  > par(mfrow = c(2, 2))
```

**Step 2: Plot First Chart**  In this first chart, we want to focus on AMZN, so we plot the AMZN line as blue, while the lines for each of the other three securities are gray lines. We then add a horizontal line using `abline()`. Note that we do not want individual x-axis and y-axis labels (Lines 3–4). Note that we are using a lot of the same variables from the earlier chart we created.

```
1  > plot(x = dt,
2  +      y = norm.prc$AMZN,
3  +      xlab = "",
4  +      ylab = "",
5  +      ylim = y.range,
6  +      type = "l",
7  +      col = "blue",
8  +      main = "AMZN")
9  > lines(x = dt, y = norm.prc$GOOG, col = "gray")
10 > lines(x = dt, y = norm.prc$AAPL, col = "gray")
11 > lines(x = dt, y = norm.prc$SPY, col = "gray")
12 > abline(h = 1)
```

**Step 3: Plot the Other Three Charts**  One approach we can use to create the next three charts is to copy-and-paste the above code three times. We then replace the lines that say AMZN to reference the subject security for that chart. For example, in the second chart, we want to replace references to AMZN with references to GOOG. We do this by replacing code in Lines 3 and 9. We then replace GOOG with AMZN in Line 10.

```
1  > # Chart 2
2  > plot(x = dt,
3  +      y = norm.prc$GOOG,
4  +      xlab = "",
5  +      ylab = "",
6  +      ylim = y.range,
7  +      type = "l",
8  +      col = "blue",
9  +      main = "GOOG")
10 > lines(x = dt, y = norm.prc$AMZN, col = "gray")
11 > lines(x = dt, y = norm.prc$AAPL, col = "gray")
12 > lines(x = dt, y = norm.prc$SPY, col = "gray")
13 > abline(h = 1)
14 >
15 > # Chart 3
16 > plot(x = dt,
17 +      y = norm.prc$AAPL,
18 +      xlab = "",
19 +      ylab = "",
20 +      ylim = y.range,
21 +      type = "l",
22 +      col = "blue",
```

```
23  +        main = "AAPL")
24  > lines(x = dt, y = norm.prc$GOOG, col = "gray")
25  > lines(x = dt, y = norm.prc$AMZN, col = "gray")
26  > lines(x = dt, y = norm.prc$SPY, col = "gray")
27  > abline(h = 1)
28  >
29  > # Chart 4
30  > plot(x = dt,
31  +        y = norm.prc$SPY,
32  +        xlab = "",
33  +        ylab = "",
34  +        ylim = y.range,
35  +        type = "l",
36  +        col = "blue",
37  +        main = "SPY")
38  > lines(x = dt, y = norm.prc$GOOG, col = "gray")
39  > lines(x = dt, y = norm.prc$AAPL, col = "gray")
40  > lines(x = dt, y = norm.prc$AMZN, col = "gray")
41  > abline(h = 1)
```

**Step 4: Create a Global Title**  Each mini-chart above has its own title, which is the ticker symbol of the security in the blue line. However, it would be more informative if we added a global title to the chart. We do this using `title()` and `main`. We should make sure we include `outer = TRUE` after entering the title of the chart. This option makes sure we apply the title to the outer margins, which we have setup above in Step 1.

```
1  > title(main = "Value of $1 Investment in AMZN, GOOG, AAPL and SPY",
2  +    outer = TRUE)
```

We return to the default one chart per page by entering `par(mfrow = c(1, 1))`.

### 1.5.3  Alternative 3—Using `ggplot` to Plot Four Mini-Charts

Figure 1.6 shows an alternative way to create these four mini-charts using `ggplot` package. `ggplot` is fairly popular, so it may be helpful to walk through an example of using that here.

**Step 1: Create a Data Frame Object of Normalized Prices**  We start by creating a data frame from the normalized price *norm.prc* dataset. We want to also create a separate date column in this dataset. I then renamed the rownames, which used to be dates if you recall from above, to a sequence of observation numbers from 1 to 1,259 (Line 3).

```
1  > date <- as.Date(index(norm.prc))
2  > data <- data.frame(date, norm.prc)
3  > rownames(data) <- seq(1, nrow(data), 1)
```

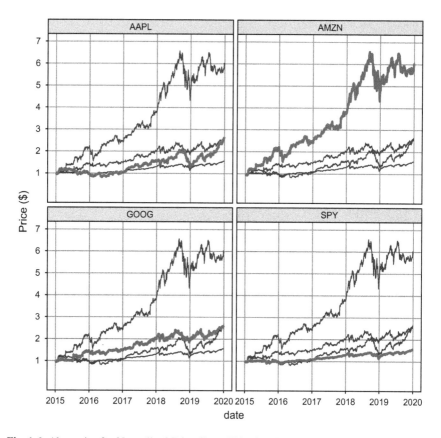

**Fig. 1.6** Alternative 3—Normalized Price Chart. This chart is an alternative presentation of the normalized price chart using `ggplot()`. This alternative to the normalized price chart looks similar to the four mini-charts in Fig. 1.5. Data source: Price data reproduced with permission of CSI ©2020. www.csidata.com

```
4   > head.tail(data)
5          date       AMZN       GOOG       AAPL        SPY
6   1 2014–12–31 1.0000000 1.0000000 1.0000000 1.0000000
7   2 2015–01–02 0.9941034 0.9969795 0.9904875 0.9994648
8   3 2015–01–05 0.9737071 0.9761967 0.9625838 0.9814148
9          date       AMZN       GOOG       AAPL        SPY
10  1257 2019–12–27 6.024811 2.575231 2.625476 1.570789
11  1258 2019–12–30 5.950991 2.545229 2.641058 1.562129
12  1259 2019–12–31 5.954052 2.546905 2.660355 1.565924
```

**Step 2: Use `tidyr()` to Pivot Data**   The dataset we labeled *data* has one column of dates and the normalized prices for each of the securities are in separate columns. We want to pivot the data so that we have three columns: one column each for the date, ticker, and normalized prices. To do this, we use `gather()` from the `tidyr` package. The result of the code in Line 2 to 3 is that the normalized prices are now stacked, so we start with the AMZN normalized prices, then the GOOG normalized prices, followed by the AAPL normalized prices, and we end with the SPY normalized prices.

```
1  > library(tidyr)
2  > pivot <- data %>%
3  +    gather(Ticker, Price, AMZN, GOOG, AAPL, SPY)
4  > head.tail(pivot)
5          date Ticker     Price
6  1 2014-12-31   AMZN 1.0000000
7  2 2015-01-02   AMZN 0.9941034
8  3 2015-01-05   AMZN 0.9737071
9          date Ticker     Price
10 5034 2019-12-27    SPY 1.570789
11 5035 2019-12-30    SPY 1.562129
12 5036 2019-12-31    SPY 1.565924
```

**Step 3: Create a Separate Variable of Ticker Names**  In the plot below, we need to create four different charts by ticker name. To make this work, we need to create a second ticker name, which we call Ticker2.

```
1  > pivot$Ticker2 <- pivot$Ticker
2  > head.tail(pivot)
3          date Ticker     Price Ticker2
4  1 2014-12-31   AMZN 1.0000000    AMZN
5  2 2015-01-02   AMZN 0.9941034    AMZN
6  3 2015-01-05   AMZN 0.9737071    AMZN
7          date Ticker     Price Ticker2
8  5034 2019-12-27    SPY 1.570789     SPY
9  5035 2019-12-30    SPY 1.562129     SPY
10 5036 2019-12-31    SPY 1.565924     SPY
```

**Step 4: Plot the Four Mini-Charts**  We use ggplot() in the ggplot2 package to create these charts, so we load that package. The scales package is needed to customize the dates in the x-axis labels (Lines 14–17). Figure 1.6 shows the output of this code. Lines 5 to 6 in the code below create the four gray lines in each of the four charts to denote the normalized prices of each of the four securities. Line 7 generates the thicker blue line to denote the security that is the focus of each chart. For example, in the AAPL chart, the thicker blue line is the AAPL normalized price. Line 8 removes the gray background of each chart and converts it to a white background. The gridlines are also converted to gray from white. Lines 9 to 11 tell R not to include a legend and what size the text of the title should be. Line 13 tells R what the text of the global chart title should be. Lines 14 to 17 relabel the x-axis from showing only 2016, 2018, and 2020 to show every year from 2015 to 2020. Lines 18 to 21 relabel the y-axis from showing $2, $4, $6 to showing labels from $0 to $7 in $1 increments. The last line generates the four mini-charts.

```
1  > library(ggplot2)
2  > library(scales)
```

```
 3  > pivot %>%
 4  +      ggplot(aes(x = date, y = Price)) +
 5  +      geom_line(data = pivot %>% dplyr::select(-Ticker),
 6  +          aes(group = Ticker2), color = "gray40") +
 7  +      geom_line(aes(color = Ticker), color = "blue", size = 1.4) +
 8  +      theme_bw() +
 9  +      theme(
10  +          legend.position = "none",
11  +          plot.title = element_text(size = 14),
12  +      ) +
13  +      ggtitle("Value of $1 Investment in AMZN, GOOG, AAPL and SPY") +
14  +      scale_x_date(labels = date_format("%Y"),
15  +          breaks = date_breaks("year"),
16  +          minor_breaks = NULL,
17  +          name =) +
18  +      scale_y_continuous(name = "Price ($)",
19  +          breaks = seq(1, 7, 1),
20  +          minor_breaks = NULL,
21  +          limits = c(0, 7)) +
22  +      facet_wrap(~Ticker)
```

## 1.6   Simple and Exponential Moving Averages

In this section, we calculate a **simple moving average** and **exponential moving average** of the AMZN close price in 2019. Both of these are used in various financial applications. For example, in Chap. 12, we show how to use the simple moving average as a technical indicator.

For example, the 20-day simple moving average (SMA) is calculated as

$$SMA_t = \frac{P_t + P_{t-1} + \cdots + P_{t-19}}{20}. \tag{1.1}$$

In comparison, the 20-day exponential moving average (EMA) is calculated as

$$EMA_t = (P_t \times k) + EMA_{t-1} \times (1 - k), \tag{1.2}$$

where $P_t$ is the price today and the smoothing factor $k = 2/(1 + 20) = 0.0952$. The 20 in the smoothing factor $k$ is the number of days in the exponential moving average. The EMA formula also requires the EMA from the day prior, so for the first EMA observation we set the 20-day EMA equal to the 20-day SMA.

Below, we will use the *data.amzn* dataset, which should still be in the R environment from our prior analyses. Otherwise, we can use `load.data()` to import the AMZN data again. In our example, we will calculate 20-day moving averages.

**Step 1: Calculate Moving Averages**  Line 1 calculates the 20-day simple moving average. We use `rollapply()`, which repeatedly applies the `mean()` function (3rd argument) over 20-day windows (2nd argument) on the AMZN close price (1st argument). Line 2 uses `EMA()` to calculate the exponential moving average on the AMZN close price (1st argument). We specify that we want a 20-day exponential moving average in the 2nd argument. As explained above, the first SMA and first EMA on January 29, 2015 are identical at $300.54.

```
1   > ma_20d <- rollapply(data.amzn$AMZN.Close, 20, mean)
2   > ema_20d <- EMA(data.amzn$AMZN.Close, n = 20)
3   > data <- cbind(data.amzn$AMZN.Close, ma_20d, ema_20d)
4   > names(data) <- c("Price", "MA_20", "EMA_20")
5   > data[18:22, ]
6                  Price    MA_20    EMA_20
7   2015-01-27 306.75       NA       NA
8   2015-01-28 303.91       NA       NA
9   2015-01-29 311.78 300.5385 300.5385
10  2015-01-30 354.53 302.7475 305.6805
11  2015-02-02 364.47 305.5450 311.2795
```

We then combine the AMZN close price, simple moving average, and exponential moving average data into one object. Since we are using a 20-day moving average, we expect that the first 19 observations would not have NAs as the value for the moving averages. This is confirmed by our output of observations 18 to 22.

Using the equations above, we can verify the SMA and EMA calculations. First, we check the first two SMA observations for January 29 and 30, 2015. Using `mean()`, we calculate the average price over the last 20 days. We can see below that the calculations match what is reported above.

```
1   > (sma1 <- mean(data$Price[1:20]))
2   [1] 300.5385
3   > (sma2 <- mean(data$Price[2:21]))
4   [1] 302.7475
```

Next, we check the first two EMA observations. As we explained above, the first EMA observation is equal to the SMA. We then use Eq. (1.2) to calculate the EMA for January 30, 2015.

```
1   > (ema1 <- sma1)
2   [1] 300.5385
3   > (ema2 <- data$Price[21] * (2 / 21) + sma1 * (1 - (2 / 21)))
4                  Price
5   2015-01-30 305.6805
```

It is oftentimes good practice to check the underlying calculations. Otherwise, we may not know if there is a glitch in the function or command that we are using. We can stop checking once we are comfortable with the results being generated.

**Step 2: Subset Data to 2019**  We subset the data to only include values for 2019. Here, we use the xts-style date subsetting. As discussed above, we could also use `subset()` to cut the data to only include observations in 2019.

```
1   > data.2019 <- data["2019-01-01/2019-12-31"]
2   > head.tail(data.2019)
3                 Price    MA_20    EMA_20
4   2019-01-02 1539.13 1558.427 1531.029
5   2019-01-03 1500.28 1544.823 1528.100
6   2019-01-04 1575.39 1540.173 1532.604
7                 Price    MA_20    EMA_20
8   2019-12-27 1869.80 1780.365 1791.791
9   2019-12-30 1846.89 1782.669 1797.039
10  2019-12-31 1847.84 1785.981 1801.877
```

**Step 3: Plot Data**  We fist create a date variable *dt*. We also create a variable *y.range* that gives the minimum and maximum values in the dataset. This allows us to chart the different objects in any order and have the y-axis range be sufficient to cover all the different objects. Otherwise, the y-axis of the chart will be around the values of the first object that we plot. In the below, we do explicitly tell R what color the AMZN price line should be, so R will default to a black color for that object. Lines 12 and 13 tell R that the moving average lines should be given different colors. The output of the plot is in Fig. 1.7.

```
1   > dt <- index(data.2019)
2   > (y.range <- range(data.2019))
3   [1] 1500.28 2020.99
4   > plot(x = dt,
5   +      y = data.2019$Price,
6   +      xlab = "Date",
7   +      ylab = "Price",
8   +      ylim = y.range,
9   +      type = "l",
10  +      lwd = 2,
11  +      main = "AMZN Price and 20-Day Moving Averages")
12  > lines(x = dt, y = data.2019$MA_20, col = "blue")
13  > lines(x = dt, y = data.2019$EMA_20, col = "red")
14  > legend("bottomright",
15  +      c("AMZN", "Simple MA", "Exponential MA"),
16  +      lwd = c(2, 2, 1),
17  +      col = c("black", "blue","red"))
```

A note when adding a legend, we can first run Lines 4 to 13 and then take a look at what area of the chart we have space to put the legend. Doing so, we can see that the bottom right portion of the chart has some space, so that is where we position the legend. Also, note that for the legend, we can order the objects in any order we desire (i.e., it does not have to follow the order that the objects are plotted). However, we need to make sure that every argument in the legend (e.g., Lines 15 to 17) should follow our selected order.

**AMZN Price and 20−Day Moving Averages**

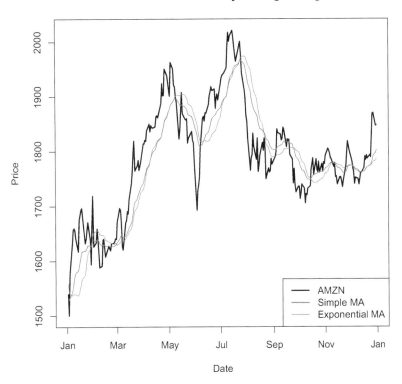

**Fig. 1.7** AMZN price and 20-day moving averages, 2019. We calculate both a 20-day simple moving average (blue line) and a 20-day exponential moving average (red line). Data source: Price data reproduced with permission of CSI ©2020. www.csidata.com

## 1.7 Volume-Weighted Average Price

We can calculate the **volume-weighted average price** (VWAP) over any time period by using the price and volume data. The idea is that, over the period, we calculate the total volume and then calculate each day's volume as a percentage of the total volume for the period. Then, we multiply each day's price by the weighted volume to calculate a daily weighted price. The VWAP is the sum of the daily weighted prices. For example, the 30-day VWAP we first calculate the total volume over the 30-day period. That is,

$$v_{total} = \sum_{t=1}^{30} = v_1 + v_2 + \cdots + v_{30}, \tag{1.3}$$

where $v_t$ is the trading volume on day $t$. Then, we calculate the VWAP as follows:

$$VWAP = \frac{v_1}{v_{total}} P_1 + \frac{v_2}{v_{total}} P_2 + \cdots + \frac{v_{30}}{v_{total}} P_{30}, \tag{1.4}$$

where $P_t$ is the price on day $t$. Below we show an example of how to calculate the 30-day VWAP for AMZN over the last 30 trading days of 2019.

**Step 1: Subset Data to Last 30 Trading Days**  We first identify the observation number of the 30th to the last trading day in the *data.amzn* dataset. Then, we create the *last.30* dataset that is a subset of *data.amzn* that only contains the close price and volume data during the last 30 trading days.

```
1  > (thirty <- nrow(data.amzn) - 30 + 1)
2  [1] 1230
3  > last.30 <- data.amzn[thirty:nrow(data.amzn), 4:5]
4  > head.tail(last.30)
5              AMZN.Close AMZN.Volume
6  2019-11-18   1752.53     2839500
7  2019-11-19   1752.79     2270800
8  2019-11-20   1745.53     2790000
9              AMZN.Close AMZN.Volume
10 2019-12-27   1869.80     6186600
11 2019-12-30   1846.89     3674700
12 2019-12-31   1847.84     2506500
```

**Step 2: Calculate the Weighted Volume**  We first calculate the total volume over the period using sum(), which gives us 92.8 million AMZN shares traded over the last 30 trading days. We then calculate the weighted volume (vol.wgt) by dividing the daily volume (AMZN.Volume) by the total volume (*tot.vol*).

```
1  > (tot.vol <- sum(last.30$AMZN.Volume))
2  [1] 92808300
3  > last.30$vol.wgt <- last.30$AMZN.Volume / tot.vol
4  > head.tail(last.30)
5              AMZN.Close AMZN.Volume     vol.wgt
6  2019-11-18   1752.53     2839500 0.03059532
7  2019-11-19   1752.79     2270800 0.02446764
8  2019-11-20   1745.53     2790000 0.03006197
9              AMZN.Close AMZN.Volume     vol.wgt
10 2019-12-27   1869.80     6186600 0.06665999
11 2019-12-30   1846.89     3674700 0.03959452
12 2019-12-31   1847.84     2506500 0.02700728
```

**Step 3: Calculated the Weighted Price**  We multiply vol.wgt and AMZN.Close on each day to get the daily weighted price (wgt.prc).

```
1  > last.30$wgt.prc <- last.30$AMZN.Close * last.30$vol.wgt
2  > head.tail(last.30)
3              AMZN.Close AMZN.Volume     vol.wgt  wgt.prc
4  2019-11-18   1752.53     2839500 0.03059532 53.61922
5  2019-11-19   1752.79     2270800 0.02446764 42.88663
6  2019-11-20   1745.53     2790000 0.03006197 52.47407
```

```
7            AMZN.Close AMZN.Volume   vol.wgt    wgt.prc
8  2019–12–27   1869.80    6186600 0.06665999 124.64085
9  2019–12–30   1846.89    3674700 0.03959452  73.12672
10 2019–12–31   1847.84    2506500 0.02700728  49.90514
```

**Step 4: Calculate the VWAP** Using sum(), we total the weighted prices in wgt.prc, which gives us a VWAP of $1,787.94. This is the 30-day VWAP for AMZN on the last trading day of 2019.

```
1  > (vwap <- sum(last.30$wgt.prc))
2  [1] 1787.941
```

Note that the code is wrapped around a parenthesis. We use that code in this book to primarily save space as it allows us to give R the instructions and output the results in one line. Without wrapping the code in parenthesis, the output of $1,787.941 will be suppressed and stored in *vwap*. If we wanted to see the value stored in *vwap*, we have to type vwap again. This two-step procedure is simplified if we add the parentheses around the code.

Note that VWAP is a more general term for a calculation that weights the prices by the volume. Thus, a specific VWAP calculation that is required for a particular analysis may be different from the one we show above. For example, VWAP can be calculated over different price frequencies. For example, in some applications, we are not only interested in the last trade price during the day but the average traded price throughout the day. To do so, we can calculate a VWAP using intraday prices and intraday volume. Although we have to use different data for that purposes, the general methodology for arriving at an intraday VWAP calculation will be similar to what we implemented above.

## 1.8   Plotting a Candlestick Chart

We now show how to use monthly prices to plot a **candlestick chart**. A candlestick chart is useful to show the range of prices for a security. We can use chartSeries() in the quantmod package to create a candlestick chart. However, we would first have to convert the data into an open-high-low-close (OHLC) object.

**Step 1: Setup Monthly Data** If we do not have quantmod pre-loaded, we can do so now using library(quantmod). We use to.monthly() to convert the daily data to monthly data. Then, to avoid any confusion with extraneous variables, we dropped the row for December 2014 (Row 1) and the adjusted close variable (Column 6).

```
1  > ohlc <- load.data("AMZN Yahoo.csv", "AMZN")
2  > ohlc <- to.monthly(ohlc)
3  > ohlc <- ohlc[-1, -6]
```

```
4   > class(ohlc)
5   [1] "xts" "zoo"
6   > head.tail(ohlc)
7            ohlc.Open ohlc.High ohlc.Low ohlc.Close ohlc.Volume
8   Jan 2015    312.58    359.50   285.25     354.53   103057100
9   Feb 2015    350.05    389.37   350.01     380.16    70846200
10  Mar 2015    380.85    388.42   365.65     372.10    55502800
11           ohlc.Open ohlc.High ohlc.Low ohlc.Close ohlc.Volume
12  Oct 2019   1746.00   1798.85  1685.06    1776.66    70360500
13  Nov 2019   1788.01   1824.69  1722.71    1800.80    52076200
14  Dec 2019   1804.40   1901.40  1735.00    1847.84    68149600
```

Here, we do not show the intermediate output of Lines 1 and 2, because the data
would be the same as what is shown above when we discussed how to convert daily
prices to monthly prices. We also used class() to show what kind of object *ohlc*
is at this point, which is an xts and zoo object. In some applications, like this one,
knowing the class of the object is important for us to know we can appropriately
apply the command that we want to use. As we show below, we want to convert the
class of the object to a quantmod.OHLC object.

**Step 2: Convert to an OHLC Object** We convert the data to an OHLC object
using the as.quantmod.OHLC(). We have to tell R what the column names
our data has. The output itself does not look different from the output in Step 1.
However, the object class has changed from xts to quantmod.OHLC (compare Line
5 in Step 1 with Line 4 in this step).

```
1   > amzn.ohlc <- as.quantmod.OHLC(ohlc,
2   +      col.names = c("Open", "High", "Low", "Close", "Volume"))
3   > class(amzn.ohlc)
4   [1] "quantmod.OHLC" "zoo"
5   > head.tail(amzn.ohlc)
6            ohlc.Open ohlc.High ohlc.Low ohlc.Close ohlc.Volume
7   Jan 2015    312.58    359.50   285.25     354.53   103057100
8   Feb 2015    350.05    389.37   350.01     380.16    70846200
9   Mar 2015    380.85    388.42   365.65     372.10    55502800
10           ohlc.Open ohlc.High ohlc.Low ohlc.Close ohlc.Volume
11  Oct 2019   1746.00   1798.85  1685.06    1776.66    70360500
12  Nov 2019   1788.01   1824.69  1722.71    1800.80    52076200
13  Dec 2019   1804.40   1901.40  1735.00    1847.84    68149600
```

**Step 3: Plot OHLC Data** We now plot the OHLC data using chartSeries().
The default theme for the plot is a black background with orange and green bars.
That theme looks better on a monitor, but for printing we may want to change
the background color. Below, we want to change the background to white, so we
changed the theme to white. This generates a chart with a white background
but maintaining the orange and green bars. Finally, we add a title using name. The
candlestick chart is shown in Fig. 1.8.

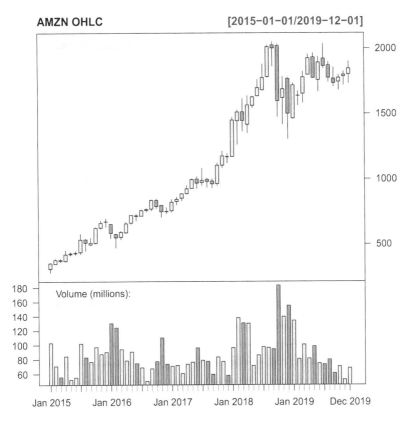

**Fig. 1.8** AMZN candlestick chart using monthly prices, 2015–2019. The background is changed from the default black background to a white background. Data source: Price data reproduced with permission of CSI ©2020. www.csidata.com

```
1   > chartSeries(amzn.ohlc,
2   +     theme = "white",
3   +     name = "AMZN OHLC")
```

**Advantages and Disadvantages of Having a Large, Active Community**

R has a very large and active community that has contributed over 16,000 packages as of June 2020, as documented on the official repository called the Comprehensive R Archive Network (CRAN). We can find all the available package on CRAN at https://cran.r-project.org/web/packages/. The large number of packages suggests that there is a decent chance that someone has already developed a package that performs the task we would like

(continued)

to undertake. Therefore, we may be able to leverage on other community members' work and save us time.

The downside of the above is that there is no standardization and quality control that covers all 16,000 packages. For example, in the above application of `chartSeries()`, we controlled the chart title by using `name`. However, when we used the standard `plot()`, we controlled the chart title using `main`. We discussed earlier using `getSymbols()` somehow does not download the last day of a date range. There was a time several years back that Yahoo Finance changed web addresses and the format of the page, so an update to the code in `getSymbols()` was necessary to get it working again.

In addition, different packages have different authors. Hence, we should expect a varying degree of styles, so we cannot expect that the structure and presentation are going to be the same.

Finally, the quality of the different packages can vary greatly. There may be some testing involved when packages are developed, but it is sometimes very difficult to find bugs especially if they arise in more obscure circumstances. If the package is popular, there is a higher likelihood that users report bugs to the authors and the authors in turn fix those bugs. This reduces the number of bugs that remains unidentified or unresolved. Less popular packages may be more susceptible to such issues. Ultimately, just like anything we get from a third party, we want to get a sense of comfort that things work reasonably well before we rely on them.

In some instances we check the functions we use in particular packages. Other times, we may also gain confidence in the reliability of the packages by looking at how many times the packages have been downloaded. For example, the `quantmod` package we used often in this chapter has been downloaded 6.4 million times from January 2015 to August 2020. As for the rest of the packages we use in this book, the downloads over that same time period range from 799,566 for the `neuralnet` package to 37.8 million for the `ggplot2` package.

## 1.9   2-Axis Price and Volume Chart

Sometimes we want to plot two different charts together. For example, if we would like to know when are the days with large stock price movements and whether those correspond with days of large volume spikes, it could be helpful to plot the price and volume of a stock in one chart. Because of the different scales of the price and volume, we cannot only use one axis. Thus, we would have to create a chart with two vertical axes but with the same horizontal axis. We show below how to implement this plot for AMZN data.

**Step 1: Obtain AMZN Price and Volume Data** We previously created the *data.amzn* dataset, which contains among other things the AMZN close price and volume data. We create a new dataset called *prc.vol* that only keeps the AMZN price and volume data from January 2015 to December 2019. This means that we delete the first observation (December 31, 2014) and keep only the fourth and fifth columns.

```
1   > prc.vol <- data.amzn[-1, c(4, 5)]
2   > head.tail(prc.vol)
3              AMZN.Close AMZN.Volume
4   2015-01-02     308.52     2783200
5   2015-01-05     302.19     2774200
6   2015-01-06     295.29     3519000
7              AMZN.Close AMZN.Volume
8   2019-12-27    1869.80     6186600
9   2019-12-30    1846.89     3674700
10  2019-12-31    1847.84     2506500
```

**Step 2: Create a Data Frame Object with Data, Price, and Volume Data** The dataset *prc.vol* is an xts object (as it is a subset of *data.amzn*), but in this application it is easier to work with this data as a data frame. As such, we convert *prc.vol* from an xts to a data frame object using `data.frame()`. Before that, we have to first create a *date* variable (Line 1) that we will then combine with the price and volume data (Line 2). Instead of combining using `c()`, we can use `data.frame()` instead. We then rename the column headers in Line 3. In Line 4, we convert the volume data into millions so that the y-axis label looks better as R may convert it to scientific notation otherwise. We also changed the row names to observation numbers because it does not make sense to have two columns of dates.

```
1   > date <- as.Date(index(prc.vol))
2   > prc.vol <- data.frame(date, prc.vol)
3   > names(prc.vol) <- c("date", "price", "volume")
4   > prc.vol$volume <- prc.vol$volume/1000000
5   > rownames(prc.vol) <- seq(1, nrow(prc.vol), 1)
6   > head.tail(prc.vol)
7          date  price volume
8   1 2015-01-02 308.52 2.7832
9   2 2015-01-05 302.19 2.7742
10  3 2015-01-06 295.29 3.5190
11          date  price volume
12  1256 2019-12-27 1869.80 6.1866
13  1257 2019-12-30 1846.89 3.6747
14  1258 2019-12-31 1847.84 2.5065
15  > range(prc.vol$price)
16  [1]   286.95 2039.51
17  > range(prc.vol$volume)
18  [1]   0.8813 23.8561
```

The above output also shows the range of prices and volumes (Lines 15 to 18). We use this information later when we generate the y-axis ranges for the price and volume data.

**Step 3: Plot the Volume Chart First** We have to choose which of the two charts we want to plot first. We are going to plot the volume as bars and price as a line. The second chart overlays on the first chart. Thus, it makes more sense for the first chart to be the volume chart, so we can overlay the price line on top of those bars. Before plotting the data, we first adjust the margins in order to make sure there is enough space around the chart to allow us to add a title and two y-axes. Thus, we use `mar = c()` to do this. The numbers denote the amount of space for the bottom, left, top, and right, respectively (Line 1). After we plot the chart, we can play around with the numbers in this line to get the chart to look the way we want it. In other words, there is nothing magical about the numbers I use in Line 1. Then, to plot the bar chart, we use `with()` first and then use the `barplot()`. We use the color red for the volume data.

```
1   > par(mar = c(5, 5, 3, 5))
2   > with(data,
3   +       barplot(prc.vol$volume,
4   +             ylim = c(0, 25),
5   +             border = NA,
6   +             ylab = "Volume (mln)",
7   +             xlab = "",
8   +             col = "red",
9   +             main = "AMZN Price and Volume \n 2015 − 2019"))
```

**Step 4: Plot the Price Chart Next** Before we add the price line, we let R know we are now adding a new chart (Line 1). Then, we use `with()` again when we use `plot()`. This allows us to add the line chart with a blue line for the price (Line 10). A key here is to not add y-axis labels (Line 11). This only prevents the y-axis labels from printing and overlapping with the y-axis labels of the bar chart. Since we do not suppress the x-axis label, we now have the date as the x-axis label. Since we do not have a y-axis label from the `plot()`, we have to manually tell R to show a y-axis label for the price. We do this in the last two lines of the code below.

```
1   > par(new = TRUE)
2   > with(data,
3   +     plot(x = prc.vol$date,
4   +           y = prc.vol$price,
5   +           ylim = c(0, 2500),
6   +           xlab = "",
7   +           ylab = "",
8   +           type = "l",
9   +           lwd = 2,
10  +           col = "blue",
11  +           yaxt = "n"))
12  > axis(side = 4)
13  > mtext(side = 4, line = 3, "Price")
```

Figure 1.9 shows the price line is superimposed on the volume bars. We can see when the volume spikes and whether there are corresponding changes in AMZN stock price and vice versa.

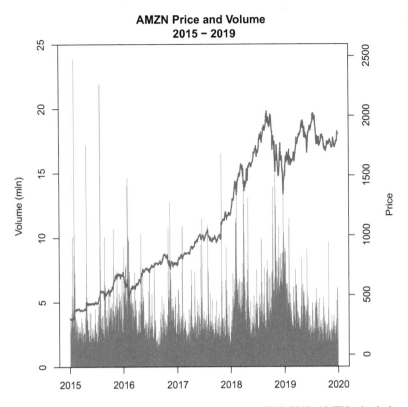

**Fig. 1.9** AMZN price and volume data in a two y-axis chart, 2015–2019. AMZN price is denoted by the blue line and AMZN trading volume is denoted by the red bars. The first y-axis on the left side reflects information for the volume, while the second y-axis on the right side reflects information for the price. Data source: Price data reproduced with permission of CSI ©2020. www. csidata.com

## 1.10  Further Reading

For more information about R commands, additional references are [4, Zuur and Meesters (2009)], [3, Teetor (2011)], and [1, Albert and Rizzo (2012)]. As we get more familiar with writing R code, an excellent text that shows how to write more efficient R code may be of interest to you, such as [2, Gillespie and Lovelace (2017)].

# References

1. Albert, J., & Rizzo, M. (2012). *R by example*. Springer.
2. Gillespie, C., & Lovelace, R. (2017). *Efficient R programming: a practical guide to smarter programming*. O'Reilly Media.
3. Teetor, P. (2011). *R cookbooks*. O'Reilly Cookbooks.
4. Zuur, A., Ieno, E., & Meesters, E. (2009). *A beginner's guide to R*. Springer.

# Chapter 2
# Individual Security Returns

In this chapter, we focus on the percentage changes in the price of a security. This is called the security's **return**. From an investments perspective, the return of a security is sometimes a more important measure than the dollar change in the price of our investment. To see why, suppose someone told you they made $500 in 1 year from investing in some stock. That information alone does not tell us whether $500 is a good return on their investment because we would need to know how much the person invested to make the $500. If the initial investment was $1000, making $500 is exceptional as that equals to a 50% return in 1 year. However, if the initial investment was $100,000, making $500 is only equal to a 0.5% return. To put this in context, the 1-year U.S. Treasury at the end of 2019 was approximately 1.5%. If a risk-free investment yields 1.5%, a risky investment making 0.5% would not be impressive.

When we describe returns as the percentage change in price, we are actually not describing the whole picture. Many securities provide intermediate cash flows, such as dividends for stocks and coupons for bonds. These intermediate cash flows could be reinvested back into the security or in some other investment. In practice, for some securities, some brokers (e.g., Schwab) allow the investor to click a box if they want dividends reinvested. Thus, reinvestment of intermediate cash flows is not merely a theoretical construct. The returns assuming the reinvestment of intermediate cash flows is called **total returns**. Because total returns are a measure of the return we receive over the entire time we hold the security, this type of return is also known as the **holding period return**. When necessary to make the distinction, we will call the return without intermediate cash flow reinvestment **price return**.

We begin this chapter by showing how to calculate price returns and total returns in Sects. 2.1 and 2.2. Then, for statistical and programming considerations, we show how to calculate logarithmic total returns in Sect. 2.3. We then show in Sect. 2.4 two ways of potentially dealing with outliers in the data: winsorization and truncation. In Sect. 2.5, we look at returns over longer time horizons and show how to cumulate daily returns into multi-day returns using both the arithmetic returns and logarithmic

C.S. Ang, *Analyzing Financial Data and Implementing Financial Models Using R*, Springer Texts in Business and Economics, https://doi.org/10.1007/978-3-030-64155-9_2

returns. Extending our discussion of weekly and monthly prices to returns, we show how to calculate **weekly returns** in Sect. 2.6 and **monthly returns** in Sect. 2.7. One common application of weekly or monthly returns is in the calculation of betas used in the cost of equity calculation, which we will demonstrate in Chap. 5. Finally, we show how to compare the performance of multiples securities using total returns in Sect. 2.8. This is similar to the normalized price charts we used earlier, but instead of using close prices we now use the adjusted close prices from Yahoo Finance. The adjusted close prices incorporate dividends, so using the changes in adjusted close prices yields a total return.

## 2.1  Price Returns

The dollar change based solely on the closing price of a security is called **capital gains** and the percentage price in the closing price of a security is its **price return**. The price return is measured over some investment horizon (e.g., 1 day, 1 month, 1 year, etc.). The length of the period depends on the application, but the return calculations should be consistent such that cumulating all the one-day returns in a given year should equal to the annual return.

The daily price return is the percentage change in the price of a security today relative to its price yesterday. That is,

$$R_t^{price} = \frac{P_t - P_{t-1}}{P_{t-1}} = \frac{P_t}{P_{t-1}} - 1, \tag{2.1}$$

where $R_t^{price}$ is the price return on day $t$, $P_t$ is the price of the security on day $t$, and $P_{t-1}$ is the price of the security the trading day before.

In this section, we will use AAPL as the example because AAPL pays dividends so we can compare the results for price returns and total returns. We now show how to calculate the daily price returns for AAPL from 2015 to 2019.

**Step 1: Import AAPL Close Price Data**  Use the load.data() function to import AAPL Yahoo.csv. Be sure that we run the pre-loaded code in Appendix B. Create a dataset *rets* that only contains the AAPL.Close.

```
 1   > data.aapl <- load.data("AAPL Yahoo.csv", "AAPL")
 2   > rets <- data.aapl$AAPL.Close
 3   > head.tail(rets)
 4                     AAPL.Close
 5   2014-12-31          110.38
 6   2015-01-02          109.33
 7   2015-01-05          106.25
 8                     AAPL.Close
 9   2019-12-27          289.80
10   2019-12-30          291.52
11   2019-12-31          293.65
```

**Step 2: Create a Variable Equal to Lag of Closing Price** For esthetic reasons, we rename AAPL.Close to price. We then use lag() with k = 1 to create a variable equal to the prior trading day's close price.

```
1  > names(rets) <- "price"
2  > rets$lag.price <- lag(rets$price, k = 1)
3  > head.tail(rets)
4              price lag.price
5  2014-12-31 110.38        NA
6  2015-01-02 109.33    110.38
7  2015-01-05 106.25    109.33
8              price lag.price
9  2019-12-27 289.80    289.91
10 2019-12-30 291.52    289.80
11 2019-12-31 293.65    291.52
```

**Step 3: Calculate Price Return** Using Eq. (2.1), we can calculate AAPL's price return. Specifically, we divide price by the lag.price and subtract 1.

```
1  > rets$price.ret <- rets$price / rets$lag.price - 1
2  > head.tail(rets)
3              price lag.price    price.ret
4  2014-12-31 110.38        NA           NA
5  2015-01-02 109.33    110.38 -0.009512548
6  2015-01-05 106.25    109.33 -0.028171608
7              price lag.price    price.ret
8  2019-12-27 289.80    289.91 -0.0003794833
9  2019-12-30 291.52    289.80  0.0059351314
10 2019-12-31 293.65    291.52  0.0073065487
```

Alternatively, we can also calculate the return using Delt(), which is in the quantmod package. As we can see below, the output from Delt() is identical to the price.ret we calculated above. Given the convenience of using Delt(), we will use Delt() going forward.

```
1  > rets2 <- Delt(rets$price)
2  > head.tail(rets2)
3             Delt.1.arithmetic
4  2014-12-31                NA
5  2015-01-02      -0.009512548
6  2015-01-05      -0.028171608
7             Delt.1.arithmetic
8  2019-12-27     -0.0003794833
9  2019-12-30      0.0059351314
10 2019-12-31      0.0073065487
```

**Step 4: Clean Up Returns Data** The output above shows that December 31, 2014 has a return of NA. We needed the December 31, 2014 price to calculate the return on January 2, 2014. Since we no longer need the December 31, 2014 price, we can delete it. To do this, inside the square brackets in Line 1, we enter −1 to the left of the comma (i.e., denoting deleting the first row). Next, we only want to keep the price.ret column in the *rets* dataset. To do this, inside the square brackets in Line 1, we enter 3 to the right of the comma (i.e., denoting the third column) .

```
 1  > rets <- rets[-1, 3]
 2  > options(scipen = 999)
 3  > head.tail(rets)
 4                     price.ret
 5  2015-01-02 -0.00951254782
 6  2015-01-05 -0.02817160838
 7  2015-01-06  0.00009413647
 8                     price.ret
 9  2019-12-27 -0.0003794833
10  2019-12-30  0.0059351314
11  2019-12-31  0.0073065487
12  > options(scipen = 0)
```

In Line 2, we used `scipen = 999` to increase the threshold before R converts the output into scientific notation. This makes the results easier to read. In the last line, we return this option back to its default value of `scipen = 0`.

---

### A Note on Stock Splits and Closing Prices from Yahoo Finance

The problem with using real world data is that it changes. In the first edition of this book, the close price on Yahoo Finance was not split-adjusted. However, as of the writing of the second edition, the close price on Yahoo Finance is now adjusted for splits. So a prior issue of using close prices from Yahoo Finance to calculate returns is no longer an issue. However, having split-adjusted close prices can bring up other issues. For example, if we wanted to calculate the market capitalization of the company, the shares outstanding that we should use should be consistent with the price we are using. Otherwise, we would calculate a completely erroneous market capitalization.

Let us take the example of Netflix (NFLX) 7-for-1 stock split on July 15, 2015. Yahoo Finance currently reports a (split-adjusted) close price of $100.37 for July 14, 2015, the day before the stock split. It was over $700 pre-split. To confuse things more, the NFLX 10-Q for the period ended June 30, 2015 (i.e., post-split) reports that NFLX has 426 million shares outstanding as of June 30, 2015. However, digging a little deeper, we know that NFLX had shares outstanding of approximately 60 million pre-split. So, NFLX market capitalization calculated using pre-split data is equal to $700 pre-split per share price multiplied by the 60 million pre-split shares outstanding. On the other hand, using post-split data, NFLX market capitalization is equal to $100 post-split price with the approximately 420 million post-split shares outstanding. Mixing and matching pre- and post-split numbers can lead to dramatically different and incorrect results.

## 2.2 Total Returns

The return to investors from holding shares of stock is not limited to changes in the price of the security. For companies that pay dividends, the shareholders holding shares prior to the ex-dividend date receive cash payments that they are able to reinvest. Therefore, the total return a shareholder can receive from a stock that pays dividends includes both the change in the price of the shares she owns as well as any income generated from the dividends and the reinvestment of those dividends on the ex-date. Returns that include dividend reinvestment are known as **holding period returns** or **total returns**.

The total return from holding a security is calculated as follows:

$$R_t = \frac{P_t + CF_t + P_{t-1}}{P_{t-1}} = \underbrace{\left[\frac{P_t}{P_{t-1}} - 1\right]}_{\text{Capital Appreciation}} + \underbrace{\frac{CF_t}{P_{t-1}}}_{\text{CF Yield}}, \tag{2.2}$$

where $CF_t$ is the cash flow (e.g., dividend) payment on day $t$ and all other terms are defined the same way as Eq. (2.1). The first term represents the capital appreciation, while the second term represents the dividend yield. The decomposition in Eq. (2.2) can also help identify the source of the return, such that investors who prefer capital gains or dividends (e.g., for tax purposes) can make a better assessment of the attractiveness of the investment for their particular investment objective. Since we will commonly use total returns, we do not add a superscript to label the total return calculation.

For a daily total return calculation, the dividend yield is zero on non-ex dividend dates. This has two implications. First, on most days, the price return and the total return are the same because we only have the changes in the capital appreciation on those dates. Second, for a non-dividend paying stock, the price return and total return are the same.

Continuing from our above example, we will use AAPL. AAPL ex-dividend dates and amount of dividends per share during 2015–2019 are as in Table 2.1.

On the ex-dividend dates, we would expect the price return and total return to deviate. The more ex-dividend dates there are during our investment horizon, the larger the deviation between cumulative price returns and cumulative total returns will be. Fortunately, Yahoo Finance already reports a variable that incorporates an adjustment for dividends. This variable is labeled the adjusted close price, which is the security's close price adjusted for both stock splits and dividends.

To see the difference in the close price and adjusted close price, we can take a look at AAPL data around November 7, 2019, the date of the stock's most last ex-dividend date in 2019. As we can see below, AAPL close price (Column 4) and adjusted close price (Column 6) are the same starting on November 7, 2019. But, prior to that date (i.e., November 6, 2019 and earlier), the close price is higher than the adjusted close price. This implies that the total return calculated using the

**Table 2.1** AAPL ex-dividend dates and dividend per share, 2015–2019

| 2015 | | 2016 | | 2017 | |
|---|---|---|---|---|---|
| Ex-date | Div/share | Ex-date | Div/share | Ex-date | Div/share |
| 02/07 | $0.47 | 02/04 | $0.52 | 02/09 | $0.57 |
| 05/07 | $0.52 | 05/05 | $0.57 | 05/11 | $0.63 |
| 08/06 | $0.52 | 08/04 | $0.57 | 08/10 | $0.63 |
| 11/05 | $0.52 | 11/03 | $0.57 | 11/10 | $0.63 |

| 2018 | | 2019 | |
|---|---|---|---|
| Ex-date | Div/share | Ex-date | Div/share |
| 02/09 | $0.63 | 02/08 | $0.73 |
| 05/11 | $0.73 | 05/10 | $0.77 |
| 08/10 | $0.73 | 08/09 | $0.77 |
| 11/08 | $0.73 | 11/07 | $0.77 |

adjusted close price must be higher than the price return calculated using the close price, which is what we expect.

```
1  > data.aapl["2019-11-04/2019-11-12"]
2              AAPL.Open AAPL.High AAPL.Low AAPL.Close AAPL.Volume AAPL.Adjusted
3  2019-11-04    257.33    257.85   255.38     257.50    25818000      256.7292
4  2019-11-05    257.05    258.19   256.32     257.13    19974400      256.3604
5  2019-11-06    256.77    257.49   255.37     257.24    18966100      256.4700
6  2019-11-07    258.74    260.35   258.11     259.43    23735100      259.4300
7  2019-11-08    258.69    260.44   256.85     260.14    17496600      260.1400
8  2019-11-11    258.30    262.47   258.28     262.20    20455300      262.2000
9  2019-11-12    261.55    262.79   260.92     261.96    21847200      261.9600
```

The ex-dividend date is important because it is the date when the stock trades without a dividend. In other words, anyone who purchased shares on November 7, 2019 would not be entitled to the dividend. Because the value of the assets on November 7, 2019 declines by the amount of the dividend paid (i.e., cash of a dollar has a market value of a dollar), we should observe a corresponding drop in the equity of AAPL of the same amount. All else equal, this reduction in equity value of AAPL will translate into a drop in the AAPL stock price.

We now continue with our AAPL example from the previous section. This time we calculate AAPL's total return from 2015 to 2019.

**Step 1: Import AAPL Adjusted Close Price Data**  Since we already imported the AAPL data earlier, we can simply create a dataset called *tot.rets* by obtaining the AAPL adjusted close price data from the *data.amzn* dataset.

```
1  > tot.rets <- data.aapl$AAPL.Adjusted
2  > head.tail(tot.rets)
3              AAPL.Adjusted
4  2014-12-31      101.41906
5  2015-01-02      100.45430
6  2015-01-05       97.62434
```

```
 7                    AAPL.Adjusted
 8   2019–12–27            289.80
 9   2019–12–30            291.52
10   2019–12–31            293.65
```

**Step 2: Create a Variable Equal to Lag of Adjusted Close Price**  To calculate the total return, we need to divide the adjusted close price today with the adjusted close price the prior trading day. Thus, using `lag()` with k = 1, we create a variable equal to the prior trading day's adjusted close price.

```
 1  > names(tot.rets) <- "adj.price"
 2  > tot.rets$lag.adjprc <- lag(tot.rets$adj.price, k = 1)
 3  > head.tail(tot.rets)
 4              adj.price lag.adjprc
 5  2014–12–31 101.41906         NA
 6  2015–01–02 100.45430   101.4191
 7  2015–01–05  97.62434   100.4543
 8              adj.price lag.adjprc
 9  2019–12–27    289.80     289.91
10  2019–12–30    291.52     289.80
11  2019–12–31    293.65     291.52
```

**Step 3: Calculate Total Return**  We calculate the total return by dividing the adjusted close price today by the adjusted close price the prior trading day. To calculate the net total return, we have to subtract 1 from the gross return.

```
 1  > tot.rets$tot.ret <- tot.rets$adj.price / tot.rets$lag.adjprc − 1
 2  > head.tail(tot.rets)
 3              adj.price lag.adjprc      tot.ret
 4  2014–12–31 101.41906         NA           NA
 5  2015–01–02 100.45430   101.4191 −0.00951261
 6  2015–01–05  97.62434   100.4543 −0.02817166
 7              adj.price lag.adjprc      tot.ret
 8  2019–12–27    289.80     289.91 −0.0003794833
 9  2019–12–30    291.52     289.80  0.0059351314
10  2019–12–31    293.65     291.52  0.0073065487
```

**Step 4: Clean Up Total Returns Data**  As we notice above, the total return for December 31, 2014 is NA. We can delete this row because we only needed the adjusted close price on this date to calculate the total return on January 2, 2015. Thus, inside the square brackets in Line 1, we enter −1 to the left of the comma. Next, since we are only interested in the total return variable, we want to only keep Column 3 of the *tot.rets* dataset. So, inside the square brackets in Line 1, we enter 3 to the right of the comma.

```
 1  > tot.rets <- tot.rets[−1, 3]
 2  > head.tail(tot.rets)
 3                   tot.ret
 4  2015–01–02 −9.512610e−03
 5  2015–01–05 −2.817166e−02
 6  2015–01–06  9.433099e−05
 7                   tot.ret
```

```
8   2019–12–27 –0.0003794833
9   2019–12–30  0.0059351314
10  2019–12–31  0.0073065487
```

We can also apply `Delt()` to **AAPL.Adjusted** to calculate the total returns.

## 2.3  Logarithmic Total Returns

The returns calculated in the preceding section using Eq. (2.2) or `Delt()` are called **arithmetic returns** or **simple returns**. In this section, we show how to calculate **logarithmic returns** or **log returns**. Log returns are used extensively in derivatives pricing, among other areas of finance. In addition, when we calculate multi-period returns later in this chapter, we show that calculating cumulative returns is relatively easier using logarithmic returns.

The logarithmic return, $r_t$, is calculated as

$$r_t = \ln\left(\frac{P_t}{P_{t-1}}\right) = \ln(1 + R_t) = \ln P_t - \ln P_{t-1}, \tag{2.3}$$

where ln is the natural logarithm operator and the rest of the variables are defined the same way as in Eq. (2.2). Therefore, we can take the difference of the log prices to calculate log returns. Also, since we calculate log returns using adjusted close prices, we will not add a subscript to the log return $r$ variable.

We now calculate the log total return for AAPL stock from January 2015 to December 2019. To apply Eq. (2.3), we apply `log()` to the adjusted close prices and then use `diff()` to take the difference.

```
1   > log.rets <- diff(log(data.aapl$AAPL.Adjusted))
2   > log.rets <- log.rets[-1, ]
3   > names(log.rets) <- "log.ret"
4   > options(scipen = 999)
5   > head.tail(log.rets)
6                      log.ret
7   2015–01–02 –0.00955814422
8   2015–01–05 –0.02857609111
9   2015–01–06  0.00009432654
10                     log.ret
11  2019–12–27 –0.0003795553
12  2019–12–30  0.0059175879
13  2019–12–31  0.0072799852
14  > options(scipen = 0)
```

Alternatively, we can also use `ROC()`, which also calculates the log returns except that the steps are less intuitive. However, as the output below shows, we get exactly the same results.

```
1   > roc.rets <- ROC(data.aapl$AAPL.Adjusted)
```

```
2   > roc.rets <- roc.rets[-1, ]
3   > names(roc.rets) <- "roc.ret"
4   > options(scipen = 999)
5   > head.tail(roc.rets)
6                          roc.ret
7   2015-01-02 -0.00955814422
8   2015-01-05 -0.02857609111
9   2015-01-06  0.00009432654
10                         roc.ret
11  2019-12-27 -0.0003795553
12  2019-12-30  0.0059175879
13  2019-12-31  0.0072799852
14  > options(scipen = 0)
```

We can now compare the arithmetic returns with log returns. As the output below shows, the differences are generally quite small on a daily basis.

```
1   > options(scipen = 999)
2   > head.tail(cbind(tot.rets$tot.ret, log.rets$log.ret))
3                      tot.ret          log.ret
4   2015-01-02 -0.00951261035 -0.00955814422
5   2015-01-05 -0.02817165617 -0.02857609111
6   2015-01-06  0.00009433099  0.00009432654
7                      tot.ret          log.ret
8   2019-12-27 -0.0003794833 -0.0003795553
9   2019-12-30  0.0059351314  0.0059175879
10  2019-12-31  0.0073065487  0.0072799852
11  > options(scipen = 0)
```

## 2.4  Winsorization and Truncation

Financial data are sometimes subject to extreme values or outliers. If we determine that these outliers are more noise than information, keeping the outliers in the data can unduly influence the outcome of the analysis and/or create problems in the interpretation of the results. There are some techniques that are used to deal with outliers, but two common approaches used in the empirical finance literature are to use **winsorization** or **truncation**.

Winsorization replaces the values greater than the $i$-th percentile and less than the $(1 - i$-th) percentile to the values at those levels. For example, if we want to winsorize at the 0.5% level, we can use quantile() to determine what is the cut-off value and then set values higher than the upper cut-off (i.e., 99.5%) to the value of the upper cut-off and set values lower than the lower cut-off (i.e., 0.5%) to the value of the lower cut-off. We show how to implement this using the tot.ret in the *tot.rets* dataset above.

**Step 1: Calculate Upper and Lower Cut-Offs**  We use quantile() to determine the upper and lower cut-off values. We then use as.numeric() to make the results numbers. As the below shows, the upper cut-off is 5.6% and the lower cut-off is −5.2%.

```
1  > (upper <- as.numeric(quantile(tot.rets, 0.995)))
2  [1] 0.05557369
3  > (lower <- as.numeric(quantile(tot.rets, 0.005)))
4  [1] −0.05225957
```

**Step 2: Winsorize the Data** We use a nested ifelse() to do this. The first ifelse() sets any total return equal to or below the lower cut-off equal to the lower cut-off total return of −5.2%. The second ifelse() sets any total return equal to or above the upper cut-off equal to the upper cut-off total return of 5.6%. If the value is in between the lower and upper cut-offs, the original value is retained. As we can see when we apply summary() to winsorize, the min and max values are equal to the lower and upper cut-off values. We can compare this to the output of summary() on the tot.ret, which has a minimum value of −0.100 and maximum value of 0.070.

```
1  > winsorize <- ifelse(tot.rets <= lower, lower,
2  +    ifelse(tot.rets >= upper, upper, tot.rets))
3  > summary(winsorize)
4        Index                    tot.ret
5   Min.    :2015−01−02   Min.    :−0.0522596
6   1st Qu.:2016−04−04    1st Qu.:−0.0058585
7   Median :2017−07−01    Median : 0.0008921
8   Mean    :2017−07−01   Mean    : 0.0010090
9   3rd Qu.:2018−09−30    3rd Qu.: 0.0089108
10  Max.    :2019−12−31   Max.    : 0.0555737
11 > summary(tot.rets$tot.ret)
12        Index                    tot.ret
13  Min.    :2015−01−02   Min.    :−0.0996073
14  1st Qu.:2016−04−04    1st Qu.:−0.0058585
15  Median :2017−07−01    Median : 0.0008921
16  Mean    :2017−07−01   Mean    : 0.0009680
17  3rd Qu.:2018−09−30    3rd Qu.: 0.0089108
18  Max.    :2019−12−31   Max.    : 0.0704215
```

An alternative to winsorization is truncation. Truncation deletes values below the lower cut-off and values above the upper cut-off. Truncation is a more systematic way of deleting observations rather than eyeballing the data and deleting observations that look like outliers.

**Step 3: Truncate Data** Using subset(), we can keep any value of tot.ret that is between the upper and lower cut-offs.

```
1  > truncate <- subset(tot.rets,
2  +    tot.rets <= upper &
3  +    tot.rets >= lower)
4  > summary(truncate)
5        Index                    tot.ret
6   Min.    :2015−01−02   Min.    :−0.0520380
7   1st Qu.:2016−04−04    1st Qu.:−0.0057517
```

```
 8   Median :2017-07-01   Median : 0.0008921
 9   Mean   :2017-07-01   Mean   : 0.0010017
10   3rd Qu.:2018-09-26   3rd Qu.: 0.0088604
11   Max.   :2019-12-31   Max.   : 0.0531672
```

There is no rule that says when we should use winsorization over truncation or vice versa. The choice depends on the analysis we would like to do. Winsorization is sometimes preferred because we still maintain the number of observations, but we are saying that we do not believe the values should be that high or that low. However, if the results change materially by the choice of winsorization or truncation, then we should make sure that we understand the source of the difference and that we are comfortable that the choice we make does not affect the reliability of our results.

## 2.5  Cumulating Multi-Day Returns

When evaluating investments, we are typically concerned with how our investment has performed over a particular time horizon. Put differently, we are interested in cumulative **multi-day returns**. We could be interested in knowing the returns of our investment over the past week or over the past month or over the past year. To fully capture the effects of being able to reinvest dividends, we should calculate daily returns and string those returns together for longer periods. Otherwise, if we simply apply Eq. (2.2) using prices at the beginning and end of the investment horizon and add the dividend, we are assuming that the dividends are received at the end of the period and no additional returns on those dividends are earned. If the assumption is that dividends were not reinvested and were kept under the mattress, then this calculation may be appropriate. However, we would argue that the more plausible alternative is for us to reinvest the dividend in a security that is of similar risk-return profile as our initial investment, and a security that satisfies that is the same security that generated those dividends. In other words, when we receive the dividend, we would reinvest that amount back into the stock. The returns of that stock going forward determine whether the reinvested dividend earned a positive or negative return.

Let us now walk through an example of cumulating multi-day returns. Suppose we are interested in knowing how much would an investment in AMZN have made through the end of 2019 if we purchased AMZN shares at the close price on December 31, 2014. We show how to implement this calculation using arithmetic returns and logarithmic returns, and show that, when consistently implemented, both types of returns yield the same result. However, some may find the programming for logarithmic returns slightly easier.

### 2.5.1  Cumulating Arithmetic Returns

To string together multiple days of arithmetic returns, we have to take the product of the daily gross returns. If $R_t$ is the **net return** on day $t$, then the **gross return** on day $t$ is equal to $1 + R_t$. In other words, for a 2-day cumulative return, we have $R_{1,2} = (1 + R_1) * (1 + R_2)$. For a 3-day cumulative return, we take $R_{1,3} = (1 + R_1) * (1 + R_2) * (1 + R_3)$. Therefore, we can generalize this calculation over a $T$-day investment horizon as

$$R_{1,T} = (1 + R_1) * (1 + R_2) * \cdots * (1 + R_T)$$

$$= \prod_{t=1}^{T}(1 + R_t). \tag{2.4}$$

**Step 1: Calculate Gross Returns** Since we have already calculated the daily arithmetic total returns above, we can use the tot.ret dataset for this step. We first calculate the gross returns by adding 1 to the values in tot.ret.

```
1   > gross.ret <- 1 + tot.rets$tot.ret
2   > head.tail(gross.ret)
3                  tot.ret
4   2015-01-02 0.9904874
5   2015-01-05 0.9718283
6   2015-01-06 1.0000943
7                  tot.ret
8   2019-12-27 0.9996205
9   2019-12-30 1.0059351
10  2019-12-31 1.0073065
```

**Step 2: Cumulate Gross Returns Daily** We can implement Eq. (2.4) in R using cumprod(), which allows us to calculate Eq. (2.4) more efficiently. Essentially, this function cumulates the gross total returns every day.

```
1   > cum.arith <- cumprod(gross.ret)
2   > head.tail(cum.arith)
3                  tot.ret
4   2015-01-02 0.9904874
5   2015-01-05 0.9625837
6   2015-01-06 0.9626745
7                  tot.ret
8   2019-12-27 2.857451
9   2019-12-30 2.874410
10  2019-12-31 2.895412
```

**Step 3: Extract Ending Value** The last value of *cum.arith* is the gross cumulative arithmetic return for AAPL from 2015 to 2019. To get to the net cumulative arithmetic return, we subtract 1 from that number. Doing so yields a cumulative arithmetic return for AAPL of 189.5% over the 5-year period from January 2015 to December 2019.

```
1  > as.numeric(cum.arith[nrow(cum.arith)]) - 1
2  [1] 1.895412
```

### 2.5.2 Cumulating Logarithmic Returns

An alternative way to calculate multi-period returns is to take the sum of the daily logarithmic returns. That is,

$$r_{1,T} = \ln((1 + R_1) * (1 + R_2) * \cdots * (1 + R_T))$$
$$= r_1 + r_2 + \cdots + r_T$$
$$= \sum_{t=1}^{T} r_t. \tag{2.5}$$

Since we already calculated the AMZN log returns in *log.rets*, calculating the cumulative log return requires one step. We only have to sum the daily log returns over the period. We do this using sum(), which yields a cumulative log return for AMZN of 106.3% over the period January 2015 to December 2019.

```
1  > (cum.log <- sum(log.rets))
2  [1] 1.063128
```

As a check on the calculations, we can convert the cumulative log return to cumulative arithmetic returns by taking the exponential. As the output below shows, the resulting figure matches the cumulative arithmetic return we calculated earlier of 189.5%.

```
1  > exp(cum.log) - 1
2  [1] 1.895412
```

Although log returns may be simpler to implement from a programming perspective, the arithmetic return has the advantage of being intuitive. When we say the arithmetic return is 189.5%, we know that if we invested $1000 at the start of the period, we would have $2895 at the end of the period. The cumulative log return of 106.3% does not have a similar intuitive explanation.

### 2.5.3 Comparing Price Return and Total Return

Using AAPL as an example and the technique to create normalized prices in Chap. 1, we create normalized values of the AAPL close price and adjusted close price. This will help us compare the performance of the arithmetic returns with total returns.

**Step 1: Calculate Normalized Close Price** In Line 1, we determine the December 31, 2014 close price of AAPL. We then divide every AAPL close price by that number to calculate the normalized price return. As we can see, a $1 investment in AAPL close price results in a value of $2.66 at the end of 2019. This indicates the price return for AAPL is 166%.

```
 1  > (first.close <- as.numeric(data.aapl$AAPL.Close[1]))
 2  [1] 110.38
 3  > prc.ret <- data.aapl$AAPL.Close / first.close
 4  > names(prc.ret) <- "Price.Ret"
 5  > head.tail(prc.ret)
 6               Price.Ret
 7  2014-12-31 1.0000000
 8  2015-01-02 0.9904875
 9  2015-01-05 0.9625838
10               Price.Ret
11  2019-12-27  2.625476
12  2019-12-30  2.641058
13  2019-12-31  2.660355
```

**Step 2: Calculate Normalized Adjusted Close Price**  We apply the same methodology to the adjusted close. This shows that a $1 invested at the adjusted close price results in a value of $2.90 at the end of 2019. This indicates the total return for AAPL is 190%.

```
 1  > (first.total <- as.numeric(data.aapl$AAPL.Adjusted[1]))
 2  [1] 101.4191
 3  > tot.ret <- data.aapl$AAPL.Adjusted / first.total
 4  > names(tot.ret) <- "Total.Ret"
 5  > head.tail(tot.ret)
 6               Total.Ret
 7  2014-12-31 1.0000000
 8  2015-01-02 0.9904874
 9  2015-01-05 0.9625837
10               Total.Ret
11  2019-12-27  2.857451
12  2019-12-30  2.874410
13  2019-12-31  2.895412
```

**Step 3: Plot the Price and Total Return**  We use plot() to create a line chart comparing the total return and price return for AAPL.

```
 1  > dt <- index(prc.ret)
 2  > (y.range <- range(prc.ret, tot.ret))
 3  [1] 0.8184454 2.8954123
 4  > plot(x = dt,
 5  +      y = tot.ret,
 6  +      xlab = "Date",
 7  +      ylab = "Normalized Price",
 8  +      type = "l",
 9  +      col = "blue",
10  +      main = "Comparing AAPL Price and Total Return")
11  > lines(x = dt, y = prc.ret, col = "darkgreen")
12  > abline(h = 1)
13  > legend("topleft",
```

```
14  +     c("Total Return", "Price Return"),
15  +     col = c("blue", "darkgreen"),
16  +     lwd = c(1, 1))
```

As Fig. 2.1 shows, the total returns are higher than the price returns. The reason for this is the reinvestment of dividends. The difference between price returns and total returns grows on each ex-dividend date. AAPL pays dividends quarterly, so over the 5-year period there are 20 ex-dividend dates. The reinvestment of those dividends is responsible for the 24% difference between total returns and price returns.

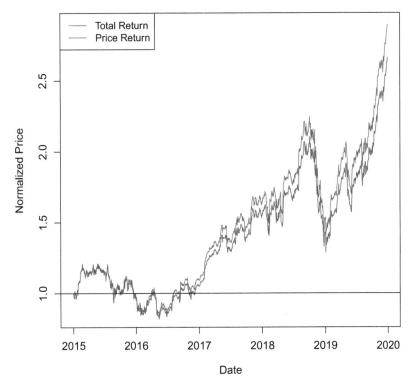

**Fig. 2.1** Comparing AAPL cumulative price returns and total returns, 2015–2019. Data source: price data reproduced with permission of CSI ©2020. www.csidata.com

## 2.6   Weekly Returns

In the previous sections, we used daily returns. However, there are applications in which we may have to use returns of lower frequency, such as **weekly returns**. A common application of weekly returns is the calculation of betas.

**Step 1: Change Daily Prices to Weekly Prices** Using to.weekly() in the xts package (which is loaded with the quantmod package), we are able to easily calculate weekly prices.

```
1  > wk <- data.aapl
2  > wk <- to.weekly(wk)
3  > head.tail(wk)
4              wk.Open wk.High wk.Low wk.Close wk.Volume wk.Adjusted
5  2015-01-02  112.82  113.13 107.35   109.33  94608000   100.45430
6  2015-01-09  108.29  113.25 104.63   112.01 283252500   102.91672
7  2015-01-16  112.60  112.80 105.20   105.99 304226600    97.38545
8              wk.Open wk.High wk.Low wk.Close wk.Volume wk.Adjusted
9  2019-12-20  277.00  282.65 276.98   279.44 183180000      279.44
10 2019-12-27  280.53  293.97 280.37   289.80  96609500      289.80
11 2019-12-31  289.46  293.68 285.22   293.65  61230000      293.65
```

**Step 2: Calculate Weekly Total Return** Since we are interested in calculating weekly total returns, we will keep the weekly adjusted close price column (Column 6).

```
1  > wk <- wk[, 6]
2  > names(wk) <- "price"
3  > head.tail(wk)
4                  price
5  2015-01-02 100.45430
6  2015-01-09 102.91672
7  2015-01-16  97.38545
8                  price
9  2019-12-20 279.44
10 2019-12-27 289.80
11 2019-12-31 293.65
```

**Step 3: Clean Up Weekly Return Dataset** We use Delt() in the  quantmod package to calculate weekly returns. Next, we clean up the dataset by removing the December 31, 2014 value (i.e., because the return value for that is NA) and rename the column header **AAPL**. The *rets.weekly* dataset contains the weekly AAPL returns. Note that since December 31, 2019 falls on a Tuesday, the last observation below is not a full week return.

```
1  > rets.weekly <- Delt(wk)
2  > rets.weekly <- rets.weekly[-1, ]
3  > names(rets.weekly) <- "AAPL"
4  > head.tail(rets.weekly)
5                      AAPL
6  2015-01-09  0.02451289
7  2015-01-16 -0.05374520
8  2015-01-23  0.06594975
```

```
 9                   AAPL
10    2019–12–20 0.01559152
11    2019–12–27 0.03707410
12    2019–12–31 0.01328505
```

> **Verify Weekly Return Calculations**
>
> It is a good habit to test our calculations to make sure that we did not make a mistake in programming. The weekly returns are Friday-to-Friday returns. Therefore, if we want to verify the weekly return for January 9, 2015, we would need the adjusted close price on January 2, 2015 (the Friday before) and on January 9, 2015. Once we figure out what the adjusted prices are on those dates, we can divide the January 9, 2015 price by the January 2, 2015 price to find the weekly return. Subtracting 1 from that amount gets us a net weekly return for the week of January 9, 2015 of 2.45%, which is the same as what we calculated above (except for the number in the 8th decimal place).
>
> ```
> 1   > data.aapl["2015–01–02/2015–01–09", 6]
> 2              AAPL.Adjusted
> 3   2015–01–02     100.45430
> 4   2015–01–05      97.62434
> 5   2015–01–06      97.63354
> 6   2015–01–07      99.00256
> 7   2015–01–08     102.80648
> 8   2015–01–09     102.91672
> 9   > 102.91672 / 100.45430 – 1
> 10  [1] 0.02451284
> ```

## 2.7   Monthly Returns

Another common return frequency used in many financial applications is **monthly returns**. The implementation is similar to the calculating weekly returns. A common application of monthly returns is the calculation of betas.

**Step 1: Change Daily Prices to Monthly Prices** Using to.monthly() in the xts package (which is loaded with the quantmod package), we are able to easily calculate monthly prices.

```
1   > mo <- data.aapl
2   > mo <- to.monthly(mo)
3   > head.tail(mo)
4            mo.Open mo.High mo.Low mo.Close  mo.Volume mo.Adjusted
5   Dec 2014  112.82  113.13 110.21   110.38   41403400    101.4191
6   Jan 2015  111.39  120.00 104.63   117.16 1305263400    107.6486
```

```
7   Feb 2015  118.05  133.60 116.08    128.46 1136535200     118.4971
8             mo.Open mo.High mo.Low mo.Close mo.Volume mo.Adjusted
9   Oct 2019  225.07  249.75 215.13    248.76  608302700     248.0154
10  Nov 2019  249.54  268.00 249.16    267.25  448331500     267.2500
11  Dec 2019  267.27  293.97 256.29    293.65  597198700     293.6500
```

**Step 2: Calculate Monthly Total Return**  Since we are interested in calculating monthly total returns, we will keep the monthly adjusted close price column (Column 6).

```
1   > mo <- mo[, 6]
2   > names(mo) <- "price"
3   > head.tail(mo)
4                price
5   Dec 2014 101.4191
6   Jan 2015 107.6486
7   Feb 2015 118.4971
8                price
9   Oct 2019 248.0154
10  Nov 2019 267.2500
11  Dec 2019 293.6500
```

**Step 3: Clean Up Monthly Return Dataset**  We use Delt() in the  quantmod package to calculate the monthly return. We then clean up the dataset by removing the December 2014 value (i.e., because the return value for that is NA) and rename the column name AAPL. The *rets.monthly* dataset contains the monthly AAPL returns.

```
1   > rets.monthly <- Delt(mo)
2   > rets.monthly <- rets.monthly[-1, ]
3   > names(rets.monthly) <- "AAPL"
4   > head.tail(rets.monthly)
5                    AAPL
6   Jan 2015   0.06142418
7   Feb 2015   0.10077668
8   Mar 2015  -0.03137180
9                    AAPL
10  Oct 2019 0.11068444
11  Nov 2019 0.07755414
12  Dec 2019 0.09878389
```

## 2.8  Comparing Performance of Multiple Securities

Total returns capture both the returns from capital appreciation and dividends. Therefore, it is a better measure of investment performance than price returns. In the last chapter, we showed how to create normalized price charts based on price returns. We will use that same technique here. However, we also show a different way of getting the same answer by cumulating the daily returns.

The *data.aapl* dataset should still be stored in memory from our previous exercises. Otherwise, we can use `load.data()` to import the AAPL price data. For these charts, we would need the data for GOOG, AMZN, and SPY as well.

```
1  > data.amzn <- load.data("AMZN Yahoo.csv", "AMZN")
2  > data.goog <- load.data("GOOG Yahoo.csv", "GOOG")
3  > data.spy <- load.data("SPY Yahoo.csv", "SPY")
```

### 2.8.1  Using Normalized Prices

**Step 1: Identify the Price on the Day of the Investment**  We assume that the investment is made on December 31, 2014. We thus create four variables to represent the adjusted close price on December 31, 2014 of each of the four securities we are plotting.

```
1  > (first.amzn <- as.numeric(data.amzn[1, 6]))
2  [1] 310.35
3  > (first.goog <- as.numeric(data.goog[1, 6]))
4  [1] 524.9587
5  > (first.aapl <- as.numeric(data.aapl[1, 6]))
6  [1] 101.4191
7  > (first.spy <- as.numeric(data.spy[1, 6]))
8  [1] 186.259
```

**Step 2: Create Dataset of Adjusted Close Prices**  We then extract the adjusted close price (Column 6) from each security's price data to create the *adj.prc* dataset. Below, we use the column name (e.g., AMZN.Adjusted) instead of the column number.

```
1  > adj.prc <- cbind(data.amzn$AMZN.Adjusted, data.goog$GOOG.Adjusted,
2  +     data.aapl$AAPL.Adjusted, data.spy$SPY.Adjusted)
3  > names(adj.prc) <- c("AMZN", "GOOG", "AAPL", "SPY")
4  > head.tail(adj.prc)
5               AMZN      GOOG       AAPL       SPY
6  2014-12-31 310.35  524.9587 101.41906 186.2590
7  2015-01-02 308.52  523.3731 100.45430 186.1593
8  2015-01-05 302.19  512.4630  97.62434 182.7974
9               AMZN      GOOG     AAPL      SPY
10 2019-12-27 1869.80 1351.89 289.80 322.86
11 2019-12-30 1846.89 1336.14 291.52 321.08
12 2019-12-31 1847.84 1337.02 293.65 321.86
```

**Step 3: Create the Dataset of Normalized Prices**  We then divide each day's price by the price on the day of the investment (i.e., price on December 31, 2014). The *norm.prc* dataset contains the normalized prices for the four securities from December 31, 2014 to December 31, 2019.

```
1  > norm.prc <- adj.prc
2  > norm.prc$AMZN <- norm.prc$AMZN / first.amzn
```

```
3   > norm.prc$GOOG <- norm.prc$GOOG / first.goog
4   > norm.prc$AAPL <- norm.prc$AAPL / first.aapl
5   > norm.prc$SPY <- norm.prc$SPY / first.spy
6   > head.tail(norm.prc)
7                    AMZN      GOOG      AAPL       SPY
8   2014-12-31 1.0000000 1.0000000 1.0000000 1.0000000
9   2015-01-02 0.9941034 0.9969795 0.9904874 0.9994647
10  2015-01-05 0.9737071 0.9761967 0.9625837 0.9814148
11                   AMZN      GOOG      AAPL       SPY
12  2019-12-27 6.024811 2.575231 2.857451 1.733392
13  2019-12-30 5.950991 2.545229 2.874410 1.723836
14  2019-12-31 5.954052 2.546905 2.895412 1.728023
```

## 2.8.2  Using Cumulative Returns

**Step 1: Create a Dataset of Gross Total Returns**  We use `Delt()` on the adjusted price of each security to calculate the net total return. Then, we add one to the net total return to get the gross total return (Lines 1–4). We then use `cbind()` to create a single dataset labeled *rets* that contains the gross total returns for each security.

```
1   > rets.amzn <- 1 + Delt(data.amzn$AMZN.Adjusted)
2   > rets.goog <- 1 + Delt(data.goog$GOOG.Adjusted)
3   > rets.aapl <- 1 + Delt(data.aapl$AAPL.Adjusted)
4   > rets.spy <- 1 + Delt(data.spy$SPY.Adjusted)
5   > rets <- cbind(rets.amzn, rets.goog, rets.aapl, rets.spy)
6   > names(rets) <- c("AMZN", "GOOG", "AAPL", "SPY")
7   > head.tail(rets)
8                    AMZN      GOOG      AAPL       SPY
9   2014-12-31        NA        NA        NA        NA
10  2015-01-02 0.9941034 0.9969795 0.9904874 0.9994647
11  2015-01-05 0.9794827 0.9791543 0.9718283 0.9819405
12                   AMZN      GOOG      AAPL       SPY
13  2019-12-27 1.0005512 0.9937445 0.9996205 0.9997522
14  2019-12-30 0.9877473 0.9883496 1.0059351 0.9944868
15  2019-12-31 1.0005144 1.0006586 1.0073065 1.0024293
```

**Step 2: Replace December 31, 2014 Value with $1**  As we can see from the output above, the values for December 31, 2014 are all NA. In prior exercises, we usually delete the December 31, 2014 observation after we calculate the normalized prices. However, for this analysis, December 31, 2014 is the day we made the $1 investment, so we can simply overwrite the NA values on this day with $1. Thus, the January 2, 2015 values apply the January 2, 2015 return to the $1 starting investment.

```
1   > rets[1, ] <- c(1, 1, 1, 1)
2   > head.tail(rets)
3                    AMZN      GOOG      AAPL       SPY
4   2014-12-31 1.0000000 1.0000000 1.0000000 1.0000000
5   2015-01-02 0.9941034 0.9969795 0.9904874 0.9994647
6   2015-01-05 0.9794827 0.9791543 0.9718283 0.9819405
7                    AMZN      GOOG      AAPL       SPY
```

```
 8   2019–12–27 1.0005512 0.9937445 0.9996205 0.9997522
 9   2019–12–30 0.9877473 0.9883496 1.0059351 0.9944868
10   2019–12–31 1.0005144 1.0006586 1.0073065 1.0024293
```

**Step 3: Cumulate the Returns of Each Security**  We now start with $1 and grow the value of our investment each day by multiplying the prior day's value by the gross return. This process can be automated in R using `cumprod()`. The *cum.rets* dataset contains the value of the investment for the four securities on each day from December 31, 2014 to December 31, 2019.

```
 1   > cum.rets <- cumprod(rets)
 2   > head.tail(cum.rets)
 3                  AMZN      GOOG      AAPL       SPY
 4   2014–12–31 1.0000000 1.0000000 1.0000000 1.0000000
 5   2015–01–02 0.9941034 0.9969795 0.9904874 0.9994647
 6   2015–01–05 0.9737071 0.9761967 0.9625837 0.9814148
 7                  AMZN      GOOG      AAPL       SPY
 8   2019–12–27 6.024811 2.575231 2.857451 1.733392
 9   2019–12–30 5.950991 2.545229 2.874410 1.723836
10   2019–12–31 5.954052 2.546905 2.895412 1.728023
```

To see that we get the same result as normalizing prices above, we can compare the December 31, 2019 values of both approaches. For convenience, we output the December 31, 2019 values under the normalized price approach. As the output below shows, whether we normalize prices or whether we cumulate returns, we end up in the same place.

```
1   > norm.prc[nrow(norm.prc), ]
2                  AMZN      GOOG      AAPL       SPY
3   2019–12–31 5.954052 2.546905 2.895412 1.728023
```

**Step 4: Plot the Cumulative Returns**  We can see how each security performed relative to the other using a line chart. AMZN performed so much better than the other securities. The output of the `plot()` is shown in Fig. 2.2.

```
 1   > plot(x = index(cum.rets),
 2   +       y = cum.rets$AMZN,
 3   +       ylim = y.range,
 4   +       xlab = "Date",
 5   +       ylab = "Value of Investment",
 6   +       type = "l",
 7   +       lwd = 3,
 8   +       main = "Value of $1 Investment in AMZN, GOOG, AAPL, and SPY")
 9   > lines(x = index(cum.rets), y = cum.rets$GOOG, col = "blue")
10   > lines(x = index(cum.rets), y = cum.rets$AAPL, col = "red")
11   > lines(x = index(cum.rets), y = cum.rets$SPY, col = "darkgreen")
12   > abline(h = 1)
13   > legend("topleft",
14   +       c("Amazon", "Google", "Apple", "S&P 500"),
15   +       col = c("black", "blue", "red","darkgreen"),
16   +       lwd = c(3, 2, 2, 1))
```

**Value of $1 Investment in AMZN, GOOG, AAPL, and SPY**

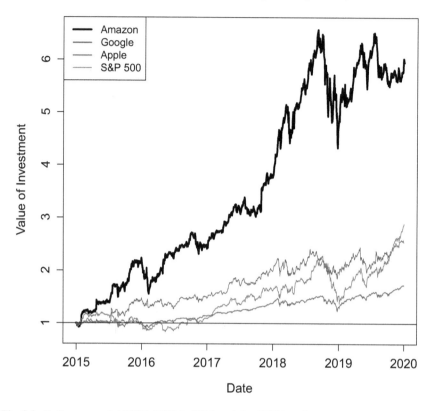

**Fig. 2.2**  Performance of AMZN, YHOO, IBM, and the SPY based on total returns, December 31, 2014–December 31, 2019. Data source: price data reproduced with permission of CSI ©2020. www.csidata.com

# Chapter 3
# Portfolio Returns

A key lesson in finance is diversification, which means we should not put all our eggs in one basket. From the perspective of portfolio management, this means that we should have more than one stock in our portfolio. In the last chapter, we looked at returns of individual securities. In this chapter, we extend the returns calculation when you have more than one security in your portfolio.

In a portfolio, some stocks will increase in value, while others will go down. Ultimately, what we should care about is the total value of our portfolio increases over time. In other words, we should resist the temptation to think that the goal is that every security in the portfolio we select should be a winner.

The **portfolio return** is the weighted-average return of the securities in the portfolio. The weight of each security in the portfolio is equal to the percentage of the total portfolio value that is invested in that security. Below, we show how portfolio returns are calculated laying out each step in the process in Sect. 3.1. For some applications, it is easier to deal with portfolio calculations using matrix algebra especially when our portfolio has a large number of securities. We do this, for example, when we discuss portfolio optimization in Chap. 7. Thus, we show in Sect. 3.2, how the portfolio return calculation is implemented using matrix algebra.

In Sect. 3.3, we then show how to construct benchmark portfolio returns, which are returns of a hypothetical portfolio that we can use to compare the performance of our portfolio. Two common approaches in benchmark portfolio construction are the use of an **equal-weighted portfolio** and a **value-weighted portfolio**. We implement both approaches assuming **quarterly rebalancing**, which is a common approach used in practice when constructing indexes. The difference between the two weighting methods is how the portfolio value is redistributed on the rebalancing dates. An equal-weighted portfolio redistributes the total value of the portfolio equally among all the securities in the portfolios on each rebalancing date. By comparison, a value-weighted portfolio redistributes the total value of the portfolio based on the market capitalization of the securities (i.e., larger firms get a bigger share of the pie) on each rebalancing date. In both approaches, the weights in each

C.S. Ang, *Analyzing Financial Data and Implementing Financial Models Using R*, Springer Texts in Business and Economics, https://doi.org/10.1007/978-3-030-64155-9_3

security are allowed to move freely depending on their performance between each rebalancing date.

Finally, we conclude this chapter by discussing time-weighted rates of return in Sect. 3.4 and money-weighted rates of return in Sect. 3.5. These are two of the more common portfolio return measures if we want to estimate returns that account for the timing of cash inflows and outflows into the portfolio.

## 3.1  Portfolio Returns the Long Way

In Chap. 2, we calculated individual security returns. We now extend this to calculating the return of a portfolio comprised of two or more securities. Mathematically, the portfolio return is equal to the weighted-average of the returns of the individual securities in the portfolio with the weights calculated as the percentage invested in that security relative to the total amount invested in the portfolio. Specifically, in an $n$-asset portfolio, we can calculate total returns as follows. First, we calculate the total investment as

$$I = \sum_{i=1}^{n} I_n$$
$$= I_1 + I_2 + \cdots + I_n, \tag{3.1}$$

where $I_i$ is the amount invested in security $i$. Next, the weight invested in each security is

$$w_i = I_i / I. \tag{3.2}$$

Thus, we have

$$\sum_{i=1}^{n} w_i = 1. \tag{3.3}$$

That is, the percentage invested in all assets must sum to 100% of the investments. This ensures that every dollar in the portfolio value is allocated to some asset. Note that if we have cash in the portfolio, we can still assign a weight as well as a return to the cash. For example, if the cash is invested in a money market mutual fund, we can use the return of that fund. Since all assets are assigned weights, the sum of those weights must still equal one (i.e., equal to 100% of the portfolio).

Finally, we can then calculate the portfolio return as

$$R_p = \sum_{i=1}^{n} w_i R_i$$

$$= w_1 R_1 + w_2 R_2 + \cdots + w_n R_n. \tag{3.4}$$

We now illustrate how to implement the portfolio return calculation assuming we invested \$50,000 in AMZN, \$30,000 in GOOG, and \$20,000 in AAPL. For this example, we assume the investment was made on December 31, 2014 and we would like to know the value of the portfolio as of December 31, 2019 if we employed a buy-and-hold strategy (i.e., we purchased the stocks on December 31, 2014 and did not sell them until December 31, 2019).

**Step 1: Import the Data for the Four Securities and Calculate Their Returns**
Using `load.data()`, we import the price data for AMZN, GOOG, and AAPL. Next, we calculate each security's returns by using `Delt()` on the adjusted price of each security. Then, we combine the returns of each security using `cbind()` and cleanup the *returns* dataset by removing the first row and renaming the return variables to **AMZN, GOOG,** and **AAPL**.

```
 1  > data.amzn <- load.data("AMZN Yahoo.csv", "AMZN")
 2  > data.goog <- load.data("GOOG Yahoo.csv", "GOOG")
 3  > data.aapl <- load.data("AAPL Yahoo.csv", "AAPL")
 4  >
 5  > rets.amzn <- Delt(data.amzn$AMZN.Adjusted)
 6  > rets.goog <- Delt(data.goog$GOOG.Adjusted)
 7  > rets.aapl <- Delt(data.aapl$AAPL.Adjusted)
 8  >
 9  > returns <- cbind(rets.amzn, rets.goog, rets.aapl)
10  > returns <- returns[-1, ]
11  > names(returns) <- c("AMZN", "GOOG", "AAPL")
12  > head.tail(returns)
13                   AMZN          GOOG          AAPL
14  2015-01-02 -0.005896623 -0.003020489 -9.512610e-03
15  2015-01-05 -0.020517267 -0.020845731 -2.817166e-02
16  2015-01-06 -0.022833293 -0.023177050  9.433099e-05
17                   AMZN          GOOG          AAPL
18  2019-12-27  0.0005511802 -0.0062555196 -0.0003794833
19  2019-12-30 -0.0122526652 -0.0116503560  0.0059351314
20  2019-12-31  0.0005143517  0.0006586174  0.0073065487
```

**Step 2: Calculate Weight of Each Security in the Portfolio** To calculate the weights, we create variables representing the dollar amount of investment in each security. We then calculate the weight of each security by calculating the percentage invested in that security divided by the total amount invested, which in this case is \$100,000. Given the amount invested in each stock, the weights would be 50% for AMZN, 30% for GOOG, and 20% for AAPL.

```
 1  > options(scipen = 999)
 2  > i.amzn <- 50000
 3  > i.goog <- 30000
 4  > i.aapl <- 20000
 5  > (i.total <- i.amzn + i.goog + i.aapl)
 6  [1] 100000
 7  >
```

```
8   > (w.amzn <- i.amzn / i.total)
9   [1] 0.5
10  > (w.goog <- i.goog / i.total)
11  [1] 0.3
12  > (w.aapl <- i.aapl / i.total)
13  [1] 0.2
```

**Step 3: Calculate Cumulative Net Return**  We calculate the cumulative net return in three steps. First, we calculate the daily gross return for each security. The gross return is one plus the net return.

```
1   > port.ret <- 1 + returns
2   > head.tail(port.ret)
3                    AMZN      GOOG      AAPL
4   2015-01-02 0.9941034 0.9969795 0.9904874
5   2015-01-05 0.9794827 0.9791543 0.9718283
6   2015-01-06 0.9771667 0.9768229 1.0000943
7                    AMZN      GOOG      AAPL
8   2019-12-27 1.0005512 0.9937445 0.9996205
9   2019-12-30 0.9877473 0.9883496 1.0059351
10  2019-12-31 1.0005144 1.0006586 1.0073065
```

Next, we use cumprod() to cumulate the gross returns on each day. Thus, the last observation in the dataset (i.e., December 31, 2019) is the cumulative gross return over the entire period.

```
1   > cum.ret <- cumprod(port.ret)
2   > head.tail(cum.ret)
3                    AMZN      GOOG      AAPL
4   2015-01-02 0.9941034 0.9969795 0.9904874
5   2015-01-05 0.9737071 0.9761967 0.9625837
6   2015-01-06 0.9514742 0.9535714 0.9626745
7                    AMZN     GOOG     AAPL
8   2019-12-27 6.024811 2.575231 2.857451
9   2019-12-30 5.950991 2.545229 2.874410
10  2019-12-31 5.954052 2.546905 2.895412
```

The third step is to calculate the cumulative net return over the period, we subtract one from the last observation above. As the output below shows, the cumulative *net* return for AMZN is 495%, GOOG is 155%, and AAPL is 190%.

```
1   > (cum.ret <- cum.ret[nrow(cum.ret)] - 1)
2                    AMZN     GOOG     AAPL
3   2019-12-31 4.954052 1.546905 1.895412
```

**Step 4: Calculate Portfolio Return**  Using Eq. (3.4), the portfolio return from December 31, 2014 to December 31, 2019 is 332%. We use as.numeric() to convert the output to a variable R considers as numeric.

```
1   > (cum.port <- as.numeric(w.amzn * cum.ret$AMZN +
2   +      w.goog * cum.ret$GOOG + w.aapl * cum.ret$AAPL))
3   [1] 3.32018
```

## 3.2 Portfolio Returns Using Matrix Algebra

As the number of securities in our portfolio grows, it becomes more cumbersome to implement Eq. (3.4). Thus, we have to use matrix algebra to make the calculations simpler. To do so, we need two vectors. Vectors are a series of numbers organized as either a row or a column. The first vector we need is a vector of weights. This is where we store the weights of the three stocks from our example above. The second vector we need is a vector of returns. This is where we store the returns of the three stocks from our example above.

Mathematically, we can define the column vector of weights as

$$\mathbf{w} = \begin{pmatrix} w_1 \\ w_2 \\ \vdots \\ w_n \end{pmatrix} \tag{3.5}$$

and the column vector of returns as

$$\mathbf{R} = \begin{pmatrix} R_1 \\ R_2 \\ \vdots \\ R_n \end{pmatrix} \tag{3.6}$$

Therefore, the portfolio return $R_p$ is equal to

$$R_p = \mathbf{w}^\mathsf{T}\mathbf{R}, \tag{3.7}$$

where $\mathsf{T}$ is the matrix transpose operator. The transpose operator converts, in this case, a column vector of weights into a row vector of weights. This step is necessary so that we can matrix multiply the two vectors. This is a technicality in matrix algebra. The takeaway from the last step above is essentially we need to, for each stock, multiply the weights by the returns and then aggregate those values.

**Step 1: Create a Row Vector of Weights** In the equation above, because we transpose the column vector of weights ($\mathbf{w}^\mathsf{T}$), we effectively use a row vector of weights. To create a row vector of weights, we use c(). We then convert that into a matrix using matrix(). The second argument in matrix() tells R that we should put all of the elements in one row.

```
1  > (weight <- c(w.amzn, w.goog, w.aapl))
2  [1] 0.5 0.3 0.2
3  > (mat.weight <- matrix(weight, 1))
4       [,1] [,2] [,3]
5  [1,]  0.5  0.3  0.2
```

**Step 2: Create a Column Vector of Returns** In the equation above, we use a column vector of returns (**R**). To create a column vector of returns, we use c (). We then convert that vector into a matrix using matrix().The second argument in matrix() is now 3, which tells R to split the data into three rows. Since we have three returns, each return gets its own row.

```
1  > (mat.returns <- matrix(cum.ret, 3))
2            [,1]
3  [1,] 4.954052
4  [2,] 1.546905
5  [3,] 1.895412
```

**Step 3: Calculate Portfolio Returns** We calculate the portfolio return by multiplying the row vector of weights by the column vector of returns using the **matrix multiplication** operator %*%. Unlike normal multiplication in which $x * y = y * x$, this may not be the case with matrix multiplication. In fact, the vectors need to be "conformable" for matrix multiplication to work. A detailed discussion of matrix multiplication is beyond the scope of this book. For our purposes, what is important is this step is necessary so we multiply the weight matrix (left side) by the return matrix (right side) to get the identical answer as to what we calculated using the manual approach earlier of 332%.

```
1  > (port.ret2 <- as.numeric(mat.weight %*% mat.returns))
2  [1] 3.32018
```

In the above, we added as.numeric() to the output. Otherwise, the output will look like a matrix instead of a number (see output below).

```
1  > (port.ret2 <- mat.weight %*% mat.returns)
2         [,1]
3  [1,] 3.32018
```

## 3.3   Constructing Benchmark Portfolio Returns

To determine how our investment is performing, we compare the returns of our investment to that of a benchmark. Selecting a benchmark is outside the scope of this book, but as a general matter the benchmark should exhibit the same risk-and-return characteristics as the subject investment. Typically, some portfolio of comparable securities are used as the benchmark. Some benchmarks are readily-available (e.g., S&P 500 ETF for large capitalization stocks). Alternatively, benchmarks can be constructed. In this section, we demonstrate how to construct an equal-weighted (EW) portfolio and value-weighted (VW) portfolio using an example of investing $1000 in AMZN, GOOG, and AAPL on December 31, 2018 and we will liquidate the holdings as of December 31, 2019.

Because we are comparing the performance of the stock to the performance of a benchmark, it is important that benchmark return we construct is realistic. For example, instead of the non-tradeable S&P 500 Index, we may want to use the SPDR S&P 500 ETF (SPY) instead if wanted a benchmark for large capitalization stocks. The latter attempts to mimic the returns of the S&P 500 Index but you could go to your brokerage account and purchase the SPY. That way, when we compare the performance of the stock to the performance of the benchmark, the investor could realistically have put their money in either security.

If we have to develop a benchmark from scratch, it is hard to find two firms that are identical. Thus, the appropriate comparison may be a combination of two or more stocks. To properly compare the performance of the subject stock and the benchmark, we need to ensure that the benchmark return is calculated in a manner that is realistic from an investment perspective as well. Otherwise, the resulting comparison may be biased. For example, mutual fund prospectuses sometimes report benchmarks that do not subtract expenses and fees. Even if we assume the reported benchmarks are appropriate, we have to make adjustments for the expenses and fees that the investor would incur had they tried to replicate the benchmark's returns. Ignoring the expenses and fees would result in benchmark returns that are overstated because we would likely be using net returns for the subject fund, which is net of expenses and fees. Thus, the comparison to the benchmark will bias the result towards showing underperformance of the subject fund.

If we have to construct our own benchmarks comprised of, say, stocks of comparable firms, then we would have to consider how to realistically create the benchmark portfolio return. One key factor is the rebalancing assumption. In terms of rebalancing, there are two issues that we would consider: rebalancing weights and rebalancing frequency. For rebalancing weights, two common approaches are equal-weighting or value-weighting. Equal-weighting assumes that the portfolio value is split evenly across the stocks in the portfolio on the rebalancing dates. Value-weighting assumes that the portfolio value is split across the stocks in the portfolio based on the market capitalization of the stocks.

For rebalancing frequency, we can assume that you only redistribute the weights at the beginning and have a buy-and-hold portfolio. The longer the holding period, the more unlikely this assumption is going to be realistic. A typical rebalancing frequency is quarterly, which means that at the end of each quarter we would redistribute the portfolio value based on the rebalancing weights assumption. For example, the S&P 500 Index uses quarterly rebalancing. In the middle of each quarter, the investments are allowed to move up and down freely, which means the weight on any given day during the quarter could be different from the weight at the start of the quarter.

Sometimes, other rules could be also used as an alternative or in addition to calendar period rebalancing. These rules would trigger rebalancing prior to the end of the quarter. For example, there could be a band around the weights applied to each stock that triggers rebalancing if the weight of the stock moves above the upper band or below the lower band. These bands could be different for each stock, as the size of the band is based on factors such as the volatility of the stock and correlation of the stock to the other stocks in the portfolio. In the example below, we ignore these additional rules and focus on calendar period rebalancing.

### 3.3.1  Quarterly Returns the Long Way

Quarterly rebalancing assumes that the value of the securities during the quarter (i.e., between rebalancing dates) is allowed to move based on how each security performs. Since our holding period for this investment is from December 31, 2018 to December 31, 2019, we will create four quarters of cumulative returns.

**Step 1: Calculate 1Q 2019 Cumulative Return**  We first subset the returns data to only include returns from January 1, 2019 to March 31, 2019 (i.e., the first quarter of 2019).

```
 1  > rets.q1 <- returns["2019-01-01/2019-03-31"]
 2  > head.tail(rets.q1)
 3                 AMZN            GOOG           AAPL
 4  2019-01-02  0.02474086  0.009887884  0.001141031
 5  2019-01-03 -0.02524152 -0.028483988 -0.099607348
 6  2019-01-04  0.05006398  0.053786157  0.042689250
 7                 AMZN            GOOG           AAPL
 8  2019-03-27 -0.010124713 -0.009792149 0.008994063
 9  2019-03-28  0.004372256 -0.003861852 0.001326494
10  2019-03-29  0.004133232  0.004125041 0.006517625
```

Next, we then calculate the gross daily returns by adding 1 to the net returns in the *rets.q1* dataset.

```
 1  > grets.q1 <- 1 + rets.q1
 2  > head.tail(grets.q1)
 3                 AMZN       GOOG       AAPL
 4  2019-01-02 1.0247409 1.009888 1.0011410
 5  2019-01-03 0.9747585 0.971516 0.9003927
 6  2019-01-04 1.0500640 1.053786 1.0426893
 7                 AMZN       GOOG       AAPL
 8  2019-03-27 0.9898753 0.9902079 1.008994
 9  2019-03-28 1.0043723 0.9961381 1.001326
10  2019-03-29 1.0041332 1.0041250 1.006518
```

Finally, because all we need is the cumulative return at the end of the quarter, we cumulate the returns during the quarter and store those returns in *crets.q1*. Note that we used `apply()` to apply `cumprod()` to each of the three columns in *grets.q1*.

```
 1  > crets.q1 <- apply(grets.q1, 2, cumprod)
 2  > (crets.q1 <- crets.q1[nrow(crets.q1), ] - 1)
 3      AMZN       GOOG       AAPL
 4  0.1856096 0.1329652 0.2093613
```

**Step 2: Calculate 2Q 2019 Cumulative Return**  The methodology followed in 2Q 2019 returns is similar to how we calculated the returns for 1Q 2019.

```
 1  > rets.q2 <- returns["2019-04-01/2019-06-30"]
 2  > head.tail(rets.q2)
 3                  AMZN            GOOG           AAPL
 4  2019-04-01  0.0187785714 0.018000353 0.006791323
 5  2019-04-02 -0.0001157326 0.005073496 0.014536706
```

```
 6    2019—04—03   0.0037045453 0.004523198 0.006854967
 7                           AMZN          GOOG          AAPL
 8    2019—06—26   0.010413804 −0.006029297  0.0216290550
 9    2019—06—27   0.003398657 −0.003509945 −0.0003002551
10    2019—06—28 −0.005592677   0.004553883 −0.0091119708
11    >
12    > grets.q2 <− 1 + rets.q2
13    > head.tail(grets.q2)
14                    AMZN      GOOG      AAPL
15    2019—04—01 1.0187786 1.018000 1.006791
16    2019—04—02 0.9998843 1.005073 1.014537
17    2019—04—03 1.0037045 1.004523 1.006855
18                    AMZN      GOOG      AAPL
19    2019—06—26 1.0104138 0.9939707 1.0216291
20    2019—06—27 1.0033987 0.9964901 0.9996997
21    2019—06—28 0.9944073 1.0045539 0.9908880
22    >
23    > crets.q2 <− apply(grets.q2, 2, cumprod)
24    > (crets.q2 <− crets.q2[nrow(crets.q2), ] − 1)
25        AMZN          GOOG          AAPL
26     0.06338902 −0.07875158   0.04597093
```

**Step 3: Calculate 3Q 2019 Cumulative Return**   The methodology followed in 3Q 2019 returns is similar to how we calculated the returns for 1Q 2019.

```
 1    > rets.q3 <− returns["2019—07—01/2019—09—30"]
 2    > head.tail(rets.q3)
 3                        AMZN          GOOG          AAPL
 4    2019—07—01 0.015082110 0.015764417 0.018340833
 5    2019—07—02 0.006305370 0.012113529 0.005854542
 6    2019—07—03 0.002424607 0.009295798 0.008286927
 7                        AMZN          GOOG          AAPL
 8    2019—09—26 −0.016111241 −0.004115461 −0.005157641
 9    2019—09—27 −0.008270884 −0.013130482 −0.004866027
10    2019—09—30   0.006062235 −0.004971036   0.023535244
11    >
12    > grets.q3 <− 1 + rets.q3
13    > head.tail(grets.q3)
14                    AMZN      GOOG      AAPL
15    2019—07—01 1.015082 1.015764 1.018341
16    2019—07—02 1.006305 1.012114 1.005855
17    2019—07—03 1.002425 1.009296 1.008287
18                    AMZN      GOOG      AAPL
19    2019—09—26 0.9838888 0.9958845 0.9948424
20    2019—09—27 0.9917291 0.9868695 0.9951340
21    2019—09—30 1.0060622 0.9950290 1.0235352
22    >
23    > crets.q3 <− apply(grets.q3, 2, cumprod)
24    > (crets.q3 <− crets.q3[nrow(crets.q3), ] − 1)
25        AMZN          GOOG          AAPL
26    −0.08328975   0.12775343   0.13591842
```

**Step 4: Calculate 4Q 2019 Cumulative Return**  The methodology followed in 4Q 2019 returns is similar to how we calculated the returns for 1Q 2019.

```
1   > rets.q4 <- returns["2019-10-01/2019-12-31"]
2   > head.tail(rets.q4)
3                     AMZN            GOOG            AAPL
4   2019-10-01 -0.0001497831 -0.011402809  0.002768250
5   2019-10-02 -0.0129173760 -0.023624572 -0.025067884
6   2019-10-03  0.0065315598  0.009518669  0.008494695
7                     AMZN            GOOG            AAPL
8   2019-12-27  0.0005511802 -0.0062555196 -0.0003794833
9   2019-12-30 -0.0122526652 -0.0116503560  0.0059351314
10  2019-12-31  0.0005143517  0.0006586174  0.0073065487
11  >
12  > grets.q4 <- 1 + rets.q4
13  > head.tail(grets.q4)
14                 AMZN       GOOG       AAPL
15  2019-10-01 0.9998502 0.9885972 1.0027682
16  2019-10-02 0.9870826 0.9763754 0.9749321
17  2019-10-03 1.0065316 1.0095187 1.0084947
18                 AMZN       GOOG       AAPL
19  2019-12-27 1.0005512 0.9937445 0.9996205
20  2019-12-30 0.9877473 0.9883496 1.0059351
21  2019-12-31 1.0005144 1.0006586 1.0073065
22  >
23  > crets.q4 <- apply(grets.q4, 2, cumprod)
24  > (crets.q4 <- crets.q4[nrow(crets.q4), ] - 1)
25        AMZN       GOOG       AAPL
26  0.06447911 0.09681708 0.31504940
```

### 3.3.2  Quarterly Returns the Shorter Way

To simplify the calculations above, we can use to.quarterly() in the quantmod package. The application of this is similar to to.weekly() and to.monthly() we discussed in Chap. 1. Here, we can start with the full price series of AMZN, GOOG, and AAPL from December 31, 2014 to December 31, 2019, calculate quarter end prices, and then calculate the quarterly returns off those prices.

**Step 1: Construct Dataset with Quarterly Prices**  For each of the three stocks, we create a dataset of the adjusted close prices. Then, we use to.quarterly() to concert the prices to quarterly prices. Looking at the AMZN example, the entry for 2014 Q4 is all $310.35. This is because we only have one observation for that quarter, which is on December 31, 2014. Thus, the open, high, low, and close are all the same for that entry. For the subsequent quarters, the open is the first close price of the month, the high is the highest close price for during the month, low is the lowest close price for the month, and close is the last close price for the month.

```
1   > prc.amzn <- data.amzn$AMZN.Adjusted
```

```
2  > qtr.amzn <- to.quarterly(prc.amzn)
3  > head.tail(qtr.amzn)
4           prc.amzn.Open prc.amzn.High prc.amzn.Low prc.amzn.Close
5  2014 Q4        310.35       310.35      310.35        310.35
6  2015 Q1        308.52       387.83      286.95        372.10
7  2015 Q2        370.26       445.99      370.26        434.09
8           prc.amzn.Open prc.amzn.High prc.amzn.Low prc.amzn.Close
9  2019 Q2       1814.19      1962.46     1692.69       1893.63
10 2019 Q3       1922.19      2020.99     1725.45       1735.91
11 2019 Q4       1735.65      1869.80     1705.51       1847.84
12 > prc.goog <- data.goog$GOOG.Adjusted
13 > qtr.goog <- to.quarterly(prc.goog)
14 > head.tail(qtr.goog)
15           prc.goog.Open prc.goog.High prc.goog.Low prc.goog.Close
16 2014 Q4      524.9587      524.9587     524.9587      524.9587
17 2015 Q1      523.3731      573.7548     491.2014      546.4996
18 2015 Q2      541.0745      563.5129     520.5100      520.5100
19           prc.goog.Open prc.goog.High prc.goog.Low prc.goog.Close
20 2019 Q2      1194.43       1287.58     1036.23        1080.91
21 2019 Q3      1097.95       1250.41     1097.95        1219.00
22 2019 Q4      1205.10       1361.17     1176.63        1337.02
23 >
24 > prc.aapl <- data.aapl$AAPL.Adjusted
25 > qtr.aapl <- to.quarterly(prc.aapl)
26 > head.tail(qtr.aapl)
27           prc.aapl.Open prc.aapl.High prc.aapl.Low prc.aapl.Close
28 2014 Q4      101.4191      101.4191    101.41906      101.4191
29 2015 Q1      100.4543      122.6850     97.38545      114.7796
30 2015 Q2      114.6136      122.7714    114.61360      116.1854
31           prc.aapl.Open prc.aapl.High prc.aapl.Low prc.aapl.Close
32 2019 Q2      189.2172      209.5103    172.1273       196.5807
33 2019 Q3      200.1861      223.2996    192.0316       223.2996
34 2019 Q4      223.9177      293.6500    218.3046       293.6500
```

**Step 2: Calculate Quarterly Returns**  From the above, we will use the quarter-end close price to calculate the returns. To do this, we use `Delt()` on the fourth column. In Line 1, we already use `cbind()`, so we can combine the quarterly returns. Line 4 then replaces the column name to the ticker symbol of the stocks. Line 5 deletes the first observation for 2014 Q4, which will have the value of NA.

```
1  > rqtr <- cbind(Delt(qtr.amzn[, 4]),
2  +      Delt(qtr.goog[, 4]),
3  +      Delt(qtr.aapl[, 4]))
4  > names(rqtr) <- c("AMZN", "GOOG", "AAPL")
5  > rqtr <- rqtr[-1, ]
6  > tail(rqtr,4)
7                AMZN        GOOG       AAPL
8  2019 Q1  0.18560959  0.13296519 0.20936127
9  2019 Q2  0.06338902 -0.07875158 0.04597093
10 2019 Q3 -0.08328975  0.12775343 0.13591842
11 2019 Q4  0.06447911  0.09681708 0.31504940
```

We output the last four observations above so that we can compare with what we calculated earlier, which I have reproduced below. As you can see, the results are identical but we are able to greatly simplify the calculations by using `to.quarterly()`.

```
1  > rbind(crets.q1, crets.q2, crets.q3, crets.q4)
2                AMZN          GOOG         AAPL
3  crets.q1  0.18560959  0.13296519 0.20936127
4  crets.q2  0.06338902 -0.07875158 0.04597093
5  crets.q3 -0.08328975  0.12775343 0.13591842
6  crets.q4  0.06447911  0.09681708 0.31504940
```

### 3.3.3  Equal-Weighted Portfolio

An equal-weighted (EW) portfolio divides the invested capital equally among the securities in the portfolio. For example, if we are investing $100 and there are two stocks in the portfolio, each stock gets a $50 investment. If we are investing $100 but there are now four stocks, then each stock gets a $25 investment. In other words, the weight for each security is $1/n$, where $n$ is the number of securities in the portfolio. As we can see, an EW portfolio would give equal weight to small firms and large firms. An example of EW indexes is the S&P 500 Equal Weight Index, which Standard & Poor's created in response to needs of the market for an "official" EW index version of the S&P 500 Index, and those provided by MSCI for developed markets, emerging markets, and all countries. These two indexes are rebalanced quarterly, which means that in between quarters the constituent weights are allowed to fluctuate based on their performance. In the examples below, we assume that the investment is made on December 31, 2018 and we want to know what the cumulative return would be at December 31, 2019.

**Step 1: Calculate the Value of the Portfolio at the End of 1Q 2019**  We first create a variable for the initial investment, *ew.i0*, which equals to $1000. Then, we grow that amount by the average return for the securities in the portfolio. Note that in an equal-weighted portfolio, the weight given to each of the three securities is 33.33%. Using `mean()` effectively applies an equal-weight to each of the three securities in our portfolio. Therefore, using `mean()` allows us to calculate the equal-weighted return easily. As the output shows, as of March 31, 2019, our EW portfolio has increased in value from $1000 to $1176.

```
1  > ew.i0 <- 1000
2  > (ew.i1 <- ew.i0 * (1 + mean(crets.q1)))
3  [1] 1175.979
```

**Step 2: Calculate the Value of the Portfolio at the End of 2Q 2019**  We then start with the $1176 portfolio value at the end of 1Q 2019 and then grow that amount by the average portfolio return for 2Q 2019. This increases the portfolio value to $1188 as of June 30, 2019.

```
1  > (ew.i2 <- ew.i1 * (1 + mean(crets.q2)))
2  [1] 1187.977
```

**Step 3: Calculate the Value of the Portfolio at the End of 3Q 2019** We then start with the $1188 portfolio value at the end of 2Q 2019 and then grow that amount by the average portfolio return for 3Q 2019. This increases the portfolio value to $1259 as of September 30, 2019.

```
1  > (ew.i3 <- ew.i2 * (1 + mean(crets.q3)))
2  [1] 1259.407
```

**Step 4: Calculate the Value of the Portfolio at the End of 4Q 2019** We then start with the $1259 portfolio value at the end of 3Q 2019 and then grow that amount by the average portfolio return for 4Q 2019. This increases the portfolio value to $1459 as of December 31, 2019.

```
1  > (ew.i4 <- ew.i3 * (1 + mean(crets.q4)))
2  [1] 1459.378
```

The above shows that our equal-weighted portfolio increased in value from $1,000 on December 31, 2018 to $1459 on December 31, 2019, a 45.9% return.

## 3.3.4 Value-Weighted Portfolio

A value-weighted (VW) portfolio invests capital in proportion to the market capitalization of the securities in the portfolio. In a VW portfolio, returns of larger firms are given more weight. Some of the major indexes are value-weighted, such as the S&P 500 Index. For our purposes, the weight of each security is set equal to the market capitalization of that security divided by the total market capitalization of all the securities in the portfolio. In addition, we only rebalance the weights at the start of each quarter using the prior quarter end's market capitalization data. The reason we use past data is because we want to avoid any look-ahead bias. To calculate the day $t$ market capitalization, we would need to know the return on day $t$ for that security. So if we use the market capitalization on day $t$ as the weights for day $t$ return, we are assuming perfect foresight as to what the end of day return would be. This is not realistic. Therefore, a more appropriate approach is to take the market capitalization the day prior to the rebalancing date, as this information would have been known to investors on the rebalancing date.

**Step 1: Calculate Weights Based on Market Cap as of December 31, 2018** We first obtain market capitalization data for AMZN, GOOG, and AAPL from http://ycharts.com. The market caps as of December 31, 2018 are $737.47 million for AMZN, $720.32 million for GOOG, and $746.08 million for AAPL, which sum to $2.2 billion. This gives us weights of 33.5% for AMZN, 32.7% for GOOG, and 33.9% for AAPL.

```
1   > mc1.amzn <- 737.47
2   > mc1.goog <- 720.32
3   > mc1.aapl <- 746.08
4   > (mc1.tot <- sum(mc1.amzn, mc1.goog, mc1.aapl))
5   [1] 2203.87
6   > (w1.amzn <- mc1.amzn / mc1.tot)
7   [1] 0.334625
8   > (w1.goog <- mc1.goog / mc1.tot)
9   [1] 0.3268432
10  > (w1.aapl <- mc1.aapl / mc1.tot)
11  [1] 0.3385318
```

**Step 2: Use Weights Calculated Above to Determine How Much Is Invested in Each Security** To determine the amount invested in each stock, we multiply the stock's weight as calculated above to the initial investment of $1000.

```
1   > vw.i0 <- 1000
2   > (vw.i0.amzn <- vw.i0 * w1.amzn)
3   [1] 334.625
4   > (vw.i0.goog <- vw.i0 * w1.goog)
5   [1] 326.8432
6   > (vw.i0.aapl <- vw.i0 * w1.aapl)
7   [1] 338.5318
```

**Step 3: Calculate Value of Portfolio as of March 31, 2019** During the quarter, we allow each investment to grow at the stock's actual return during that quarter. For example, AMZN started the quarter at $334.63 and ended the quarter at $396.73. Summing the end of quarter values for all three securities yields a portfolio value as of March 31, 2019 of $1176.

```
1   > (vw.i1.amzn <- vw.i0.amzn * (1 + crets.q1[1]))
2       AMZN
3   396.7346
4   > (vw.i1.goog <- vw.i0.goog * (1 + crets.q1[2]))
5       GOOG
6   370.302
7   > (vw.i1.aapl <- vw.i0.aapl * (1 + crets.q1[3]))
8       AAPL
9   409.4072
10  > (vw.i1 <- sum(vw.i1.amzn, vw.i1.goog, vw.i1.aapl))
11  [1] 1176.444
```

**Step 4: Calculate Value of Portfolio as of June 30, 2019** We use the three steps above to calculate the value of the portfolio as of June 30, 2019. We start by obtaining market capitalization data as of March 31, 2019 for each security. This yields a total market capitalization of $2.6 billion. Instead of calculating the weights separately, we include the weight in each investment in Lines 7, 10, and 13. In those lines, we multiply the portfolio value as of March 31, 2019 of $1176 by the market cap of the security divided by the total market cap of $2.6 billion. We then grow that investment amount by the cumulative return for each security during 2Q 2019. This yields a portfolio value as of June 30, 2019 of $1191.

```
1   > mc2.amzn <- 876.22
2   > mc2.goog <- 815.67
3   > mc2.aapl <- 895.67
4   > (mc2.tot <- sum(mc2.amzn, mc2.goog, mc2.aapl))
5   [1] 2587.56
6   >
7   > (vw.i2.amzn <- vw.i1 * (mc2.amzn / mc2.tot) * (1 + crets.q2[1]))
8        AMZN
9   423.6294
10  > (vw.i2.goog <- vw.i1 * (mc2.goog / mc2.tot) * (1 + crets.q2[2]))
11       GOOG
12  341.6426
13  > (vw.i2.aapl <- vw.i1 * (mc2.aapl / mc2.tot) * (1 + crets.q2[3]))
14     AAPL
15  425.94
16  > (vw.i2 <- sum(vw.i2.amzn, vw.i2.goog, vw.i2.aapl))
17  [1] 1191.212
```

**Step 5: Calculate Value of Portfolio as of September 30, 2019** We follow the approach in Step 4 to calculate the portfolio value as of September 30, 2019, which yields a portfolio value of \$1256.

```
1   > mc3.amzn <- 939.29
2   > mc3.goog <- 750.42
3   > mc3.aapl <- 910.64
4   > (mc3.tot <- sum(mc3.amzn, mc3.goog, mc3.aapl))
5   [1] 2600.35
6   >
7   > (vw.i3.amzn <- vw.i2 * (mc3.amzn / mc3.tot) * (1 + crets.q3[1]))
8        AMZN
9   394.4473
10  > (vw.i3.goog <- vw.i2 * (mc3.goog / mc3.tot) * (1 + crets.q3[2]))
11       GOOG
12  387.6821
13  > (vw.i3.aapl <- vw.i2 * (mc3.aapl / mc3.tot) * (1 + crets.q3[3]))
14       AAPL
15  473.8612
16  > (vw.i3 <- sum(vw.i3.amzn, vw.i3.goog, vw.i3.aapl))
17  [1] 1255.991
```

**Step 6: Calculate Value of Portfolio as of December 31, 2019** We follow the approach in Step 4 to calculate the portfolio value as of December 31, 2019, which yields a portfolio value of \$1466.

```
1   > mc4.amzn <- 859.28
2   > mc4.goog <- 842.21
3   > mc4.aapl <- 995.15
4   > (mc4.tot <- sum(mc4.amzn, mc4.goog, mc4.aapl))
5   [1] 2696.64
6   >
7   > (vw.i4.amzn <- vw.i3 * (mc4.amzn / mc4.tot) * (1 + crets.q4[1]))
8        AMZN
9   426.0252
```

```
10   > (vw.i4.goog <- vw.i3 * (mc4.goog / mc4.tot) * (1 + crets.q4[2]))
11       GOOG
12   430.2472
13   > (vw.i4.aapl <- vw.i3 * (mc4.aapl / mc4.tot) * (1 + crets.q4[3]))
14       AAPL
15   609.5285
16   > (vw.i4 <- sum(vw.i4.amzn, vw.i4.goog, vw.i4.aapl))
17   [1] 1465.801
```

The above shows that our value-weighted portfolio increased in value from $1000 on December 31, 2018 to $1466 on December 31, 2019, a 46.6% return.

The above shows the basic calculation, but the steps above can be used when we add or subtract stocks in the portfolio. Essentially, on the rebalancing date, we would re-allocate the total portfolio value to the stocks in the portfolio. If we moved from three to four stocks, for example, in the equal-weight portfolio we would invest one-fourth of the portfolio value in each stock. For the value-weighted portfolio, the market capitalization of the fourth stock is added to get the total market capitalization of the four stocks and the investment is then distributed accordingly.

### 3.3.5  Daily Portfolio Returns

Sometimes we would like to obtain the daily values of the portfolio so we can plot how our investment performed on a daily basis. This has the advantage of seeing whether there were interim movements that could be of interest to us. We start with the EW portfolio and then discuss the VW portfolio.

**Step 1: Calculate EW Daily Values of Each Security in Portfolio for 1Q**  We begin by dividing our initial investment *ew.i0* of $1000 by 3, the number of securities in the portfolio. This means we divide the total portfolio value equally across the three securities with each security getting $333.33. We then multiply this number to the cumulative gross return of each security on each day. The cumulative gross return of each security on each day is calculated by using  cumprod() on the 1Q gross return dataset *grets.q1*.

```
 1   > ew.val1 <- ew.i0 / 3 * cumprod(grets.q1)
 2   > head.tail(ew.val1)
 3                    AMZN     GOOG     AAPL
 4   2019-01-02 341.5803 336.6293 333.7137
 5   2019-01-03 332.9583 327.0407 300.4733
 6   2019-01-04 349.6275 344.6310 313.3003
 7                    AMZN     GOOG     AAPL
 8   2019-03-27 391.8631 377.5617 399.9795
 9   2019-03-28 393.5765 376.1036 400.5100
10   2019-03-29 395.2032 377.6551 403.1204
```

**Step 2: Calculate Daily Total EW Portfolio Value for 1Q** We then create the column called tot, which equals the sum of the daily values of the investments in each of the three securities. We calculate the total by applying rowSums() to *ew.val1* dataset. As we can see, the ending total EW portfolio value as of March 29, 2019 is equal to what we got earlier, which is stored in the variable *ew.i1*.

```
1  > ew.val1$tot <- rowSums(ew.val1)
2  > head.tail(ew.val1)
3                   AMZN     GOOG     AAPL      tot
4  2019-01-02 341.5803 336.6293 333.7137 1011.9233
5  2019-01-03 332.9583 327.0407 300.4733  960.4724
6  2019-01-04 349.6275 344.6310 313.3003 1007.5588
7                   AMZN     GOOG     AAPL      tot
8  2019-03-27 391.8631 377.5617 399.9795 1169.404
9  2019-03-28 393.5765 376.1036 400.5100 1170.190
10 2019-03-29 395.2032 377.6551 403.1204 1175.979
11 >
12 > ew.i1
13 [1] 1175.979
```

**Step 3: Repeat the Same Procedure For 2Q to 4Q** We repeat Steps 1 and 2 above for 2Q to 4Q. We find that the EW portfolio ending value at the end of the year is \$1459.38, which is identical to the ending value we calculated earlier and stored in the variable *ew.i4*.

```
1  > ew.val2 <- as.numeric(ew.val1[nrow(ew.val1), 4]) / 3 * cumprod(grets.q2)
2  > ew.val2$tot <- rowSums(ew.val2)
3  > head.tail(ew.val2)
4                   AMZN     GOOG     AAPL      tot
5  2019-04-01 399.3540 399.0489 394.6550 1193.058
6  2019-04-02 399.3077 401.0735 400.3920 1200.773
7  2019-04-03 400.7870 402.8876 403.1367 1206.811
8                   AMZN     GOOG     AAPL      tot
9  2019-06-26 417.7655 360.7520 413.9078 1192.425
10 2019-06-27 419.1853 359.4858 413.7836 1192.455
11 2019-06-28 416.8409 361.1228 410.0132 1187.977
12 >
13 > ew.val3 <- as.numeric(ew.val2[nrow(ew.val2), 4]) / 3 * cumprod(grets.q3)
14 > ew.val3$tot <- rowSums(ew.val3)
15 > head.tail(ew.val3)
16                  AMZN     GOOG     AAPL      tot
17 2019-07-01 401.9647 402.2349 403.2551 1207.455
18 2019-07-02 404.4993 407.1074 405.6160 1217.223
19 2019-07-03 405.4800 410.8918 408.9773 1225.349
20                  AMZN     GOOG     AAPL      tot
21 2019-09-26 363.8320 454.7843 441.6208 1260.237
22 2019-09-27 360.8228 448.8128 439.4719 1249.107
23 2019-09-30 363.0102 446.5817 449.8150 1259.407
24 >
25 > ew.val4 <- as.numeric(ew.val3[nrow(ew.val3), 4]) / 3 * cumprod(grets.q4)
26 > ew.val4$tot <- rowSums(ew.val4)
27 > head.tail(ew.val4)
28                  AMZN     GOOG     AAPL      tot
```

```
29   2019-10-01 419.7394 415.0154 420.9644 1255.719
30   2019-10-02 414.3175 405.2108 410.4117 1229.940
31   2019-10-03 417.0236 409.0679 413.8980 1239.990
32                   AMZN       GOOG      AAPL      tot
33   2019-12-27 452.1815 465.5673 544.8228 1462.572
34   2019-12-30 446.6410 460.1433 548.0564 1454.841
35   2019-12-31 446.8708 460.4463 552.0608 1459.378
36   >
37   > ew.i4
38   [1] 1459.378
```

**Step 4: Combine the Quarterly Portfolio Values** We want to stack each of the quarterly datasets on top of each other. To do this, we use `rbind()`.

```
1    > ew.port <- rbind(ew.val1, ew.val2, ew.val3, ew.val4)
2    > head.tail(ew.port)
3                    AMZN       GOOG      AAPL       tot
4    2019-01-02 341.5803 336.6293 333.7137 1011.9233
5    2019-01-03 332.9583 327.0407 300.4733  960.4724
6    2019-01-04 349.6275 344.6310 313.3003 1007.5588
7                    AMZN       GOOG      AAPL       tot
8    2019-12-27 452.1815 465.5673 544.8228 1462.572
9    2019-12-30 446.6410 460.1433 548.0564 1454.841
10   2019-12-31 446.8708 460.4463 552.0608 1459.378
```

**Step 5: Calculate VW Daily Values of Each Security in Portfolio for 1Q** We create a *vw.val1* dataset that is equal to the cumulative daily gross returns. To do so, we apply `cumprod()` on the daily gross return datasets *grets.q1*. Then, we overwrite each security's gross return by the product of the security's gross return and the security's weight for that quarter, which we have previously calculated. These weights for the first quarter are stored in the variables *vw.i0.amzn*, *vw.i0.goog*, and *vw.i0.appl*. The zero denotes that these weights are calculated using the market capitalization prior to the 1st quarter. As we can see below, the weights applicable for the second quarter are based on the market capitalization of the securities at the end of the 1st quarter, so the index will be 1 instead of 0.

```
1    > vw.val1 <- cumprod(grets.q1)
2    > vw.val1$AMZN <- vw.val1$AMZN * vw.i0.amzn
3    > vw.val1$GOOG <- vw.val1$GOOG * vw.i0.goog
4    > vw.val1$AAPL <- vw.val1$AAPL * vw.i0.aapl
5    > head.tail(vw.val1)
6                    AMZN       GOOG      AAPL
7    2019-01-02 342.9039 330.0750 338.9180
8    2019-01-03 334.2485 320.6732 305.1593
9    2019-01-04 350.9823 337.9209 318.1863
10                   AMZN       GOOG      AAPL
11   2019-03-27 393.3816 370.2105 406.2173
12   2019-03-28 395.1016 368.7808 406.7561
13   2019-03-29 396.7346 370.3020 409.4072
```

**Step 6: Calculate Daily Total EW Portfolio Value for 1Q** We then create the column called tot, which equals the sum of the daily values of the investments in each of the three securities. We calculate the total by applying `rowSums()` to *vw.val1* dataset. As we can see, the ending total VW portfolio value as of March 29, 2019 is equal to what we got earlier, which is stored in the variable *vw.i1*.

```
1   > vw.val1$tot <- rowSums(vw.val1)
2   > head.tail(vw.val1)
3                  AMZN      GOOG      AAPL       tot
4   2019-01-02 342.9039 330.0750 338.9180 1011.897
5   2019-01-03 334.2485 320.6732 305.1593  960.081
6   2019-01-04 350.9823 337.9209 318.1863 1007.090
7                  AMZN      GOOG      AAPL       tot
8   2019-03-27 393.3816 370.2105 406.2173 1169.809
9   2019-03-28 395.1016 368.7808 406.7561 1170.638
10  2019-03-29 396.7346 370.3020 409.4072 1176.444
11  >
12  > vw.i1
13  [1] 1176.444
```

**Step 7: Repeat the Same Procedure For 2Q–4Q**   We repeat Steps 1 and 2 above for 2Q–4Q. We find that the EW portfolio ending value at the end of the year is $1465.80, which is identical to the ending value we calculated earlier and stored in the variable *vw.i4*.

```
1   > vw.val2 <- cumprod(grets.q2)
2   > (vw.i1tot <- as.numeric(vw.val1[nrow(vw.val1), 4]))
3   [1] 1176.444
4   > vw.val2$AMZN <- vw.val2$AMZN * vw.i1tot * (mc2.amzn / mc2.tot)
5   > vw.val2$GOOG <- vw.val2$GOOG * vw.i1tot * (mc2.goog / mc2.tot)
6   > vw.val2$AAPL <- vw.val2$AAPL * vw.i1tot * (mc2.aapl / mc2.tot)
7   > vw.val2$tot <- rowSums(vw.val2)
8   > head.tail(vw.val2)
9                  AMZN      GOOG      AAPL       tot
10  2019-04-01 405.8576 377.5228 409.9853 1193.366
11  2019-04-02 405.8107 379.4382 415.9451 1201.194
12  2019-04-03 407.3140 381.1544 418.7964 1207.265
13                 AMZN      GOOG      AAPL       tot
14  2019-06-26 424.5690 341.2918 429.9859 1195.847
15  2019-06-27 426.0120 340.0938 429.8568 1195.963
16  2019-06-28 423.6294 341.6426 425.9400 1191.212
17  >
18  > vw.val3 <- cumprod(grets.q3)
19  > (vw.i2tot <- as.numeric(vw.val2[nrow(vw.val2), 4]))
20  [1] 1191.212
21  > vw.val3$AMZN <- vw.val3$AMZN * vw.i2tot * (mc3.amzn / mc3.tot)
22  > vw.val3$GOOG <- vw.val3$GOOG * vw.i2tot * (mc3.goog / mc3.tot)
23  > vw.val3$AAPL <- vw.val3$AAPL * vw.i2tot * (mc3.aapl / mc3.tot)
24  > vw.val3$tot <- rowSums(vw.val3)
25  > head.tail(vw.val3)
26                 AMZN      GOOG      AAPL       tot
27  2019-07-01 436.7753 349.1842 424.8123 1210.772
28  2019-07-02 439.5294 353.4141 427.2994 1220.243
29  2019-07-03 440.5951 356.6994 430.8404 1228.135
30                 AMZN      GOOG      AAPL       tot
```

```
31   2019–09–26 395.3403 394.8029 465.2290 1255.372
32   2019–09–27 392.0705 389.6190 462.9652 1244.655
33   2019–09–30 394.4473 387.6821 473.8612 1255.991
34   >
35   > vw.val4 <− cumprod(grets.q4)
36   > (vw.i3tot <− as.numeric(vw.val3[nrow(vw.val3), 4]))
37   [1] 1255.991
38   > vw.val4$AMZN <− vw.val4$AMZN * vw.i3tot * (mc4.amzn / mc4.tot)
39   > vw.val4$GOOG <− vw.val4$GOOG * vw.i3tot * (mc4.goog / mc4.tot)
40   > vw.val4$AAPL <− vw.val4$AAPL * vw.i3tot * (mc4.aapl / mc4.tot)
41   > vw.val4$tot <− rowSums(vw.val4)
42   > head.tail(vw.val4)
43                  AMZN      GOOG      AAPL       tot
44   2019–10–01 400.1595 387.7959 464.7855 1252.741
45   2019–10–02 394.9904 378.6344 453.1343 1226.759
46   2019–10–03 397.5703 382.2385 456.9835 1236.792
47                  AMZN      GOOG      AAPL       tot
48   2019–12–27 431.0882 435.0323 601.5371 1467.658
49   2019–12–30 425.8062 429.9640 605.1073 1460.877
50   2019–12–31 426.0252 430.2472 609.5285 1465.801
51   >
52   > vw.i4
53   [1] 1465.801
```

**Step 8: Combine the Quarterly Portfolio Values** We want to stack each of the quarterly datasets on top of each other. To do this, we use `rbind()`.

```
1    > vw.port <− rbind(vw.val1, vw.val2, vw.val3, vw.val4)
2    > head.tail(vw.port)
3                  AMZN      GOOG      AAPL       tot
4    2019–01–02 342.9039 330.0750 338.9180 1011.897
5    2019–01–03 334.2485 320.6732 305.1593  960.081
6    2019–01–04 350.9823 337.9209 318.1863 1007.090
7                  AMZN      GOOG      AAPL       tot
8    2019–12–27 431.0882 435.0323 601.5371 1467.658
9    2019–12–30 425.8062 429.9640 605.1073 1460.877
10   2019–12–31 426.0252 430.2472 609.5285 1465.801
```

**Step 9: Combine EW and VW Daily Portfolio Values** Using `cbind()`, we combine the tot columns of *ew.port* and *vw.port* datasets.

```
1    > portfolios <− cbind(ew.port$tot, vw.port$tot)
2    > names(portfolios) <− c("EW", "VW")
3    > head.tail(portfolios)
4                     EW        VW
5    2019–01–02 1011.9233 1011.897
6    2019–01–03  960.4724  960.081
7    2019–01–04 1007.5588 1007.090
8                     EW        VW
9    2019–12–27 1462.572 1467.658
10   2019–12–30 1454.841 1460.877
11   2019–12–31 1459.378 1465.801
```

To visually see the weights of the various securities in the portfolio during each quarter, we can create four pie charts. Line 1 tells R that we are creating four charts on a page laid out as $2 \times 2$. Lines 2–3 calculate the weights of each security during the first quarter. We use round() to round the numbers to one decimal place. Lines 3–5 paste the ticker and weight of each security. Lines 9–12 use pie() to create a pie chart for the first quarter weights.

```
1   > par(mfrow = c(2,2))
2   > (Q1.pie.values <- round((c(mc1.amzn, mc1.goog, mc1.aapl) /
3   +     mc1.tot) * 100, digits = 1))
4   [1] 33.5 32.7 33.9
5   > (Q1.pie.labels <- c(paste("AMZN (", Q1.pie.values[1], "%)", sep = ""),
6   +     paste("GOOG (", Q1.pie.values[2],"%)", sep = ""),
7   +     paste("AAPL (", Q1.pie.values[3], "%)", sep = "")))
8   [1] "AMZN (33.5%)" "GOOG (32.7%)" "AAPL (33.9%)"
9   > pie(Q1.pie.values,
10  +     labels = Q1.pie.labels,
11  +     col = c("black","blue","red"),
12  +     main = "Q1 Value Weighting")
```

We then repeat the same process to create the pie charts for second through fourth quarter weights.

```
1   > (Q2.pie.values <- round((c(mc2.amzn, mc2.goog, mc2.aapl) /
2   +     mc2.tot) * 100, digits = 1))
3   [1] 33.9 31.5 34.6
4   > (Q2.pie.labels <- c(paste("AMZN (", Q2.pie.values[1], "%)", sep = ""),
5   +     paste("GOOG (", Q2.pie.values[2],"%)", sep = ""),
6   +     paste("AAPL (", Q2.pie.values[3], "%)", sep = "")))
7   [1] "AMZN (33.9%)" "GOOG (31.5%)" "AAPL (34.6%)"
8   > pie(Q2.pie.values,
9   +     labels = Q2.pie.labels,
10  +     col = c("black","blue","red"),
11  +     main = "Q2 Value Weighting")
12  >
13  > (Q3.pie.values <- round((c(mc3.amzn, mc3.goog, mc3.aapl) /
14  +     mc3.tot) * 100, digits = 1))
15  [1] 36.1 28.9 35.0
16  > (Q3.pie.labels <- c(paste("AMZN (", Q3.pie.values[1], "%)", sep = ""),
17  +     paste("GOOG (", Q3.pie.values[2],"%)", sep = ""),
18  +     paste("AAPL (", Q3.pie.values[3], "%)", sep = "")))
19  [1] "AMZN (36.1%)" "GOOG (28.9%)" "AAPL (35%)"
20  > pie(Q3.pie.values,
21  +     labels = Q3.pie.labels,
22  +     col = c("black","blue","red"),
23  +     main = "Q3 Value Weighting")
24  >
25  > (Q4.pie.values <- round((c(mc4.amzn, mc4.goog, mc4.aapl) /
26  +     mc4.tot) * 100, digits = 1))
27  [1] 31.9 31.2 36.9
28  > (Q4.pie.labels <- c(paste("AMZN (", Q4.pie.values[1], "%)", sep = ""),
29  +     paste("GOOG (", Q4.pie.values[2],"%)", sep = ""),
30  +     paste("AAPL (", Q4.pie.values[3], "%)", sep = "")))
31  [1] "AMZN (31.9%)" "GOOG (31.2%)" "AAPL (36.9%)"
```

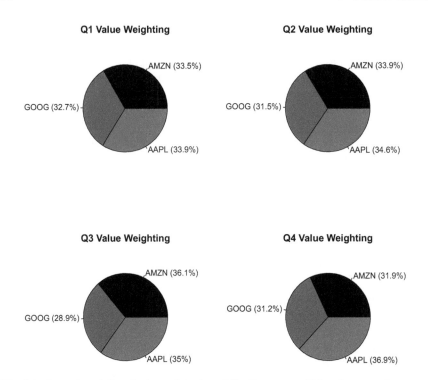

**Fig. 3.1** Quarterly weights of value-weighted portfolio. Data source: Yahoo finance and company SEC filings

```
32   > pie(Q4.pie.values,
33   +      labels = Q4.pie.labels,
34   +      col = c("black","blue","red"),
35   +      main = "Q4 Value Weighting")
36   >
37   > par(mfrow = c(1, 1))
```

Figure 3.1 shows the output of the above code.

In Appendix D, we will leverage on the value-weighted portfolio we created above to generate a hypothetical portfolio that we will use later on. This may be a good time to visit the appendix to construct those hypothetical returns.

## 3.4   Time-Weighted Rate of Return

The **time-weighted rate of return** (TWRR) strings together returns over various time increments. For the TWRR calculation, the market value of our portfolio and cash inflows and outflows can occur on various dates over the holding period.

We first define a holding period return $HPR_t$ as

$$HPR_t = \frac{V_t + C_t - V_{t-1}}{V_{t-1}}, \tag{3.8}$$

where $V_t$ is the market value of the portfolio at time $t$ and $C_t$ is the portfolio cash inflow or outflow at time $t$. For example, if the portfolio receives funds, $C_t$ will be a positive cash flow. Conversely, if there is a withdrawal of funds from the portfolio, $C_t$ will be a negative cash flow. For each date in which we calculate the market value and date in which a cash inflow or outflow occurs, we calculate the holding period return. Then, we string together the various holding period returns over the investment period. In other words,

$$TWRR = \prod_{t=1}^{T}(1 + HPR_t)$$

$$= (1 + HPR_1) * (1 + HPR_2) * \cdots * (1 + HPR_T). \tag{3.9}$$

In our example, we will calculate the time-weighted rate of return given the market value of the portfolio and cash flows reported in Table 3.1.

**Step 1: Enter the Data**  We input the data into R manually. We start with three column vectors: date, market value of the portfolio, and cash flows of the portfolio. We then use cbind() to combine these three vectors.

```
1   > dates <- as.Date(c("2018–12–31", "2019–03–31", "2019–06–30",
2   +                      "2019–07–31", "2019–09–30", "2019–12–31"))
3   > mv <- c(2000000, 1950000, 2000000, 2220000, 2400000, 2500000)
4   > cf <- c(0, 0, 0, 20000, 0, –5000)
5   > cbind(data.frame(dates), mv, cf)
6           dates      mv    cf
7   1 2018–12–31 2000000     0
8   2 2019–03–31 1950000     0
9   3 2019–06–30 2000000     0
10  4 2019–07–31 2220000 20000
11  5 2019–09–30 2400000     0
12  6 2019–12–31 2500000 –5000
```

**Step 2: Calculate the Holding Period Return**  The holding period return is calculated as $(MV_t - MV_{t-1} + CF_t)/MV_{t-1}$. We use a for-loop to recursively do the holding period return calculation for each time interval.

**Table 3.1** Hypothetical portfolio market values and cash flows for use in the time-weighted rate of return example

| Date | Value | Cash flow |
|---|---|---|
| 12/31/2018 | $2,000,000 | $0 |
| 03/31/2019 | $1,950,000 | $0 |
| 06/30/2019 | $2,000,000 | $0 |
| 07/31/2019 | $2,220,000 | $20,000 |
| 09/30/2019 | $2,400,000 | $0 |
| 12/31/2019 | $2,500,000 | –$5000 |

```
1  > hpr <- rep(0, length(cf))
2  > for (i in (2 : (length(cf)))){
3  +    hpr[i] <- (mv[i] − mv[i −1] + cf[i]) / mv[i − 1]
4  + }
5  > cbind(data.frame(dates), mv, cf, hpr)
6         dates       mv      cf        hpr
7  1 2018−12−31 2000000       0  0.00000000
8  2 2019−03−31 1950000       0 −0.02500000
9  3 2019−06−30 2000000       0  0.02564103
10 4 2019−07−31 2220000   20000  0.12000000
11 5 2019−09−30 2400000       0  0.08108108
12 6 2019−12−31 2500000   −5000  0.03958333
```

**Step 3: Calculate the Cumulative Portfolio Return**  We begin with calculating the gross periodic return by adding 1 to the holding period return. We then cumulate the gross returns using cumprod(). The last value in *cum.ret* is the cumulative portfolio return over the period, which in our example is 25.9%. The 25.9% is the time-weighted rate of return of our portfolio.

```
1  > (gross.ret <- 1 + hpr)
2  [1] 1.000000 0.975000 1.025641 1.120000 1.081081 1.039583
3  > (cum.ret <- cumprod(gross.ret))
4  [1] 1.000000 0.975000 1.000000 1.120000 1.210811 1.258739
5  > cum.ret[length(cf)] − 1
6  [1] 0.2587387
```

## 3.5  Money-Weighted Rate of Return

The **money-weighted rate of return** (MWRR) is the **internal rate of return** (IRR) of the portfolio. The IRR is the discount rate that would set the net present value of a project (i.e., present value of cash flows less investment cost) equal to zero. Analogously, the MWRR is the rate of return that would set the present value of the portfolio cash flows equal to the initial value. That is,

$$
0 = \sum_{t=0}^{T} \frac{C_t}{(1 + MWRR)^t}
$$

$$
= C_0 + \frac{C_1}{(1 + MWRR)} + \frac{C_2}{(1 + MWRR)^2} + \cdots + \frac{C_T}{(1 + MWRR)^T}, \quad (3.10)
$$

where $C_t$ is the cash flow at time $t$. We have to decide what sign we will give cash inflows and outflows. For example, if cash inflows to the portfolio are viewed as an investment by the investor, then $C_0$ is negative while subsequent cash inflows would be negative as well but cash outflows (i.e., distributions from the portfolio to the investor) are positive. According to Descartes' Rule of Signs, there can be as many different MWRRs (i.e., roots) as there are sign changes. So to have at least one

MWRR, we need to have at least one sign change (i.e., an investment amount that is a negative cash flow and distributions going forward that are positive cash flows). However, the more sign changes there are (i.e., mix of inflows and outflows), the more possible MWRRs (i.e., roots) there could be.

We can also re-write the above equation by letting $a = 1/(1 + MWRR)$ such that

$$0 = C_0 + aC_1 + a^2 C_2 + \cdots + a^n C_n. \qquad (3.11)$$

The above equation shows us that the formula for the MWRR is a polynomial of degree $T$, which would have to be solved numerically (e.g., by trial-and-error or iteratively).

In many IRR calculations, the cash flows are assumed to come at regular time intervals (e.g., monthly or annual). However, in the context of portfolios, the cash flows could come at any time. Thus, we need to create two functions to make the MWRR calculation work. First, we create a function that calculates the net present value (NPV) of the cash flows that calculates the present value based on the date the cash flows arrive. Second, we create the MWRR function that uses this NPV function.

**Step 1: Create Present Value Function**  The present value function we create takes each cash flow that occurs on different dates, and then discounts those to the present. This function takes on three arguments: vector of cash flows, vector of dates when those cash flows occur, and a discount rate $r$. Line 2 creates a vector of how many years from the valuation date the cash flows will occur. Line 3 calculates a discount factor for each of those cash flow dates. Line 4 calculates the present value of each cash flow by multiplying the future value of the cash flow by the present value factor. Line 5 calculates the present value of all the cash flow streams.

```
1  > pv <- function(cf, dates, r) {
2  +    t <- as.numeric((dates - dates[1]) / 365)
3  +    pv_factor <- (1 + r)^-t
4  +    pv <- cf * pv_factor
5  +    value <- sum(pv)
6  +    return(value)
7  + }
```

**Step 2: Create MWRR Function**  The mwrr() function takes on three arguments: vector of cash flows, vector of dates when those cash flows occur, and a guess as to what the rate of return should be. Because we iterate to find the solution, having a good guess (i.e., the last argument) helps in getting to the correct solution. In instances when there are multiple possible roots, picking the right guess helps us get to the correct MWRR. Line 2 tells R by how much to change the guess under each iteration. Line 3 determines the tolerance level of the error before we deem the solution acceptable. The idea here is that we are doing trial-and-error to get to the rate of return. The value of *tol* determines how far our value should be before we stop iterating. Line 4 is the starting value of the current rate of return that is passed on through the loop. Line 5 sets the iteration counter to 0. We will use this counter to see how many iterations R took to get to the solution.

```
1   > mwrr <- function(cf, dates, guess) {
2   +    delta.x <- 0.01
3   +    tol <- 0.0000001
4   +    cur.x <- guess
5   +    iter <- 0
6   +    for (i in 1:1000) {
7   +       fx <- pv(cf, dates, cur.x)
8   +       cur.x.delta <- cur.x - delta.x
9   +       fx.delta <- pv(cf, dates, cur.x.delta)
10  +       dx <- (fx - fx.delta) / delta.x
11  +       cur.x <- cur.x - (fx / dx)
12  +       iter <- iter + 1
13  +       cat("At iteration", iter, "MWRR equals", cur.x, "\n")
14  +       if (abs(fx) < tol) break
15  +    }
16  + }
```

In the for-loop, we currently run 1000 iterations. That should be generally sufficient. Line 7 uses pv() to calculate the value of the cash flows using our guess of the rate of return *cur.x*. Line 8 subtracts the increment of *delta.x* to the current guess, and then re-runs pv() using that new guess in Line 9. The difference between the present value in Lines 7 and 9 are evaluated in Line 10. Adjustments to the rate of return guess are made in Line 11 and we repeat the calculation as long as the net present value in *fx* is greater than the tolerance (Line 14). Line 12 adds 1 to the number of iterations if we have to repeat the loop. That way, when the answer is reported in Line 13, we will know the correct iteration number as well.

Now that we have the mwrr() function setup, we can go through a couple of examples. In our first example, suppose you paid $100,000 on December 31, 2018 and the value of the portfolio ends up at $120,000 on December 31, 2019. Without doing any calculations, we know the rate of return on this portfolio is 20% per year. Now supposed that when we ran the mwrr() function, we assumed the guess was a 10% rate of return. The function took seven iterations but it ended up with the correct answer of 20% after seven iterations. Although at iteration 4 it appears like we have reached our solution, since our tolerance level is quite small (i.e., we want a more precise solution) the calculation continues to iterate.

```
1    > cf <- c(-100000, 120000)
2    > dates <- as.Date(c("2018-12-31", "2019-12-31"))
3    > mwrr(cf, dates, 0.1)
4    At iteration 1 MWRR equals 0.1908333
5    At iteration 2 MWRR equals 0.1998536
6    At iteration 3 MWRR equals 0.1999988
7    At iteration 4 MWRR equals 0.2
8    At iteration 5 MWRR equals 0.2
9    At iteration 6 MWRR equals 0.2
10   At iteration 7 MWRR equals 0.2
```

Let us now look at our second example, which now considers three cash flows. On December 31, 2018, we invested $200,000 in the portfolio. We took a distribution on June 30, 2019 of $20,000. By December 31, 2019, the portfolio value ends up at $220,000. Calculating the MWRR for this set of cash flows is now a little harder than the MWRR in our first example. Using mwrr() and a guess of 20%, we found the MWRR of the portfolio of 21.0% in seven iterations.

```
 1   > cf <- c(-200000, 20000, 220000)
 2   > dates <- as.Date(c("2018-12-31", "2019-06-30", "2019-12-31"))
 3   > mwrr(cf, dates, 0.2)
 4   At iteration 1 MWRR equals 0.2099248
 5   At iteration 2 MWRR equals 0.210089
 6   At iteration 3 MWRR equals 0.2100903
 7   At iteration 4 MWRR equals 0.2100903
 8   At iteration 5 MWRR equals 0.2100903
 9   At iteration 6 MWRR equals 0.2100903
10   At iteration 7 MWRR equals 0.2100903
```

So, what is the deal with the guess? The guess allows us to get to the solution sooner. For example, if instead of a guess of 20%, we said the guess was 80%. Given that we know the correct answer is 21.0%, we know that 80% is much farther from 21.0% than 20% is. We thus would expect the mwrr() function to take more iterations to find the right solution. As the output below shows, it took the mwrr() function nine iterations to get to the same solution of 21.0%. When the cash flows are more complicated and the guess is further away from the correct answer, it will take the function more iterations to get to the solution.

```
 1   > mwrr(cf, dates, 0.8)
 2   At iteration 1 MWRR equals -0.06770725
 3   At iteration 2 MWRR equals 0.1446252
 4   At iteration 3 MWRR equals 0.2060536
 5   At iteration 4 MWRR equals 0.2100441
 6   At iteration 5 MWRR equals 0.21009
 7   At iteration 6 MWRR equals 0.2100903
 8   At iteration 7 MWRR equals 0.2100903
 9   At iteration 8 MWRR equals 0.2100903
10   At iteration 9 MWRR equals 0.2100903
```

## 3.6 Further Reading

Calculating returns is at the heart of most investment applications. We have only touched upon basic return concepts in the previous chapter and in this chapter. A very good discussion of these and other return calculations can be found in [1]. We also discussed equal-weighted indexes from S&P and MSCI. More information on those indexes can be found in the following documents: S&P 500 Equal Weight Index Methodology, August 2016 and MSCI Equal Weight Indexes Methodology, September 2014.

We discussed how to value-weight individual portfolios and we used the amount we invested in each security to calculate the weight. However, when off-the-shelf indexes, such as the S&P 500 Index, state they are value-weighted, those indexes do not mean that a specific amount is invested in each security in the index. Thus, to value-weight, these indexes need a measure of the total value of each component of the index. It may be natural to use the market capitalization of each stock, which is equal to the stock's price multiplied by its total shares outstanding. However, some of the outstanding shares are not really available to trade. This is because some of the firm's shares are held by investors that could not trade their shares or do not intend to sell their shares, such as those held by insiders or long term holders with a strategic purpose for not trading those shares. Thus, these shares are often removed from the shares outstanding to arrive at the stock's **free float**. Using free float is viewed as increasing investability because only shares that could be bought and sold are included in the calculation. The float-adjusted market capitalization approach is used by major index providers like S&P and FTSE Russell for their indexes.

## Reference

1. Maginn, J., Tuttle, D., Pinto, J., & McLeavey, D. (2007). *Managing investment portfolios: A dynamic process* (3rd edn.). Hoboken: Wiley.

# Chapter 4
# Risk

Most investments inherently contain some level of risk. As regular individual investors, we are unlikely to take advantage of any mispricing in the market and create arbitrage profits. This is because many institutional investors have dedicated a considerable amount of resources to identify and profit from these opportunities the moment they become available. By doing so, such opportunities are quickly eliminated. Therefore, any mispricing that is profitable would unlikely reach the hands of regular individual investors.

The closest instrument to a truly risk-free security is a short maturity US Treasury security. This is because the likelihood of the US government defaulting on very short-term obligations is extremely low and other risks, such as inflation risk and liquidity risk, are also negligible for very short maturities. However, being default risk-free, we expect low returns from US Treasuries relative to comparable securities that expose investors to default risk.

By extension to other securities, to induce investors to put their money in riskier securities, the risky security must offer the prospect of higher returns. This means that we should expect higher returns from a particular investment only if we are willing to take on more risk. This is known as the risk-return trade-off, which we discuss in Sect. 4.1.

As a concept, risk is fairly easy to grasp. We know that larger swings in price are riskier than smaller swings in price. We know that more frequent price changes are riskier than less frequent price changes. We know that putting your money in a savings account is less risky than investing in stocks. However, a measure that quantifies all attributes of risk still eludes us. Thus, the measures of risk we use today, including those we discuss in this chapter, are all imperfect measures of risk.

There is some appeal to a quantitative measure of risk. In the late 1950s, Harry Markowitz gave us what is still the most common measure of risk we use today, which is the **variance** of a security's price. The variance or, its positive square root, **standard deviation** is a measure of how far a security's return deviates from its average during the period. Variance does not care whether the deviation from the

© The Author(s), under exclusive license to Springer Nature Switzerland AG 2021      103
C.S. Ang, *Analyzing Financial Data and Implementing Financial Models Using R*,
Springer Texts in Business and Economics,
https://doi.org/10.1007/978-3-030-64155-9_4

average is a positive deviation or a negative deviation. Both are treated as risk. We discuss these in the individual security context in Sect. 4.2 and portfolio context in Sect. 4.3.

Because of the actual or perceived deficiencies of variance as a measure of risk, many other measures of risk were developed and are frequently used in practice. Some focus on measuring loss or **downside risk**, such as Value-at-Risk (VaR) and Expected Shortfall (also known as conditional VaR or tail VaR). We discuss these measures in Sects. 4.4 and 4.5, respectively. In Sect. 4.6, we walk through the calculation of several measures of risk that are modifications of the close price-to-close price calculation that is applied when calculating the variance. Examples of these are Parkinson, Garman–Klass, Rogers–Satchell–Yoon, and Yang and Zhang. These measures have been shown to be several times more efficient than close-to-close volatility.

## 4.1   Risk-Return Trade-Off

The main trade-off we have to consider when making investments is between risk and return. You can expect to get a higher expected return from a security only if you are willing to take on more risk. To see why, let us first take a look at how stocks and bonds have performed over the last 5 years. For purposes of our analysis, we compare the performance of the S&P 500 ETF (SPY) and the SPDR Bloomberg Barclays 1–3 Month T-Bill ETF (BIL) over the 5-year period from 2015 to 2019.

**Step 1: Import SPY and BIL Data**  If load.data() and head.tail() have not been loaded, please refer to Appendix B to show how to pre-load these two functions. We then import SPY Yahoo.csv and BIL Yahoo.csv into R.

```
1  > data.spy <- load.data("SPY Yahoo.csv", "SPY")
2  > head.tail(data.spy)
3              SPY.Open SPY.High SPY.Low SPY.Close SPY.Volume SPY.Adjusted
4  2014-12-31    207.99   208.19  205.39    205.54  130333800      186.2590
5  2015-01-02    206.38   206.88  204.18    205.43  121465900      186.1593
6  2015-01-05    204.17   204.37  201.35    201.72  169632600      182.7974
7              SPY.Open SPY.High SPY.Low SPY.Close SPY.Volume SPY.Adjusted
8  2019-12-27    323.74   323.80  322.28    322.86   42528800      322.86
9  2019-12-30    322.95   323.10  320.55    321.08   49729100      321.08
10 2019-12-31    320.53   322.13  320.15    321.86   57077300      321.86
11 >
12 > data.bil <- load.data("BIL Yahoo.csv", "BIL")
13 > head.tail(data.bil)
14              BIL.Open BIL.High BIL.Low BIL.Close BIL.Volume BIL.Adjusted
15 2014-12-31     91.48    91.48   91.46     91.48     348100      87.49528
16 2015-01-02     91.46    91.48   91.46     91.46     341000      87.47615
17 2015-01-05     91.46    91.48   91.46     91.48    2518300      87.49528
18              BIL.Open BIL.High BIL.Low BIL.Close BIL.Volume BIL.Adjusted
19 2019-12-27     91.42    91.43   91.42     91.42    1325600      91.42
```

| 20 | 2019–12–30 | 91.44 | 91.44 | 91.43 | 91.43 | 1381300 | 91.43 |
|----|------------|-------|-------|-------|-------|---------|-------|
| 21 | 2019–12–31 | 91.45 | 91.45 | 91.43 | 91.43 | 1406200 | 91.43 |

**Step 2: Calculate Returns for SPY and BIL**  To calculate returns, we use `Delt()` on the adjusted close prices. In Line 3, we combine *ret.spy* and *ret.bil* using `cbind()`. We also rename the column names to SPY and BIL to make the names more meaningful.

```
1  > ret.spy <- Delt(data.spy$SPY.Adjusted)
2  > ret.bil <- Delt(data.bil$BIL.Adjusted)
3  > rets <- cbind(ret.spy, ret.bil)
4  > names(rets) <- c("SPY", "BIL")
5  > head.tail(rets)
6                      SPY             BIL
7  2014–12–31          NA              NA
8  2015–01–02 −0.0005352814 −0.0002185947
9  2015–01–05 −0.0180595459  0.0002186425
10                     SPY             BIL
11 2019–12–27 −0.0002477767 −0.0001093952
12 2019–12–30 −0.0055132196  0.0001094071
13 2019–12–31  0.0024292950  0.0000000000
```

**Step 3: Calculate Cumulative Return of SPY and BIL**  Note that the return on December 31, 2014 for both SPY and BIL is NA. Usually we will clean up this dataset and delete that row. However, thinking ahead to what we want to do, which is to create a chart that compares the performance of SPY and BIL assuming we had invested at the close of December 31, 2014, we will want to have the normalized $1 value on December 31, 2014. Thus, we hard code 0 for SPY and BIL in Line 1 because, in Line 3, we calculate the daily gross return that will add 1 to this value. Hence, we will end up with a value of 1 for SPY and 1 for BIL as of December 31, 2014 after implementing Line 3. We then use `cumprod()` to cumulate the daily gross return. The resulting values in the *cum.ret* dataset can be interpreted as the value of a $1 investment in SPY and BIL on December 31, 2014. By December 31, 2019, we can see that the $1 investment in SPY is now worth $1.73 and the $1 investment in BIL is now worth $1.04.

```
1  > (rets[1, ] <- c(0, 0))
2  [1] 0 0
3  > gross.ret <- 1 + rets
4  > cum.ret <- cumprod(gross.ret)
5  > head.tail(cum.ret)
6                SPY       BIL
7  2014–12–31 1.0000000 1.0000000
8  2015–01–02 0.9994647 0.9997814
9  2015–01–05 0.9814148 1.0000000
10               SPY       BIL
11 2019–12–27 1.733392 1.044856
12 2019–12–30 1.723836 1.044971
13 2019–12–31 1.728023 1.044971
```

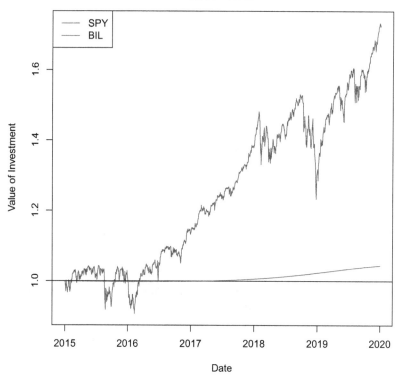

**Fig. 4.1** Difference in returns from investing in stocks versus investing in bonds, 2015–2019. Data source: price data reproduced with permission of CSI ©2020. www.csidata.com

**Step 4: Plot the Normalized Price Chart** From the last step, we know that stocks significantly outperformed bonds over the period 2015 to 2019. However, it is still interesting to see the data visually. As Fig. 4.1 shows, there were periods in 2015 in which an investment in SPY would have underperformed an investment in BIL, but beginning mid-2016 SPY began to substantially outperform BIL. This pattern of outperformance by stocks over bonds holds for even longer time periods.

```
 1  > plot(x = index(cum.ret),
 2  +      y = cum.ret$SPY,
 3  +      xlab = "Date",
 4  +      ylab = "Value of Investment",
 5  +      type = "l",
 6  +      col = "darkgreen",
 7  +      main = "Value of $1 Invested in the S&P 500 index and T-Bills
 8  + 2015 - 2019")
 9  > lines(x = index(cum.ret),
10  +       y = cum.ret$BIL,
```

```
11  +     col = "blue")
12  > abline(h = 1)
13  > legend("topleft",
14  +     c("SPY", "BIL"),
15  +     lty = 1,
16  +     col = c("darkgreen", "blue"))
```

Given how much stocks outperformed bonds, why then would investors bother putting money in bonds? The reason for this is that an investment in stocks is much riskier than an investment in bonds. To see this, we can plot the returns of SPY and BIL.

**Step 5: Plot Daily Returns of SPY and BIL**  We use `plot()` to show a simple chart of the returns of SPY and BIL.

```
1  > plot(rets$SPY,
2  +     col = "red",
3  +     main = "Volatility of S&P 500 Index and T-BIlls")
4  > lines(rets$BIL,
5  +     col = "darkgreen")
```

Figure 4.2 shows the volatility of stock and bond returns. The chart demonstrates how much more volatile stock returns are compared to bond returns. Volatility in this case is reflected in the large swings in the equity returns (red) compared to the relatively flat, close to zero, bond volatility line (green).

## 4.2 Individual Security Risk

**Risk** or **volatility** is something that is difficult to quantitatively measure accurately. Just like how we described Fig. 4.2, we can qualitatively say that a stock's return is volatile, but it is sometimes better to be able to quantify such a statement. It is common for investors to use **variance** or, its positive square root, **standard deviation** as the measure of risk.

The variance of an asset's return, $\sigma^2$, is

$$\sigma^2 = \frac{1}{T-1} \sum_{t=1}^{T} (R_t - \bar{R})^2, \tag{4.1}$$

where $R_t$ is the return of the asset on day $t$ for $t = 1, \ldots, T$ and $\bar{R}$ is the average return of the asset over days 1 to $T$. The standard deviation is equal to the square root of the variance and is denoted by $\sigma$. This calculation technically calculates the sample variance and sample standard deviation. For ease of exposition, going forward, we drop the term sample when describing variance and standard deviation.

What Eq. (4.1) tells us about variance as a measure of volatility is that variance captures deviations from the average. This deviation is squared because, from the

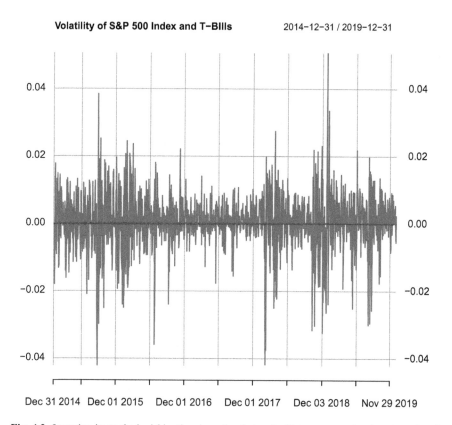

**Fig. 4.2** Investing in stocks is riskier than investing in bonds. Data source: price data reproduced with permission of CSI ©2020. www.csidata.com

perspective of variance, both positive and negative deviations from the mean are considered risk. Had we not squared the deviations from the mean, positive and negative deviations could offset each other.

### 4.2.1 Standard Deviation and Variance

We now calculate the variance and standard deviation of SPY and BIL returns over the period 2015 to 2019.

**Step 1: Create a Combined Return Dataset** Using cbind(), we combine the SPY and BIL returns into one dataset. Note that we removed the observation for December 31, 2014 because we only needed that to calculate the return on the first trading day of 2015.

```
1  > returns <- cbind(ret.spy[-1, ], ret.bil[-1, ])
2  > names(returns) <- c("SPY", "BIL")
3  > head.tail(returns)
4                      SPY            BIL
5  2015-01-02 -0.0005352814 -0.0002185947
6  2015-01-05 -0.0180595459  0.0002186425
7  2015-01-06 -0.0094190191 -0.0002185947
8                      SPY            BIL
9  2019-12-27 -0.0002477767 -0.0001093952
10 2019-12-30 -0.0055132196  0.0001094071
11 2019-12-31  0.0024292950  0.0000000000
```

**Step 2: Calculate SPY and BIL Standard Deviation**  We use sd() to calculate the standard deviation of the returns.

```
1  > (sd.spy <- sd(returns$SPY))
2  [1] 0.008455189
3  > (sd.bil <- sd(returns$BIL))
4  [1] 0.0001624406
```

**Step 3: Calculate SPY and BIL Variance**  We use var() to calculate the variance of the returns directly from the *returns* dataset. Alternative, we can get to the same answer by squaring the standard deviation. Note that we used scipen = 999 because the BIL variance is small (as we would expect given that it is a proxy for the risk-free rate) and the output would convert to scientific notation. A higher scipen value increases the threshold before R converts values to scientific notation.

```
1  > options(scipen = 999)
2  >
3  > (var.spy <- as.numeric(var(returns$SPY)))
4  [1] 0.00007149022
5  > sd.spy^2
6  [1] 0.00007149022
7  >
8  > (var.bil <- as.numeric(var(returns$BIL)))
9  [1] 0.00000002638694
10 > sd.bil^2
11 [1] 0.00000002638694
12 >
13 > options(scipen = 0)
```

Note that the variance and standard deviation change over time. Below, we show the standard deviation of the returns for each year from 2015 to 2019.

**Step 1: Add Year Variable to Returns Dataset**  In order to separate out the data by year, we need to first create a *Year* variable. Note that the *Year* dataset has a capital Y. This is because when we use cbind() in Line 7, the dataset name becomes the variable name.

```
1  > Year <- as.Date(index(returns))
2  > Year <- format(Year, "%Y")
3  > head.tail(Year)
4  [1] "2015" "2015" "2015"
5  [1] "2019" "2019" "2019"
```

```
 6   >
 7   > returns <- cbind(returns, Year)
 8   > head.tail(returns)
 9                    SPY            BIL Year
10   2015-01-02 -0.0005352814 -0.0002185947 2015
11   2015-01-05 -0.0180595459  0.0002186425 2015
12   2015-01-06 -0.0094190191 -0.0002185947 2015
13                    SPY            BIL Year
14   2019-12-27 -0.0002477767 -0.0001093952 2019
15   2019-12-30 -0.0055132196  0.0001094071 2019
16   2019-12-31  0.0024292950  0.0000000000 2019
```

**Step 2: Calculate Standard Deviation for 2015**  We use subset() to keep only returns in 2015. We then use apply() to apply the function sd() in the columns of the dataset *returns*. However, since Column 3 in returns is the Year column, we delete that column when calculating the standard deviation.

```
 1   > ret.2015 <- subset(returns, returns$Year == "2015")
 2   > head.tail(ret.2015)
 3                    SPY            BIL Year
 4   2015-01-02 -0.0005352814 -0.0002185947 2015
 5   2015-01-05 -0.0180595459  0.0002186425 2015
 6   2015-01-06 -0.0094190191 -0.0002185947 2015
 7                    SPY            BIL Year
 8   2015-12-29  0.010671977  0.0000000000 2015
 9   2015-12-30 -0.007087809  0.0002186300 2015
10   2015-12-31 -0.010003283 -0.0002185822 2015
11   >
12   > (sd.2015 <- apply(ret.2015[, -3], 2, sd))
13           SPY           BIL
14   0.0097211644 0.0001656093
```

**Step 3: Calculate Standard Deviation for 2016–2019**  We use the same methodology in Step 2 to calculate the standard deviation for each year from 2016 to 2019.

```
 1   > ret.2016 <- subset(returns, Year == "2016")
 2   > (sd.2016 <- apply(ret.2016[, -3], 2, sd))
 3           SPY           BIL
 4   0.0082278382 0.0001849057
 5   >
 6   > ret.2017 <- subset(returns, Year == "2017")
 7   > (sd.2017 <- apply(ret.2017[, -3], 2, sd))
 8           SPY           BIL
 9   0.0042482157 0.0001925367
10   >
11   > ret.2018 <- subset(returns, Year == "2018")
12   > (sd.2018 <- apply(ret.2018[, -3], 2, sd))
13           SPY           BIL
14   0.0107321803 0.0001169287
15   >
16   > ret.2019 <- subset(returns, Year == "2019")
17   > (sd.2019 <- apply(ret.2019[, -3], 2, sd))
18           SPY           BIL
```

19    0.007881128 0.000119230

**Step 4: Plot the Annual Standard Deviation** We first combine the five datasets that contain the standard deviation for each year. Since we are stacking the datasets on top of each other, we use `rbind()`.

```
1  > sd.all <- rbind(sd.2015, sd.2016, sd.2017, sd.2018, sd.2019)
2  > rownames(sd.all) <- seq(2015, 2019, 1)
3  > sd.all
4            SPY         BIL
5  2015 0.009721164 0.0001656093
6  2016 0.008227838 0.0001849057
7  2017 0.004248216 0.0001925367
8  2018 0.010732180 0.0001169287
9  2019 0.007881128 0.0001192300
```

The standard deviations reported above are daily standard deviations. Since we typically want to report the annualized standard deviation, we multiply the dataset *sd.all* by $\sqrt{252}$. The 252 inside the square root sign denotes the typical number of trading days in a calendar year.

```
1  > (sd.all <- sd.all * sqrt(252))
2            SPY         BIL
3  2015 0.15431870 0.002628966
4  2016 0.13061288 0.002935287
5  2017 0.06743833 0.003056425
6  2018 0.17036808 0.001856185
7  2019 0.12510903 0.001892717
```

We then chart the data using a bar plot (Fig. 4.3) in which the two bars are beside each other. To do this, we first transpose the data using `t()`. Then, we use `barplot()` with the argument `beside = TRUE`, which tells R not to stack the bars.

```
1  > (t.sd <- t(sd.all))
2            2015        2016        2017        2018        2019
3  SPY 0.154318700 0.130612881 0.067438334 0.170368081 0.125109035
4  BIL 0.002628966 0.002935287 0.003056425 0.001856185 0.001892717
5  > barplot(t.sd,
6  +     beside = TRUE,
7  +     main = "Annualized Standard Deviation of SPY and BIL Returns
8  +     2015 to 2019",
9  +     ylim = c(0, 0.2),
10 +     col = c("blue", "red"),
11 +     border = c(0, 0),
12 +     legend.text = c("SPY", "BIL"))
```

## 4.3  Portfolio Risk

The previous section discusses how to calculate risk for an individual security. However, most investments are done in the context of a portfolio of securities. Markowitz [7] showed that it is the **covariance** of the assets in the portfolio that

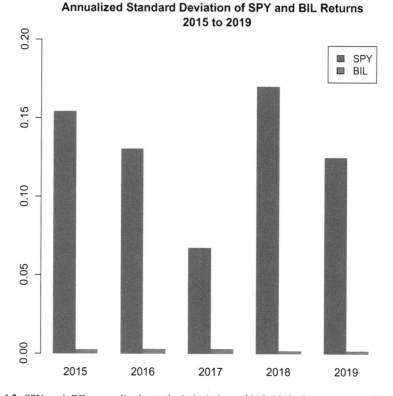

**Fig. 4.3** SPY and BIL annualized standard deviations, 2015–2019. Data source: price data reproduced with permission of CSI ©2020. www.csidata.com

is important when assessing risk in the context of a portfolio. In particular, when we add assets that are not perfectly correlated with the securities in our portfolio, the overall risk of the portfolio decreases. In fact, Reilly and Brown [9] state that using 12–18 well-selected stocks can yield 90% of the maximum benefits of diversification. Therefore, a portfolio's risk is generally lower than the weighted average of the standard deviations of each of the securities in the portfolio.

### 4.3.1   Two Assets Using Manual Approach

In the case of a two-asset portfolio, portfolio risk is calculated as

$$\sigma_p^2 = w_1^2 \sigma_1^2 + w_2^2 \sigma_2^2 + 2\sigma_{1,2} w_1 w_2, \tag{4.2}$$

where $w_i$ is the weight of security $i$ in the portfolio, $\sigma_i$ is the volatility of security $i$, and $\sigma_{1,2}$ is the covariance between the returns of securities 1 and 2. Note that $\sigma_{1,2} = \rho_{1,2}\sigma_1\sigma_2$, so in some instances Eq. (4.2) may be shown using the correlation coefficient term ($\rho_{1,2}$) instead, but we have to add the product of the standard deviation of the two assets ($\sigma_1\sigma_2$).

Let us work through an example of calculating the portfolio risk for a \$10,000 portfolio that invested \$6000 in SPY and \$4000 in BIL.

**Step 1: Determine Weights of Securities in Portfolio** This above investment amounts mean that the weights for our securities would be 60% SPY and 40% BIL.

```
1  > w.spy <- 0.6
2  > w.bil <- 0.4
```

**Step 2: Calculate Annualized Standard Deviation and Covariance for SPY and BIL** We have the daily standard deviations, *sd.spy* and *sd.bil*, calculated earlier. We can annualize those standard deviations by multiplying them by $\sqrt{252}$ (Lines 1 and 3). In Line 6, we then use cov() to calculate the covariance of SPY and BIL. Covariance is a measure of how the returns of SPY and BIL move together. To get to the annualized covariance, we multiply the daily covariance by 252 (Line 6).

```
1  > (annual.sd.spy <- sd.spy * sqrt(252))
2  [1] 0.134222
3  > (annual.sd.bil <- sd.bil * sqrt(252))
4  [1] 0.002578664
5  > options(scipen = 999)
6  > (covar <- as.numeric(cov(returns$SPY, returns$BIL) * 252))
7  [1] 0.0000009189221
8  > options(scipen = 0)
```

**Step 3: Calculate Portfolio Risk** Using Eq. (4.2), we calculate the portfolio variance (Lines 1 and 2). Taking the square root of the portfolio variance gives us the portfolio standard deviation of 8.1% (Line 4).

```
1  > (port.var <- w.spy^2 * annual.sd.spy^2 +
2  +      w.bil^2 * annual.sd.bil^2 + 2 * covar * w.spy * w.bil)
3  [1] 0.006487098
4  > (port.sd <- sqrt(port.var))
5  [1] 0.08054252
```

**Verifying Relationship Between Covariance and Correlation**

Mathematically, we know $\sigma_{1,2} = \rho_{1,2}\sigma_1\sigma_2$, where $\sigma_{1,2}$ is the covariance of assets 1 and 2, $\rho_{1,2}$ is the correlation of assets 1 and 2, $\sigma_1$ is the standard deviation of asset 1, and $\sigma_2$ is the standard deviation of asset 2. Thus, to verify the relationship between covariance and correlation, we first calculate the correlation between SPY and BIL. We then multiply the correlation by

(continued)

the annualized standard deviation of SPY and BIL. As the output shows, the covariance we calculate in Line 5 is identical to the covariance we calculate in Line 7 of Step 2 above.

```
1  > (correl <- as.numeric(cor(returns$SPY, returns$BIL)))
2  [1] 0.002654974
3  > options(scipen = 999)
4  > correl * annual.sd.spy * annual.sd.bil
5  [1] 0.0000009189221
6  > options(scipen = 0)
```

### 4.3.2  Two Assets Using Matrix Algebra

We have three terms in Eq. (4.2), which is the formula to calculate the portfolio risk for two assets. The number of terms required to calculate portfolio risk grows with the number of securities we add to our portfolio because the covariance term (e.g., the third term in Eq. (4.2)) has to be calculated for each pair of securities in the portfolio. For $n$ securities, we have $n$ variance terms and $n(n-1)$ covariance terms. As such, if we had 10 securities, we would have 10 variance terms and 90 [= (10 * 9)] covariance terms. Since most portfolios are constructed with more than two securities, manually laying out and calculating each set of covariance terms become extremely cumbersome and could get quite confusing to implement. However, we can use matrix algebra to make this process simpler.

Consider a portfolio with $n$ assets. Let the vector of weights be

$$\mathbf{w} = (w_1 \quad w_2 \quad \cdots \quad w_n),\qquad(4.3)$$

and the **variance–covariance matrix** be

$$\Sigma = \begin{pmatrix} \sigma_1^2 & \sigma_{1,2} & \cdots & \sigma_{1,n} \\ \sigma_{2,1} & \sigma_2^2 & \cdots & \sigma_{2,n} \\ \vdots & \vdots & \ddots & \vdots \\ \sigma_{n,1} & \sigma_{n,2} & \cdots & \sigma_n^2 \end{pmatrix}.\qquad(4.4)$$

Thus, portfolio variance equals

$$\sigma_P^2 = \mathbf{w}\Sigma\mathbf{w}^\mathsf{T}.\qquad(4.5)$$

The $\mathsf{T}$ superscript denotes the transposition of a matrix, which in our cases flips the row vector of weights into a column vector of weights.

Let us now show how we calculate the portfolio risk for the same two-asset portfolio in our earlier example but using matrix algebra.

**Step 1: Create Vector of Weights** The weight vector is created by using `matrix()`. This corresponds to the first term in Eq. (4.5) of **w**. The third term in the equation is the transposed version of **w**.

```
1  > (weight <- matrix(c(w.spy, w.bil), 1))
2        [,1] [,2]
3  [1,]  0.6  0.4
```

**Step 2: Construct Variance–Covariance Matrix** We first create a matrix of the returns. We use the *returns* dataset but drop the **Year** column (Column 3). We then use `cov()` to generate the variance–covariance matrix. We annualize the variances and covariances by multiplying the variance–covariance matrix by 252.

```
1  > mat.rets <- as.matrix(returns[, -3])
2  > head.tail(mat.rets)
3                     SPY              BIL
4  2015-01-02 -0.0005352814 -0.0002185947
5  2015-01-05 -0.0180595459  0.0002186425
6  2015-01-06 -0.0094190191 -0.0002185947
7                     SPY              BIL
8  2019-12-27 -0.0002477767 -0.0001093952
9  2019-12-30 -0.0055132196  0.0001094071
10 2019-12-31  0.0024292950  0.0000000000
11 > options(scipen = 999)
12 > (vcov <- cov(mat.rets) * 252)
13                 SPY              BIL
14 SPY 0.0180155364219 0.0000009189221
15 BIL 0.0000009189221 0.0000066495084
16 > options(scipen = 0)
```

**Step 3: Calculate Portfolio Risk** Using Eq. (4.5), we calculate the portfolio variance (Line 1). We then take the square root of the portfolio variance to arrive at the portfolio standard deviation of 8.1% (Line 4). This is the exact portfolio standard deviation that we calculated in the manual approach.

```
1  > (mat.port.var <- weight %*% vcov %*% t(weight))
2             [,1]
3  [1,] 0.006487098
4  > (mat.port.sd <- sqrt(mat.port.var))
5            [,1]
6  [1,] 0.08054252
```

### 4.3.3  Multiple Assets

In this section, we show how to extend the above technique to calculate portfolio risk using matrix algebra for multiple assets. This will not be the only time we will use matrix algebra in a portfolio context. We will use this technique again when we discuss mean–variance portfolio optimization in Chap. 7 . We continue from our previous example but add AMZN and GOOG to the portfolio. For our example, we assume that our $10,000 portfolio is now invested $3000 in AMZN, $3000 in GOOG, $2000 in SPY, and $2000 in BIL.

**Step 1: Calculate Individual Security Returns** Since we already have *ret.spy* and *ret.bil* in the R environment, we only need to calculate the returns for AMZN and GOOG. Then, we combine the returns in the *rets* dataset.

```
1  > data.amzn <- load.data("AMZN Yahoo.csv", "AMZN")
2  > data.goog <- load.data("GOOG Yahoo.csv", "GOOG")
3  >
4  > ret.amzn <- Delt(data.amzn$AMZN.Adjusted)
5  > ret.goog <- Delt(data.goog$GOOG.Adjusted)
6  >
7  > rets <- cbind(ret.amzn[-1, ], ret.goog[-1, ], ret.spy[-1, ], ret.bil[-1, ])
8  > names(rets) <- c("AMZN", "GOOG", "AAPL", "SPY")
9  > head.tail(rets)
10                   AMZN            GOOG           AAPL              SPY
11 2015-01-02 -0.005896623 -0.003020489 -0.0005352814 -0.0002185947
12 2015-01-05 -0.020517267 -0.020845731 -0.0180595459  0.0002186425
13 2015-01-06 -0.022833293 -0.023177050 -0.0094190191 -0.0002185947
14                   AMZN            GOOG           AAPL              SPY
15 2019-12-27  0.0005511802 -0.0062555196 -0.0002477767 -0.0001093952
16 2019-12-30 -0.0122526652 -0.0116503560 -0.0055132196  0.0001094071
17 2019-12-31  0.0005143517  0.0006586174  0.0024292950  0.0000000000
```

**Step 2: Create Vector of Weights** The investment amounts in the different securities indicate a weight of 30% for AMZN, 30% for GOOG, 20% for SPY, and 20% for BIL.

```
1  > (weight <- matrix(weight, 1))
2       [,1] [,2] [,3] [,4]
3  [1,]  0.3  0.3  0.2  0.2
```

**Step 3: Construct Variance–Covariance Matrix** We first create a matrix of the returns by using  matrix() on the *rets* dataset. Then, we use cov() to generate the variance–covariance matrix. We multiply the variance–covariance matrix by 252 to annualize the variances and covariances.

```
1  > mat.rets <- as.matrix(rets)
2  > head.tail(mat.rets)
3                   AMZN            GOOG           AAPL              SPY
4  2015-01-02 -0.005896623 -0.003020489 -0.0005352814 -0.0002185947
5  2015-01-05 -0.020517267 -0.020845731 -0.0180595459  0.0002186425
6  2015-01-06 -0.022833293 -0.023177050 -0.0094190191 -0.0002185947
7                   AMZN            GOOG           AAPL              SPY
8  2019-12-27  0.0005511802 -0.0062555196 -0.0002477767 -0.0001093952
9  2019-12-30 -0.0122526652 -0.0116503560 -0.0055132196  0.0001094071
10 2019-12-31  0.0005143517  0.0006586174  0.0024292950  0.0000000000
11 > options(scipen = 999)
```

```
12  > (vcov <- cov(mat.rets) * 252)
13                    AMZN            GOOG            AAPL            SPY
14  AMZN   0.085500156136  0.044974252237  0.0237802740394 -0.0000070028676
15  GOOG   0.044974252237  0.057580505313  0.0214485761771 -0.0000048924342
16  AAPL   0.023780274039  0.021448576177  0.0180155364219  0.0000009189221
17  SPY   -0.000007002868 -0.000004892434  0.0000009189221  0.0000066495084
18  > options(scipen = 0)
```

**Step 4: Calculate Portfolio Risk** Using Eq. (4.5), we calculate the portfolio variance. We then take the square root of the portfolio variance to arrive at the portfolio standard deviation of 16.5%.

```
1  > (mat.port.var <- weight %*% vcov %*% t(weight))
2            [,1]
3  [1,] 0.02711962
4  > (mat.port.sd <- sqrt(mat.port.var))
5            [,1]
6  [1,] 0.1646804
```

**Benefit of Diversification**

The portfolio standard deviation of 16.5% is lower than the weighted average standard deviation of the four securities in the portfolio of 18.7%. In fact, when assets that are not perfectly correlated are added to a portfolio, there is at least some diversification benefit.

```
1  > (annual.sd.amzn <- sd(rets$AMZN) * sqrt(252))
2  [1] 0.2924041
3  > (annual.sd.goog <- sd(rets$GOOG) * sqrt(252))
4  [1] 0.2399594
5  > (sd.vector <- c(annual.sd.amzn, annual.sd.goog,
6  +    annual.sd.spy, annual.sd.bil))
7  [1] 0.292404097 0.239959383 0.134221967 0.002578664
8  > (sd.mat <- as.matrix(sd.vector, 1))
9            [,1]
10  [1,] 0.292404097
11  [2,] 0.239959383
12  [3,] 0.134221967
13  [4,] 0.002578664
14  > weight %*% sd.mat
15            [,1]
16  [1,] 0.1870692
```

## 4.4   Value-at-Risk

A popular measure of the risk of loss in a portfolio is **Value-at-Risk** or **VaR**. VaR
measures the loss in our portfolio over a pre-specified time horizon, assuming some
level of probability. For example, we will calculate as an example the 1 and 5%
1-Day VaR on December 31, 2019 of a Hypothetical Portfolio (see Appendix D
for construction of hypothetical portfolio). In this section, we will use the market
convention of representing VaR as a positive number and using the significance level
(e.g., 1 or 5%) instead of confidence level (e.g., 99 or 95%). We also implement two
types of VaR calculations: Gaussian VaR and Historical VaR. The former assumes
that the data follow a normal or Gaussian distribution, while the latter uses the
distributional properties of the actual data. As such, Historical VaR requires more
data to implement and Gaussian VaR requires less data to implement.

For our example, we assume that we made an investment in a hypothetical
portfolio on December 31, 2018 of \$1,000,000. From the prior chapter, we know
the value of the portfolio would have increased to \$1,465,801 as of December 31,
2019. We will calculate the 1-Day VaR on this date based on a 1 and 5% significance
level.

### 4.4.1   Gaussian VaR

One of the simplest approaches to estimate VaR is to assume that the portfolio
returns follow a normal distribution. Hence, this approach is called **Gaussian VaR**.
Because of the distributional assumption, we only need at least 1 year of daily
returns data, which is approximately 252 observations.

**Step 1: Import Portfolio Returns Data** We use `read.csv()` to import the
return data for the hypothetical portfolio into R. We have to construct a variable
that is recognized by R as a date. We then create a workable data frame object with
date and portfolio return variables.

```
1  > returns <- read.csv("Hypothetical Portfolio (Daily).csv", header = TRUE)
2  > head.tail(returns)
3     X        Date      Port.Ret
4   1 1 2019-01-02  0.01189697
5   2 2 2019-01-03 -0.05120679
6   3 3 2019-01-04  0.04896317
7        X        Date      Port.Ret
8    250 250 2019-12-27 -0.001856224
9    251 251 2019-12-30 -0.004619646
10   252 252 2019-12-31  0.003370195
11  >
12  > Date <- as.Date(returns$Date)
13  > Date <- data.frame(Date)
14  > head.tail(Date)
15          Date
```

```
16   1 2019–01–02
17   2 2019–01–03
18   3 2019–01–04
19              Date
20   250 2019–12–27
21   251 2019–12–30
22   252 2019–12–31
23   > str(Date)
24   'data.frame': 252 obs. of  1 variable:
25    $ Date: Date, format: "2019–01–02" "2019–01–03" "2019–01–04" ...
26   >
27   > Port.Ret <- returns[, 3]
28   > rets <- cbind(Date, Port.Ret)
29   > head.tail(rets)
30              Date      Port.Ret
31   1 2019–01–02  0.01189697
32   2 2019–01–03 –0.05120679
33   3 2019–01–04  0.04896317
34              Date      Port.Ret
35   250 2019–12–27 –0.001856224
36   251 2019–12–30 –0.004619646
37   252 2019–12–31  0.003370195
38   > str(rets)
39   'data.frame': 252 obs. of  2 variables:
40    $ Date    : Date, format: "2019–01–02" "2019–01–03" "2019–01–04" ...
41    $ Port.Ret: num  0.0119 –0.0512 0.049 0.0105 0.0143 ...
```

**Step 2: Calculate the Mean and Standard Deviation of the Portfolio Returns**
We calculate the average or mean portfolio return using mean(), and we calculate
the standard deviation of returns using sd(). The portfolio mean return is 0.16%.
In practice, some assume that because we are calculating VaR over a short horizon,
the mean return is zero. In our example, we do not make this assumption but we can
see from the output below that our calculated mean is small. The standard deviation
of portfolio returns is 1.3%.

```
1   > (port.ret <- mean(rets$Port.Ret))
2   [1] 0.001604052
3   > (port.sd  <- sd(rets$Port.Ret))
4   [1] 0.01308291
```

**Step 3: Calculate 1 and 5% VaR**  The VaR is calculated as follows:

$$VaR_\alpha = -(\mu - \sigma * Z_\alpha)I, \tag{4.6}$$

where $\alpha$ is the significance level of the VaR, $\mu$ is the portfolio average return, $\sigma$
is the standard deviation of the portfolio returns, $Z_\alpha$ is the $z$-score based on the
VaR significance level, and $I$ is the current portfolio value. The $z$-score is calculated
using qnorm(), which returns the inverse cumulative density function.  qnorm()
takes as an argument the desired significance level (e.g., 1 or 5%). As such, for the
1% Gaussian VaR, we use qnorm(0.01), and for the 5% Gaussian VaR, we use
qnorm(0.05).

```
1  > (var01.gauss <- (port.ret + qnorm(0.01) * port.sd) * 1465801)
2  [1] -42261.03
3  > (var01.gauss <- abs(var01.gauss))
4  [1] 42261.03
5  >
6  > (var05.gauss <- (port.ret + qnorm(0.05) * port.sd) * 1465801)
7  [1] -29192.05
8  > (var05.gauss <- abs(var05.gauss))
9  [1] 29192.05
```

What then is the interpretation of the above result? The above means that there is a 1% (5%) chance that our portfolio loses more than \$42,261 (\$29,192) over the next day.

### 4.4.2  Historical VaR

**Historical VaR** uses a mix of current weights in the portfolio and a simulation of historical returns of the securities in the portfolio to construct a simulated series of portfolio profits and losses. Given its use of historical returns data, Historical VaR works better when we have lots of data. Typically, 3–5 years of data is recommended for this approach. We use 5 years in our illustration below.

**Step 1: Import 5 Years of Data for AMZN, GOOG, and AAPL** We import the data for the three securities using `load.data()`. We then calculate the daily total returns for each security from 2015 to 2019.

```
1   > data.amzn <- load.data("AMZN Yahoo.csv", "AMZN")
2   > ret.amzn <- Delt(data.amzn$AMZN.Adjusted)
3   > ret.amzn <- ret.amzn[-1, ]
4   > names(ret.amzn) <- "AMZN"
5   > head.tail(ret.amzn)
6                         AMZN
7   2015-01-02 -0.005896623
8   2015-01-05 -0.020517267
9   2015-01-06 -0.022833293
10                        AMZN
11  2019-12-27  0.0005511802
12  2019-12-30 -0.0122526652
13  2019-12-31  0.0005143517
14  >
15  > data.goog <- load.data("GOOG Yahoo.csv", "GOOG")
16  > ret.goog <- Delt(data.goog$GOOG.Adjusted)
17  > ret.goog <- ret.goog[-1, ]
18  > names(ret.goog) <- "GOOG"
19  > head.tail(ret.goog)
20                        GOOG
21  2015-01-02 -0.003020489
22  2015-01-05 -0.020845731
23  2015-01-06 -0.023177050
24                        GOOG
```

```
25   2019-12-27 -0.0062555196
26   2019-12-30 -0.0116503560
27   2019-12-31  0.0006586174
28   >
29   > data.aapl <- load.data("AAPL Yahoo.csv", "AAPL")
30   > ret.aapl <- Delt(data.aapl$AAPL.Adjusted)
31   > ret.aapl <- ret.aapl[-1, ]
32   > names(ret.aapl) <- "AAPL"
33   > head.tail(ret.aapl)
34                     AAPL
35   2015-01-02 -0.00951261035
36   2015-01-05 -0.02817165617
37   2015-01-06  0.00009433099
38                     AAPL
39   2019-12-27 -0.0003794833
40   2019-12-30  0.0059351314
41   2019-12-31  0.0073065487
```

**Step 2: Combine Returns in One Dataset**   Using `cbind()`, we combine the three
return series into one dataset labeled *ret.data*.

```
1    > ret.data <- cbind(ret.amzn, ret.goog, ret.aapl)
2    > head.tail(ret.data)
3                     AMZN          GOOG         AAPL
4    2015-01-02 -0.005896623 -0.003020489 -0.00951261035
5    2015-01-05 -0.020517267 -0.020845731 -0.02817165617
6    2015-01-06 -0.022833293 -0.023177050  0.00009433099
7                     AMZN          GOOG         AAPL
8    2019-12-27  0.0005511802 -0.0062555196 -0.0003794833
9    2019-12-30 -0.0122526652 -0.0116503560  0.0059351314
10   2019-12-31  0.0005143517  0.0006586174  0.0073065487
```

**Step 3: Identify the Percentage Invested in Each Security as of December 31,
2019**   From the exercise of constructing the portfolio in Appendix D, we know that
the weights of the securities are 31.8% for AMZN, 31.3% for GOOG, and 36.9%
for AAPL. We then use these weights to disaggregate the $1,465,801 portfolio value
to how much is invested in AMZN, GOOG, and AAPL.

```
1    > w.amzn <- 0.3186484
2    > w.goog <- 0.3123183
3    > w.aapl <- 0.3690333
4    > value <- 1465801
5    > (v.amzn <- w.amzn * value)
6    [1] 467075.1
7    > (v.goog <- w.goog * value)
8    [1] 457796.5
9    > (v.aapl <- w.aapl * value)
10   [1] 540929.4
```

**Step 4: Calculate Simulated Portfolio Return Applying Current Security
Values to Historical Security Returns**   We assume that the current value of each
security in the portfolio remains frozen over our VaR time horizon (e.g., 1 day).

We then apply each daily return combination for AMZN, GOOG, and AAPL to the current value of each security. That is, the portfolio profit and loss $\pi$ are defined as follows:

$$\pi_t = V^{AMZN} R_t^{AMZN} + V^{GOOG} R_t^{GOOG} + V^{AAPL} R_t^{AAPL}, \tag{4.7}$$

where $V$ denotes the value of the security in the portfolio and $R_t$ is the return of the security on day $t$. The variable name generated by this step is AMZN.Ret, which is misleading, so we rename this variable to PnL, which is more fitting.

```
1   > sim.pnl <- v.amzn * ret.data$AMZN +
2   +     v.goog * ret.data$GOOG +
3   +     v.aapl * ret.data$AAPL
4   > names(sim.pnl) <- "PnL"
5   > head.tail(sim.pnl)
6                    PnL
7   2015-01-02  -9282.586
8   2015-01-05 -34365.084
9   2015-01-06 -21224.209
10                   PnL
11  2019-12-27  -2811.586
12  2019-12-30  -7845.920
13  2019-12-31   4494.080
```

**Step 5: Calculate Appropriate Quantile for 1 and 5% VaR**  To find the VaR of significance level $\alpha$, we use quantile(). The first argument is the *negative* of the simulated portfolio profit and loss, and the second argument is the $1 - \alpha$ confidence level.

```
1   > (var01.hist <- as.numeric(quantile(-sim.pnl$PnL, 0.99)))
2   [1] 58408.56
3   > (var05.hist <- as.numeric(quantile(-sim.pnl$PnL, 0.95)))
4   [1] 32777.87
```

Note that the higher the significance level (i.e., 5% vs. 1%) or the lower the confidence level (i.e., 95% vs. 99%), the lower the VaR amount. To see why this is the case, suppose we were asked to give our best estimate of a number that we are sure we would not exceed. The higher the estimate of this number we come up with, the higher the likelihood that we will not exceed it. Accordingly, we should expect to have a higher number at the 99% confidence level than at the 95% confidence level.

**Step 6: Plot the VaR in Relation to Profit and Loss Density**  Sometimes it may be easier for us to visualize the data by plotting (i) the density of the simulated portfolio profits and losses, (ii) the normal distribution of profits and losses based on the mean and standard deviation of the simulated portfolio profits and losses, and (iii) our estimates of the 1 and 5% 1-Day Historical VaR. Below, we first plot (i) and (iii), and then we add (ii).

To get the density of the simulated portfolio profits and losses, we use density().

```
1   > (ret.d <- density(sim.pnl$PnL))
2
3   Call:
4    density.default(x = sim.pnl$PnL)
5
6   Data: sim.pnl$PnL (1258 obs.); Bandwidth 'bw' = 3217
7
8           x                 y
9    Min.    :-88361   Min.    :0.000000001125
10   1st Qu.:-35891    1st Qu.:0.000000216745
11   Median : 16578    Median :0.000000933908
12   Mean   : 16578    Mean    :0.000004760038
13   3rd Qu.: 69047    3rd Qu.:0.000005316302
14   Max.    :121516   Max.    :0.000028759710
```

We can now plot (i) and (iii). Note that we add two vertical lines using abline() (Lines 7 and 8). The two lines denote our estimates of VaR, so we can visually see from the density of the simulated portfolio profits and losses, where the Historical VaR estimates are.

```
1   > plot(ret.d,
2   +     xlab = "Profit & Loss",
3   +     ylab = "",
4   +     yaxt = "n",
5   +     main = "Density of Simulated Portfolio P&L Over Three Years
6   +     And 1% and 5% 1-Day Historical Value-at-Risk (VaR)")
7   > abline(v = -quantile(-sim.pnl$PnL, 0.99), col = "red")
8   > abline(v = -quantile(-sim.pnl$PnL, 0.95), col = "blue")
```

Next, we now turn to number (ii) on the list, which is the plot of the normal distribution based on the mean and standard deviation of the simulated portfolio profits and losses. First, we create a sequence of 1000 numbers between the smallest and largest simulated profits and losses. This creates bounds for the normal density plot that follows.

```
1   > x <- seq(min(sim.pnl$PnL), max(sim.pnl$PnL), length = 1000)
2   > head.tail(x)
3   [1] -78710.10 -78519.33 -78328.56
4   [1] 111484.1 111674.9 111865.7
```

Then, we use dnorm() to provide the values of the probability density function for the normal distribution. dnorm() takes on three arguments. The first argument uses the x vector we created above. The second argument is the mean of the simulated portfolio profits and losses, which we calculate using mean(). The third argument is the standard deviation of the simulated portfolio profits and losses, which we calculate using sd(). This creates a normal density based on the mean and standard deviation of the simulated portfolio profits and losses.

```
1   > y <- dnorm(x, mean = mean(sim.pnl$PnL), sd = sd(sim.pnl$PnL))
2   > head.tail(y)
3   [1] 0.000000006530747 0.000000006783921 0.000000007046273
4   [1] 0.000000000006207750 0.000000000005892707 0.000000000005593146
```

We then use `lines()` to add the normal density plot as a green bell curve onto the plot.

```
1  > lines(x, y, type = "l", col = "darkgreen")
```

Finally, we add a legend on the top-right portion of the chart, so the legend does not obstruct the other elements of the chart.

```
1  > legend("topright",
2  +     c("Simulated P&L Distribution","Normal Distribution",
3  +        "1% 1-Day VaR","5% 1-Day VaR"),
4  +        col = c("black", "darkgreen", "red", "blue"),
5  +        lty = 1)
```

Figure 4.4 shows the results of our plot. The chart shows that the distribution of simulated profits and losses is more peaked than what a normal distribution with the same mean and standard deviation would show. This means that there are more observations that are around the mean or average profits of $1127 in the simulated profits and losses.

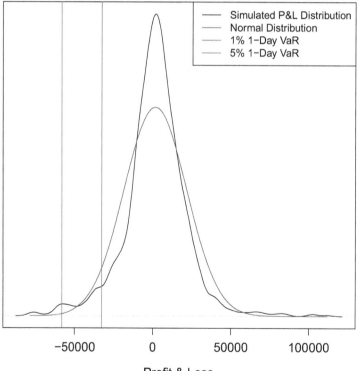

**Density of Simulated Portfolio P&L Over Three Years
And 1% and 5% 1-Day Historical Value−at−Risk (VaR)**

**Fig. 4.4** Density of simulated portfolio profits and losses and Historical Value-at-Risk

> **Using the Square Root of $T$ Rule to**
> **Scale 1-Day VaR To a $T$-Day VaR**
>
> Assuming the returns are independent and identically distributed (i.i.d), we can calculate a $T$-Day VaR by multiplying the 1-Day VaR by $\sqrt{T}$. For example, our 1% 1-Day Historical VaR was $58,408. Assuming returns are i.i.d., the 1% 10-Day Historical VaR is equal to $184,704.

```
1  > as.numeric(quantile(-sim.pnl$PnL, 0.99) * sqrt(10))
2  [1] 184704.1
```

## 4.5 Expected Shortfall

VaR has been criticized for not being able to fully capture how large the loss we should expect if the loss exceeds VaR or what is known as **tail risk**. That is, VaR does not allow us to answer the question, "if losses exceed VaR, how bad should we expect that loss to be?" The answer to this question is called **expected shortfall** (ES). ES is known by many other names, such as **tail VaR** and **tail loss**. Some texts may call our ES measure **expected tail loss** and use the term expected shortfall to denote the average loss that exceeds a benchmark VaR. We stay with our definition of ES but bear in mind some texts may not have the same definition of ES. To parallel our VaR discussion, we show two methods of calculating ES: Gaussian ES and Historical ES. Similar to the difference between Gaussian VaR and Historical VaR, the difference between Gaussian ES and Historical ES is the assumption of a specific distribution or a lack thereof.

### 4.5.1 Gaussian ES

We can also calculate ES by assuming a normal distribution. This is known as a **Gaussian ES**. The calculation of Gaussian ES is

$$\mu + \frac{\sigma * \phi(\Phi(\alpha))}{\alpha}, \tag{4.8}$$

where $\phi(\cdot)$ is the height of the normal probability density function and is implemented in R using `dnorm()`, while $\Phi(\cdot)$ is the inverse normal cumulative density function and is implemented in R using `qnorm()`. $\mu$ is the mean portfolio return, $\sigma$ is the standard deviation of the portfolio returns, and $\alpha$ is the significance level.

```
1  > (es01.gauss <- 1465801 * (port.ret +
```

```
2  +     port.sd * (dnorm(qnorm(.01)) / 0.01)))
3  [1] 53461.89
4  > (es05.gauss <- 1465801 * (port.ret +
5  +     port.sd * (dnorm(qnorm(.05)) / 0.05)))
6  [1] 41907.76
```

The result means that there is a 1% (5%) chance our losses exceed VaR, but when it does, we expect that, on average, we will lose $53,462 ($41,908).

### 4.5.2  Historical ES

The Historical ES is calculated by taking the average portfolio loss that falls short of the Historical VaR estimate.

**Step 1: Identify Historical VaR Limit for Portfolio**  Let us first create variables that hold the historical 1 and 5% VaR estimates. These are the same values we calculated above, but we are taking the negative of the VaR. We will use this value to determine which P&Ls are less than the VaR limit.

```
1  > (var01.limit <- -var01.hist)
2  [1] -58408.56
3  > (var05.limit <- -var05.hist)
4  [1] -32777.87
```

**Step 2: Identify Simulated Portfolio Losses in Excess of VaR**  To calculate ES, we have to find the average of the losses exceeding a loss of $58,409 for the 1% 1-Day Historical ES and $32,778 for the 5% 1-Day Historical ES. As such, the next step would be for us to use the P&L of our simulated portfolio to identify those losses. To do this, we use dummy variables to indicate what returns fall short of the threshold for 1 and 5% Historical ES.

```
1  > es.pnl <- sim.pnl$PnL
2  > es.pnl$d01 <- ifelse(es.pnl$PnL < var01.limit, 1, 0)
3  > es.pnl$d05 <- ifelse(es.pnl$PnL < var05.limit, 1, 0)
4  > head.tail(es.pnl)
5                      PnL d01 d05
6  2015-01-02  -9282.586   0   0
7  2015-01-05 -34365.084   0   1
8  2015-01-06 -21224.209   0   0
9                      PnL d01 d05
10 2019-12-27  -2811.586   0   0
11 2019-12-30  -7845.920   0   0
12 2019-12-31   4494.080   0   0
```

**Step 3: Extract Portfolio Losses in Excess of VaR**  Now that we have dummy variables that identify the days on which the P&L falls short of the Historical VaR estimate, we then extract those observations into a data object for the 1% Historical ES *shortfall01* and 5% Historical ES *shortfall05*. We have 13 observations for the 1% Historical ES, and we have 63 observations for the 5% Historical ES.

```
1   > shortfall01 <- subset(es.pnl[, 1:2], d01 == 1)
2   > dim(shortfall01)
3   [1] 13  2
4   > head.tail(shortfall01)
5                  PnL d01
6   2015-08-21 -76694.06   1
7   2015-08-24 -59973.70   1
8   2016-02-05 -59976.79   1
9                  PnL d01
10  2019-01-03 -78710.10   1
11  2019-05-13 -60747.39   1
12  2019-08-05 -59180.06   1
13  >
14  > shortfall05 <- subset(es.pnl[, c(1, 3)], d05 == 1)
15  > dim(shortfall05)
16  [1] 63  2
17  > head.tail(shortfall05)
18                  PnL d05
19  2015-01-05 -34365.08   1
20  2015-01-27 -37510.83   1
21  2015-03-10 -33493.70   1
22                  PnL d05
23  2019-08-05 -59180.06   1
24  2019-08-14 -44426.01   1
25  2019-08-23 -53964.10   1
```

**Step 4: Compute Average of Losses in Excess of VaR** To get the ES, we take
the average of the returns is each of these data objects. We find that the 1% 1-Day
Historical ES is equal to $66,174. This result means that we have a 1% probability
that the loss in our portfolio would exceed the VaR, but, when it does, we expect
that, on average, we would lose $66,174. Similarly, the 5% 1-Day Historical ES is
equal to $47,269. This result means that we have a 5% probability that the loss in
our portfolio would exceed the VaR, but, when it does, we expect that, on average,
we would lose $47,269.

```
1   > (es01.hist <- -mean(shortfall01$PnL))
2   [1] 66173.53
3   > (es05.hist <- -mean(shortfall05$PnL))
4   [1] 47269
```

### 4.5.3 Comparing VaR and ES

We expect that the ES values to be bigger losses than VaR because, by definition,
the ES is the mean or average loss when losses exceed VaR. This can easily be seen
when we combine the results into one table. We use a combination of rbind()
and cbind() to bring together all the resulting calculations of Historical and
Gaussian VaR and ES. The intermediate results look quite messy with variable
names and row labels that are not too meaningful. As such, we rename the variable

names (Line 3–4) to make them easy to understand and consistent. We also rename
the row names (Line 5) to conform to the convention of using significance levels
and, to avoid confusion, include the time horizon in our VaR and ES calculations.

```
1  > combo <- rbind(cbind(var01.hist, es01.hist, var01.gauss, es01.gauss),
2  +       cbind(var05.hist, es05.hist, var05.gauss, es05.gauss))
3  > colnames(combo) <- c("VaR Historical","ES Historical","VaR Gaussian",
4  +    "ES Gaussian")
5  > rownames(combo) <- c("1% 1-Day", "5% 1-Day")
6  > combo
7          VaR Historical ES Historical VaR Gaussian ES Gaussian
8  1% 1-Day       58408.56      66173.53     42261.03    53461.89
9  5% 1-Day       32777.87      47269.00     29192.05    41907.76
```

Looking at the output above, we can easily confirm that the ES is larger than the
VaR.

## 4.6  Alternative Risk Measures

In this section, we will discuss alternative risk measures beyond close-to-close
volatility. The advantage of using close-to-close volatility is that it only requires
us to look at closing prices, but we have to use many observations to get a good
estimate of volatility. Using a large number of observations entails getting a long
historical series. The earlier part of such long historical data may be less relevant to
measure volatility today. Therefore, we may want to consider alternative measures
of volatility that are more efficient than close-to-close volatility by utilizing the
open, high, and low prices during the trading day in addition to the close price. The
measures we will discuss are the Parkinson, Garman–Klass, Rogers–Satchell–Yoon,
and Yang-Zhou. Parkinson uses high and low prices, while the remaining alternative
volatility measures all use open, high, low, and close prices.

### 4.6.1  Parkinson

The Parkinson volatility measure uses the stock's high and low price of the day.
Specifically,

$$\sigma_{\text{Parkinson}} = \sqrt{\frac{1}{4T \ln 2} \sum_{t=1}^{T} \ln\left(\frac{h_t}{l_t}\right)^2}, \tag{4.9}$$

where $T$ is the number of days in the sample period, $h_t$ is the high price on day $t$,
and $l_T$ is the low price on day $t$. We now demonstrate how to implement this in R.
We apply this to AMZN data we have been using as our example in this book.

**Step 1: Import AMZN High and Low Data**  Since we are doing completely new analyses, we can also clear the R environment at this point using `rm(list = ls())` and load the pre-loaded functions in Appendix B. After loading the AMZN data, we then delete the first observation (December 31, 2014) and keep Columns 2 (high price) and 3 (low price).

```
1  > data.amzn <- load.data("AMZN Yahoo.csv", "AMZN")
2  > parkinson <- data.amzn[-1, 2:3]
3  > head.tail(parkinson)
4             AMZN.High AMZN.Low
5  2015-01-02    314.75   306.96
6  2015-01-05    308.38   300.85
7  2015-01-06    303.00   292.38
8             AMZN.High AMZN.Low
9  2019-12-27   1901.40  1866.01
10 2019-12-30   1884.00  1840.62
11 2019-12-31   1853.26  1832.23
```

**Step 2: Calculate the Terms in the Parkinson Formula**  We first calculate the term $\ln(h_t / l_t)$ and then take the square of that value.

```
1  > parkinson$log.hi.low <- log(parkinson$AMZN.High / parkinson$AMZN.Low)
2  > parkinson$log.square <- (parkinson$log.hi.low)^2
3  > head.tail(parkinson)
4             AMZN.High AMZN.Low log.hi.low   log.square
5  2015-01-02    314.75   306.96 0.02506126 0.0006280666
6  2015-01-05    308.38   300.85 0.02472098 0.0006111270
7  2015-01-06    303.00   292.38 0.03567846 0.0012729527
8             AMZN.High AMZN.Low log.hi.low   log.square
9  2019-12-27   1901.40  1866.01 0.01878800 0.0003529890
10 2019-12-30   1884.00  1840.62 0.02329471 0.0005426434
11 2019-12-31   1853.26  1832.23 0.01141246 0.0001302443
```

**Step 3: Sum the Values of *log.square***  Using `sum()`, we total the values under the column labeled *log.square*.

```
1  > (parkinson.sum <- sum(parkinson$log.square))
2  [1] 0.6898327
```

**Step 4: Calculate the Daily Parkinson Volatility Measure**  We can now combine the various components and calculate the daily Parkinson volatility measure. The output below shows that the daily Parkinson volatility is 1.4%.

```
1  > (parkinson.vol <- sqrt(1 /
2  +    (4 * nrow(parkinson) * log(2)) * parkinson.sum))
3  [1] 0.01406335
```

**Step 5: Calculate the Annualized Parkinson Volatility**  Using the square root of time rule, we convert daily volatility into annual volatility by multiplying the daily volatility by the square root of 252. The output below shows that the annualized Parkinson volatility is 22.3%.

```
1  > (annual.parkinson.vol <- parkinson.vol * sqrt(252))
2  [1] 0.2232488
```

## 4.6.2   Garman–Klass

The Garman–Klass volatility measure can be viewed as an extension of the
Parkinson volatility measure that includes opening and closing prices. The Garman–
Klass volatility is calculated as follows:

$$\sigma_{\text{Garman--Klass}} = \sqrt{\frac{1}{2T}\sum_{t=1}^{T}\ln\left(\frac{h_t}{l_t}\right)^2 - \frac{2\ln 2 - 1}{T}\ln\left(\frac{c_t}{o_t}\right)^2},\qquad(4.10)$$

where $o_t$ is the open price on day $t$ and all other terms are defined in the same
manner as in Eq. (4.9). We now demonstrate how to calculate the Garman–Klass
volatility measure using our Amazon data.

**Step 1: Import AMZN Open, High, Low, and Close Price Data**  Since we already
imported the data when we calculated the Parkinson risk measure, we can simply
extract the appropriate columns from *data.amzn*. In addition, since we are not
calculating returns, we can delete the first observation (December 31, 2014).

```
1  > gk <- data.amzn[-1, 1:4]
2  > head.tail(gk)
3              AMZN.Open AMZN.High AMZN.Low AMZN.Close
4  2015-01-02    312.58    314.75   306.96     308.52
5  2015-01-05    307.01    308.38   300.85     302.19
6  2015-01-06    302.24    303.00   292.38     295.29
7              AMZN.Open AMZN.High AMZN.Low AMZN.Close
8  2019-12-27   1882.92   1901.40  1866.01    1869.80
9  2019-12-30   1874.00   1884.00  1840.62    1846.89
10 2019-12-31   1842.00   1853.26  1832.23    1847.84
```

**Step 2: Calculate the First Term**  We calculate the first term inside the square root
in Eq. (4.10). Notice also that the summation term is equal to the *parkinson.sum* we
calculated above. Using that same variable can help us save some time coding.

```
1  > (gk1 <- (1 / (2 * nrow(gk))) * parkinson.sum)
2  [1] 0.0002741783
```

**Step 3: Calculate the Second Term**  We now calculate the second term inside the
square root in Eq. (4.10). Because the resulting value in this step is small, we use
the `scipen = 999` option to increase R's threshold before converting the values
to scientific notation (Line 1) but return this option to its default value at the end of
this step (Line 5).

```
1  > options(scipen = 999)
```

```
2  > (gk2 <- ((2 * log(2) - 1) / nrow(gk)) *
3  +   sum(log(gk$AMZN.Close / gk$AMZN.Open)^2))
4  [1] 0.00007911461
5  > options(scipen = 0)
```

**Step 4: Calculate the Daily Garman–Klass Volatility** Subtracting the second term from the first and taking the square root of the difference give us the daily Garman–Klass risk measure. The output below shows the Garman–Klass volatility is 1.40%.

```
1  > (gk.vol <- sqrt(gk1 - gk2))
2  [1] 0.01396652
```

**Step 5: Annualize the Volatility** To get the annualized volatility, we multiply the daily volatility by the square root of 252. The output shows that the Garman–Klass volatility is equal to 22.2%.

```
1  > (annual.gk.vol <- gk.vol * sqrt(252))
2  [1] 0.2217117
```

### 4.6.3   Rogers, Satchell, and Yoon

The prior risk measures we have discussed all assume that the mean return is zero. In contrast, the Rogers, Satchell, and Yoon (RSY) volatility properly measures the volatility of securities with an average return that is not zero. The equation for the RSY volatility is

$$\sigma_{\text{RSY}} = \sqrt{\frac{1}{T} \sum_{t=1}^{T} \left( \ln\left(\frac{h_t}{c_t}\right) \ln\left(\frac{h_t}{o_t}\right) + \ln\left(\frac{l_t}{c_t}\right) \ln\left(\frac{l_t}{o_t}\right) \right)}, \qquad (4.11)$$

where all the variables are defined in the same way as they were in the Parkinson and Garman–Klass formulas. We now demonstrate how to implement the RSY volatility measure on our Amazon data.

**Step 1: Obtain Open, High, Low, and Close Data** This is similar to Step 1 of the Garman–Klass procedure.

```
1  > rsy <- data.amzn[-1, 1:4]
2  > head.tail(rsy)
3               AMZN.Open AMZN.High AMZN.Low AMZN.Close
4  2015-01-02     312.58    314.75   306.96     308.52
5  2015-01-05     307.01    308.38   300.85     302.19
6  2015-01-06     302.24    303.00   292.38     295.29
7               AMZN.Open AMZN.High AMZN.Low AMZN.Close
8  2019-12-27    1882.92   1901.40  1866.01    1869.80
9  2019-12-30    1874.00   1884.00  1840.62    1846.89
10 2019-12-31    1842.00   1853.26  1832.23    1847.84
```

**Step 2: Calculate the Product of First Two Log Terms**  We first calculate each of the log terms *rsy1* and  *rsy2* and then take their product *rsy12*.

```
 1  > rsy1 <- as.numeric(log(rsy$AMZN.High / rsy$AMZN.Close))
 2  > head.tail(rsy1)
 3  [1] 0.01999204 0.02027684 0.02577485
 4  [1] 0.016758970 0.019894025 0.002928885
 5  >
 6  > rsy2 <- as.numeric(log(rsy$AMZN.High / rsy$AMZN.Open))
 7  > head.tail(rsy2)
 8  [1] 0.006918278 0.004452452 0.002511435
 9  [1] 0.009766683 0.005321992 0.006094318
10  >
11  > options(scipen = 999)
12  > rsy12 <- rsy1 * rsy2
13  > head.tail(rsy12)
14  [1] 0.00013831048 0.00009028166 0.00006473186
15  [1] 0.00016367955 0.00010587585 0.00001784956
16  > options(scipen = 0)
```

**Step 3: Calculate the Product of Last Two Log Terms**  Similar to Step 2, we first calculate each of the log terms  *rsy3* and *rsy4* and then take their product *rsy34*.

```
 1  > rsy3 <- as.numeric(log(rsy$AMZN.Low / rsy$AMZN.Close))
 2  > head.tail(rsy3)
 3  [1] -0.005069219 -0.004444144 -0.009903612
 4  [1] -0.002029033 -0.003400683 -0.008483578
 5  >
 6  > rsy4 <- as.numeric(log(rsy$AMZN.Low / rsy$AMZN.Open))
 7  > head.tail(rsy4)
 8  [1] -0.01814298 -0.02026853 -0.03316703
 9  [1] -0.009021320 -0.017972715 -0.005318145
10  >
11  > options(scipen = 999)
12  > rsy34 <- rsy3 * rsy4
13  > head.tail(rsy34)
14  [1] 0.00009197073 0.00009007626 0.00032847336
15  [1] 0.00001830455 0.00006111950 0.00004511689
16  > options(scipen = 0)
```

**Step 4: Calculate the RSY Volatility Measure**  We now bring together all the terms and calculate the annualized risk measure. We first add *rsy12* and *rsy34* and sum the totals. Next, we divide the value by the number of observations in *rsy*. Then, we take the square root of this value. This results in a daily RSY volatility of 1.40%.

```
 1  > (rsy.vol <- sqrt((1 / nrow(rsy)) * sum((rsy12 + rsy34))))
 2  [1] 0.01395756
```

**Step 5: Annualize the RSY Volatility Measure**  We multiply the daily RSY volatility by the square root of 252. As the output shows, the RSY volatility measure also shows an annualized RSY volatility of 22.2%.

```
 1  > (annual.rsy.vol <- rsy.vol * sqrt(252))
 2  [1] 0.2215693
```

### 4.6.4   Yang and Zhang

Yang and Zhang developed a volatility measure that handles both opening jumps and drift. We can think of the Yang–Zhang volatility as the sum of the overnight (i.e., volatility from the prior day's close to today's open) and a weighted average of the RSY volatility and the day's open-to-close volatility. The following Yang–Zhang volatility formula looks slightly more involved than the others but becomes clearer when we break the calculations down into pieces.

$$\sigma_{\text{Yang–Zhang}} = \sqrt{\sigma^2_{\text{overnight vol}} + k\sigma^2_{\text{open-to-close vol}} + (1-k)\sigma^2_{\text{RSY}}}, \qquad (4.12)$$

where

$$\sigma^2_{\text{overnight vol}} = \frac{1}{T-1} \sum_{t=1}^{T} \left( \ln\left(\frac{o_t}{c_{t-1}}\right) - \text{Avg} \ln\left(\frac{o_t}{c_{t-1}}\right) \right)^2,$$

$$\sigma^2_{\text{open-to-close vol}} = \frac{1}{T-1} \sum_{t=1}^{T} \left( \ln\left(\frac{c_t}{o_t}\right) - \text{Avg} \ln\left(\frac{c_t}{o_t}\right) \right)^2, \text{ and}$$

$$k = \frac{\alpha - 1}{\alpha + \frac{T+1}{T-1}}.$$

Below we demonstrate how to calculate the Yang–Zhang volatility measure on AMZN stock data.

**Step 1: Import AMZN Open, High, Low, and Close Data and Create Variable for Prior Day's Close Price**  We import Columns 1 to 4 from *data.amzn*. The high and low prices are used to compute the RSY volatility measure (i.e., part of the third term in Eq. (4.12)). Since we have already calculated the RSY volatility measure previously, we can just call in *rsy.vol* instead of recalculating the number. Also, we need the lag close price, so we use Lag() to construct values in a column labeled Lag.Close that takes on the prior day's close price. We also delete the first observation (December 31, 2014).

```
1   > yz <- data.amzn[, 1:4]
2   > yz$Lag.Close <- Lag(yz$AMZN.Close, k = 1)
3   > yz <- yz[-1, ]
4   > head.tail(yz)
5                AMZN.Open AMZN.High AMZN.Low AMZN.Close Lag.Close
6   2015-01-02    312.58    314.75    306.96    308.52    310.35
7   2015-01-05    307.01    308.38    300.85    302.19    308.52
8   2015-01-06    302.24    303.00    292.38    295.29    302.19
9                AMZN.Open AMZN.High AMZN.Low AMZN.Close Lag.Close
10  2019-12-27   1882.92   1901.40   1866.01   1869.80   1868.77
11  2019-12-30   1874.00   1884.00   1840.62   1846.89   1869.80
12  2019-12-31   1842.00   1853.26   1832.23   1847.84   1846.89
```

**Step 2: Calculate the First Term in the Yang–Zhang Equation**  We first calculate
the mean of the variable because we need to subtract this number from each
observation. Then, we calculate the entire first term.

```
1  > (yz1.mean <- mean(log(yz$AMZN.Open / yz$Lag.Close)))
2  [1] 0.00163107
3  > (yz1 <- 1 / (nrow(yz) - 1) * sum((
4  +    log(yz$AMZN.Open / yz$Lag.Close) - yz1.mean)^2))
5  [1] 0.0001711657
```

**Step 3: Calculate the Second Term in the Yang–Zhang Equation**  We first
calculate the mean of the variable because we need to subtract this number from
each observation. Then, we calculate the entire second term.

```
1  > (yz2.mean <- mean(log(yz$AMZN.Close / yz$AMZN.Open)))
2  [1] -0.0002128886
3  >
4  > (yz2 <- 1 / (nrow(yz) - 1) * sum((
5  +    log(yz$AMZN.Close / yz$AMZN.Open) - yz2.mean)^2))
6  [1] 0.0002049215
```

**Step 5: Calculate $k$**  In their paper, Yang and Zhang [11] suggest that the value of
$\alpha$ in practice should be 1.34. We follow that assumption here.

```
1  > (k <- 0.34 / (1.34 + (nrow(yz) + 1) / (nrow(yz) - 1)))
2  [1] 0.1452004
```

**Step 6: Calculate the Annualized Yang–Zhang Volatility**  We combine the terms
and calculate the annualized Yang–Zhang volatility measure. The output below
shows the annualized Yang and Zhang volatility is 30.4%.

```
1  > (annual.yz.vol <- sqrt(yz1 + k * yz2 +
2  +    (1 - k)*rsy.vol^2) * sqrt(252))
3  [1] 0.3042969
```

### 4.6.5  Comparing the Risk Measures

Before we can compare these different risk measures, we have to first calculate the
close-to-close volatility for this same time period. We use log returns in doing the
calculation, which we do in Line 1. We then calculate the daily standard deviation
of the return using sd() and annualize the volatility by multiplying by $\sqrt{252}$.

```
1  > rets <- diff(log(data.amzn$AMZN.Adjusted))
2  > rets <- rets[-1, ]
3  > names(rets) <- "AMZN"
4  > head.tail(rets)
5                      AMZN
```

```
 6   2015–01–02 −0.005914077
 7   2015–01–05 −0.020730670
 8   2015–01–06 −0.023098010
 9                        AMZN
10   2019–12–27  0.0005510283
11   2019–12–30 −0.0123283480
12   2019–12–31  0.0005142195
13   > (sd <- sd(rets))
14   [1] 0.01825005
15   > (annual.sd <- sd * sqrt(252))
16   [1] 0.2897105
```

Now, we can create a table that compares the five different risk measures. As the
output shows, the Yang and Zhang volatility measure yields the highest volatility
(30.4%) and then followed by the close-to-close volatility or standard deviation
(29.0%). The Parkinson, Garman–Klass, and RSY volatilities are very similar at
approximately 22%.

```
 1   > vols <- rbind(annual.sd, annual.parkinson.vol,
 2   +    annual.gk.vol, annual.rsy.vol, annual.yz.vol)
 3   > rownames(vols) <- c("Close–to–Close", "Parkinson",
 4   +    "Garman–Klass", "Rogers et al", "Yang–Zhang")
 5   > colnames(vols)<-c("Volatility")
 6   > vols
 7                  Volatility
 8   Close–to–Close  0.2897105
 9   Parkinson       0.2232488
10   Garman–Klass    0.2217117
11   Rogers et al    0.2215693
12   Yang–Zhang      0.3042969
```

## 4.7   Further Reading

Many investment textbooks (e.g., Bodie et al. [3] and Reilly and Brown [9]) have
more detailed discussions of standard deviation/variance as well as portfolio risk.
Excellent advanced discussions of volatility can be found in [4] and [8]. Jorion [6]
and Gregoriou [5] provide excellent and readable treatments of Value-at-Risk. In
addition, a comprehensive discussion of Value-at-Risk and expected shortfall can
be found in [1]. Bennett and Gil [2] and Sinclair [10] have a very good discussion
of various other risk measures.

# References

1. Alexander, C. (2009). *Value at risk models, market risk analysis* (vol. 4). Hoboken: Wiley.
2. Bennet, C., & Gil, M. (2012). Measuring historical volatility. Santander Equity Derivatives Report.
3. Bodie, Z., Kane, A., & Marcus, A. (2012). *Essentials of investments* (9th edn.). New York: McGraw-Hill/Irwin.
4. Derman, E., & Miller, M. (2016). *The volatility smile.* Hoboken: Wiley.
5. Gregoriou, G. (2009). *The VaR modeling handbook.* New York: McGraw-Hill.
6. Jorion, P. (2006). *Value-at-risk: The new benchmark for managing financial risk* (3rd edn.). New York: McGraw-Hill.
7. Markowitz, H. (1952). Portfolio selection. *The Journal of Finance, 7,* 77–91.
8. Rebonato, R. (1999). *Volatility and correlation.* Hoboken: Wiley.
9. Reilly, F., & Brown, K. (2002). *Investment analysis & portfolio management*, 10th edn. Boston: South-Western Cengage Learning.
10. Sinclair, E. (2008). *Volatility trading.* Hoboken: Wiley.
11. Yang, D., & Zhang, Q. (2000). Drift-independent volatility estimation based on high, low, open, and close prices. *Journal of Business, 73,* 477–491.

# Chapter 5
# Factor Models

In finance, only risk factors that cannot be diversified away are priced in the discount rate. There are many models that have been developed to identify such factors, but virtually all of those models do not have a theoretical basis. The one model that has a theoretical basis and is commonly used in many real world applications is the **Capital Asset Pricing Model** (CAPM) by [25, Sharpe (1964)], which we discuss in Sect. 5.1. The CAPM is a single factor model with market risk as the lone factor. The CAPM allows us to provide a factor loading on market risk for the subject firm. In other words, the CAPM allows us to determine how sensitive a particular stock is to market movements. For example, defensive stocks are expected to move less or possibly in the opposite direction as the market movement.

The CAPM is the model taught in virtually all finance textbooks and courses. However, since its development in the 1960s, many studies have found that the CAPM does not perform well in empirical testing. However, it is unclear whether these results are a result of the CAPM's failure or our inability to come up with the appropriate market proxy. The "market" in the CAPM is supposed to capture all assets, but from a practical perspective we are unable to find a proxy that reflects that. This argument was first introduced in [24, Roll (1977)] and is known as **Roll's Critique**. Moreover, the CAPM is still the most common approach used by investors and finance professionals compared to any other factor model.

For practical applications, a simpler model similar to the CAPM is often used called the **market model**. We discuss the market model in Sect. 5.2. Like the CAPM, the market model is a single factor model. The primary difference is that the market model does not impose a theoretical restriction on what the "market" should be and, thus, any broad-based stock market index, such as the S&P 500 Index or MSCI World Index, is often used.

One of the critical components of the CAPM is beta, which is the sensitivity of the stock's return to the market's return. Because of the importance of beta in many applications, such as in the calculation of the discount rate, we show how to perform a couple of analyses to test the robustness of the beta estimate. In Sect. 5.3,

© The Author(s), under exclusive license to Springer Nature Switzerland AG 2021     137
C.S. Ang, *Analyzing Financial Data and Implementing Financial Models Using R*,
Springer Texts in Business and Economics,
https://doi.org/10.1007/978-3-030-64155-9_5

we show how to perform an analysis of betas over time. We show that betas are not stationary. That is, we cannot assume the beta in one period will remain the same in a different period. Moreover, even for betas during the same period, the beta could change depending on how we estimate the returns over that period. For example, in Sect. 5.4, we show choosing which day of the week to calculate weekly returns may matter. Thus, the typical Friday-to-Friday return may lead to biased results in some situations, but we cannot tell beforehand unless we perform the calculation. Although we only show how the analysis is performed using weekly returns, this reference day issue is also relevant when using monthly returns. The typical case is to calculate monthly returns using end-of-the-month prices, but we can also test the sensitivity of how or if the beta changes when, say, we calculate returns based on the middle of the month.

Because of the poor performance of the CAPM empirically, different models have been developed that include multiple factors. These are known as **multi-factor models**. These models include additional factors that help explain more of the variation in expected stock returns. However, most of these multi-factor models are an empirical or statistical result (i.e., they are not based on theory). A well-known example of this is the three factor model developed by Eugene Fama and Kenneth French (FF Model). In [15, Fama and French (1993)], Fama and French found that there were large return premium associated with size and value. As a proxy for size, Fama and French used the difference in returns between small and large capitalization stocks. As a proxy for value, they used the difference in returns of high and low book-to-market (i.e., value and growth stocks) stocks. Although based on the statistical results, the Fama–French model is the prevalent factor model used in academia. We discuss the Fama–French model in Sect. 5.5. An issue with models based on statistics is that using different time periods or including other data may change the results. For example, using more recent and broader data, Fama and French found in their 2012 and 2017 articles that there was either no size premium or the size premium is close to zero, which implies that only the result for the value premium remains robust.

Implementing these factor models requires the use of regression analysis. In turn, regression analysis is based on a number of assumptions. We show below how to test three of the key assumptions. One key assumption is that the variances are constant (i.e., homoskedastic). In Sect. 5.6, we show how to implement a test to identify heteroskedasticity and show how to generate heteroskedasticity-consistent standard errors to correct the issue. Another assumption is that the residuals are normally distributed. We go through several tests for non-normality in Sect. 5.7. Finally, the residuals are assumed to be not serially correlated. We show how to test for autocorrelation in the residuals in Sect. 5.8.

One of the ways factor models are used in practice and in academia is through a technique called an **event study**. We discuss this in Sect. 5.9. Event studies are used to determine whether publicly disclosed new information had an effect on the value of the firm. From an academic perspective, much of what we know about corporate finance is a result of conducting event studies. The popularity of event studies has now crossed over to other fields, such as the law. In securities litigation, for example, expert witnesses testifying on whether certain disclosures affected the stock price are generally required to use an event study in virtually all situations.

There is a proliferation of factors in the academic literature, so choosing the best factors in a factor model may not be simple and straightforward. In general, we have to use a combination of qualitative and quantitative approaches. There is no bright-line test to make this determination, and the most appropriate method likely depends on the facts and circumstances. In Sect. 5.10, we go through an example of how we can use a quantitative approach to select the best set of factors among a larger set of factors with "best" being defined using statistical criteria. This could complement qualitative analysis of which regressors should be used in a particular model.

## 5.1  CAPM

The most commonly used factor model by investors and practitioners is the Capital Asset Pricing Model (CAPM) by [25, Sharpe (1964)]. Berk and van Binsbergen [6] used mutual fund flows from a sample of 4,275 mutual funds covering the period January 1977 to March 2011. This study focused on the behavior of mutual fund investors, which covers a great majority of households with an annual income of over $100,000. The authors found the CAPM was the model that was most consistent with how mutual fund investors set their required rate of return. Pinto et al. [23] surveyed professional equity analysts who are members of the CFA Institute. The equity analysts in their sample spent a majority of their time evaluating individual securities for purposes of making investment recommendations or portfolio decisions. The authors found that 68% of the 1,436 equity analysts who responded to their survey used the CAPM. Graham and Harvey [17] surveyed Fortune 500 CFOs and financial executives from 4,440 firms who were members of the Financial Executives Institute. Among other things, the survey investigated how the respondents made capital budgeting (i.e., investment) decisions. Based on the survey responses, the authors found that over 70% of their survey respondents always or almost always used the CAPM. Because of this, we discuss the CAPM first.

The value of a company's stock is equal to the present value of the stream of potential dividend payments (i.e., free cash flow) shareholders expect to receive from holding the stock. Since these potential dividend payments are expected to arrive at different points in time in the future, we have to discount those potential dividend payments by an appropriate risk-adjusted discount rate. The CAPM gives us the appropriate risk-adjusted discount rate to use to discount these equity cash flows. From a mathematical perspective, the formula for the CAPM is

$$R_i = R_f + \beta_i (R_m - R_f), \tag{5.1}$$

where $R_i$ is the return on asset $i$, $R_f$ is the return on the risk-free asset, $R_m$ is the return on the market proxy, and $\beta_i$ is the sensitivity of asset $i$ to the overall market.

**CAPM Regression** Although the CAPM equation can often be expressed as Eq. (5.1), empirical tests of the CAPM typically convert the formula into its **excess return** form. The term "excess" means the return over the risk-free rate of both the subject security's return and the market return. That is, the CAPM in excess return form is as follows:

$$R_i - R_f = \alpha + \beta_i(R_m - R_f), \tag{5.2}$$

where all the variables are defined in the same way as in Eq. (5.1). The $\alpha$ and $\beta_i$ in Eq. (5.2) are estimated using an OLS regression. CAPM restricts $\alpha = 0$ in theory, but in practice this is typically not done.

To perform the CAPM regression, we need to choose (i) the length of the estimation period, (ii) the frequency of the returns data, and (iii) the risk-free rate used in the calculation. This could potentially give us a large number of combinations and could potentially lead to a large number of variations in the results. We have to use our judgment given the facts and circumstances. In our implementation, for (i) and (ii), we follow the methodology used on Yahoo Finance, which is to use five years of monthly returns or 60 monthly return observations for both the subject security's return and the market return. For (iii), we use the 3-Month Treasury Bill rate.

**Step 1: Import Portfolio Returns Data** First, we import the portfolio returns data into R. In practice, we will likely have kept track of our portfolio's return, and for implementing the CAPM, we would then import that data. To mimic this for our example, we constructed a hypothetical portfolio and saved the monthly returns of that portfolio as Hypothetical Portfolio (monthly).csv. The details of how this portfolio is constructed are described in Appendix C. We import the CSV file containing our portfolio returns using the read.csv() command. Using dim(), we see that there are 60 observations in this dataset. The number to the right tells us the dataset contains two columns.

```
1  > data.port <- read.csv("Hypothetical Portfolio (Monthly).csv")
2  > data.port <- data.port[, -1]
3  > dim(data.port)
4  [1] 60  2
5  > head.tail(data.port)
6         Date      Port.Ret
7  1 Jan 2015  0.07306837
8  2 Feb 2015  0.07258170
9  3 Mar 2015 -0.02373268
10        Date      Port.Ret
11 58 Oct 2019 0.05596117
12 59 Nov 2019 0.04224452
13 60 Dec 2019 0.04982447
```

**Step 2: Import SPY Data** We do this using load.data(). We then use to.monthly() to convert the data into monthly prices. Then, using Delt() (Line 13), we calculate the returns based on the adjusted close prices. Next, we

clean up the data by removing the first observation (December 2014) and renaming
the variable market (Lines 14 and 15). Using dim() (Line 16), we can see that we
have 60 market returns.

```
1  > SPY <- load.data("SPY Yahoo.csv", "SPY")
2  > market <- to.monthly(SPY)
3  > head.tail(market)
4            SPY.Open SPY.High SPY.Low SPY.Close SPY.Volume SPY.Adjusted
5  Dec 2014   207.99   208.19  205.39    205.54  130333800     186.2590
6  Jan 2015   206.38   206.88  198.55    199.45 3183506000     180.7403
7  Feb 2015   200.05   212.24  197.86    210.66 1901638100     190.8988
8            SPY.Open SPY.High SPY.Low SPY.Close SPY.Volume SPY.Adjusted
9  Oct 2019   297.74   304.55  284.82    303.33 1386748300     301.8459
10 Nov 2019   304.92   315.48  304.74    314.31 1037123500     312.7722
11 Dec 2019   314.59   323.80  307.13    321.86 1285175800     321.8600
12 >
13 > market <- Delt(market$SPY.Adjusted)
14 > market <- market[-1, ]
15 > names(market) <- "Market"
16 > dim(market)
17 [1] 60  1
18 > head.tail(market)
19               Market
20 Jan 2015 -0.02962929
21 Feb 2015  0.05620466
22 Mar 2015 -0.01570571
23               Market
24 Oct 2019 0.02210462
25 Nov 2019 0.03619827
26 Dec 2019 0.02905545
```

**Step 3: Import Risk-Free Rate Data** For the risk-free rate, we import the 3-
month Treasury data (DGS3MO) from the FRED database. We save the data to
a file labeled **DGS3MO.csv** and import the file into R using read.csv(). The
DGS3MO data are daily yields, but it includes non-trading days. Details on how
to download this data are found in Chap. 9 Sect. 9.1. For example, Line 5 below
is the data for January 1, 2015, which is a non-trading day, and the value under
the DGS3MO column is a dot (.). In Line 8, we use subset() to remove all
observations with a dot for value. In this line, we tell R to keep observations in
which **DGS3MO** is not equal to a dot. The "not equal to" is symbolized by an
exclamation point (!) followed by an equal sign (=).

```
1  > raw.rf <- read.csv("DGS3MO.csv", header = TRUE)
2  > head(raw.rf, 3)
3           DATE DGS3MO
4  1 2014-12-31   0.04
5  2 2015-01-01      .
6  3 2015-01-02   0.02
7  >
8  > raw.rf <- subset(raw.rf, DGS3MO != ".")
9  > str(raw.rf)
10 'data.frame': 1251 obs. of 2 variables:
```

```
11    $ DATE  : Factor w/ 1305 levels "2014-12-31","2015-01-01",..: 1 3 4 5 6 7 ...
12    $ DGS3MO: Factor w/ 223 levels ".","0.00","0.01",..: 6 4 5 5 5 5 4 ...
13  > head.tail(raw.rf)
14          DATE DGS3MO
15   1 2014-12-31   0.04
16   3 2015-01-02   0.02
17   4 2015-01-05   0.03
18          DATE DGS3MO
19   1303 2019-12-27   1.57
20   1304 2019-12-30   1.57
21   1305 2019-12-31   1.55
```

**Step 3a: Create Date Variable** We will later convert this data into an xts object. As such, we need a variable that R recognizes as a date. Unfortunately, the DATE column (Column 1) in the dataset is considered as a Factor (i.e., categorical variable) by R. So we have to create a separate variable that is considered by R as a Date variable.

```
1   > Date <- as.Date(raw.rf$DATE, "%Y-%m-%d")
2   > str(Date)
3    Date[1:1251], format: "2014-12-31" "2015-01-02" "2015-01-05" "2015-01-06" ...
4   > head.tail(Date)
5   [1] "2014-12-31" "2015-01-02" "2015-01-05"
6   [1] "2019-12-27" "2019-12-30" "2019-12-31"
```

**Step 3b: Create Numeric DGS3MO Variable** Similar to the issue with the date, R considers the values under the DGS3MO column (Column 2 in *raw.rf*) a Factor. To be useful to us, we would like this to be a numeric variable. We use a combination of `as.numeric()` and `as.character()` to convert the data into what R would recognize as numeric.

```
1   > DGS3MO <- as.numeric(as.character(raw.rf$DGS3MO))
2   > str(DGS3MO)
3    num [1:1251] 0.04 0.02 0.03 0.03 0.03 0.03 0.02 0.03 0.03 0.04 ...
4   > head.tail(DGS3MO)
5   [1] 0.04 0.02 0.03
6   [1] 1.57 1.57 1.55
```

**Step 3c: Combine Datasets and Convert into xts Object** We use `cbind()` to combine the *Date* and *DGS3MO* vectors. Most of the dataset and vector labels in this book are in lowercase. However, when these vectors are combined, the name of the vector becomes the variable name. Thus, to save time and lines of code, we can name the vector the way we want the variable names to appear. Next, after combining the two vectors into *rf*, we use `xts()` to convert the dataset into an xts object.

```
1   > rf <- cbind(data.frame(Date), data.frame(DGS3MO))
2   > head.tail(rf)
3         Date DGS3MO
4   1 2014-12-31   0.04
5   2 2015-01-02   0.02
```

```
 6   3 2015–01–05    0.03
 7                 Date DGS3MO
 8    1249 2019–12–27    1.57
 9    1250 2019–12–30    1.57
10    1251 2019–12–31    1.55
11   >
12   > rf <- xts(rf$DGS3MO, order.by = rf$Date)
13   > names(rf) <- "DGS3MO"
14   > head.tail(rf)
15                DGS3MO
16    2014–12–31    0.04
17    2015–01–02    0.02
18    2015–01–05    0.03
19                DGS3MO
20    2019–12–27    1.57
21    2019–12–30    1.57
22    2019–12–31    1.55
```

**Step 3d: Convert into Monthly Prices** Now that *rf* is an xts object, we can now convert the data into monthly prices using to.monthly().

```
 1   > rf <- to.monthly(rf)
 2   > head.tail(rf)
 3             rf.Open rf.High rf.Low rf.Close
 4   Dec 2014    0.04    0.04   0.04     0.04
 5   Jan 2015    0.02    0.04   0.02     0.02
 6   Feb 2015    0.02    0.03   0.01     0.02
 7             rf.Open rf.High rf.Low rf.Close
 8   Oct 2019    1.82    1.82   1.54     1.54
 9   Nov 2019    1.52    1.62   1.52     1.59
10   Dec 2019    1.60    1.60   1.53     1.55
```

**Step 3e: Calculate the Monthly Yield** We will use the open price for each month as the yield for that month. The yield reported is an annual yield, so we convert the annual yield into a monthly yield assuming geometric averaging. In other words, if we compound the monthly yield 12 times, we will get to the annual yield. Then, we subset the data to only include data from January 2015 to December 2019. Because the monthly yields are small, we use scipen = 999 to increase the threshold before R converts the values into scientific notation. It is sometimes easier to read numbers in decimals than in scientific notation.

```
 1   > options(scipen = 999)
 2   > rf <- (1 + rf$rf.Open / 100)^(1 / 12) - 1
 3   > head.tail(rf)
 4                   rf.Open
 5   Dec 2014 0.00003332722
 6   Jan 2015 0.00001666514
 7   Feb 2015 0.00001666514
 8                   rf.Open
 9   Oct 2019 0.001504160
10   Nov 2019 0.001257927
11   Dec 2019 0.001323654
```

```
12  >
13  > rf.sub <- subset(rf,
14  +     index(rf) >= as.yearmon("Jan 2015") &
15  +     index(rf) <= as.yearmon("Dec 2019"))
16  > head.tail(rf.sub)
17                    rf.Open
18  Jan 2015 0.00001666514
19  Feb 2015 0.00001666514
20  Mar 2015 0.00001666514
21                    rf.Open
22  Oct 2019 0.001504160
23  Nov 2019 0.001257927
24  Dec 2019 0.001323654
25  > options(scipen = 0)
```

**Step 4: Combine into One Dataset** We combine the portfolio, market, and risk-free rate data into one dataset using `cbind()`. During the process, we rename the variables as well.

```
1   > combo <- cbind(market, rf.sub, data.port$Port.Ret)
2   > names(combo) <- c("Market", "Risk.Free", "Port.Ret")
3   > head.tail(combo)
4                Market    Risk.Free      Port.Ret
5   Jan 2015 -0.02962929 1.666514e-05  0.07306837
6   Feb 2015  0.05620466 1.666514e-05  0.07258170
7   Mar 2015 -0.01570571 1.666514e-05 -0.02373268
8                Market    Risk.Free      Port.Ret
9   Oct 2019 0.02210462 0.001504160 0.05596117
10  Nov 2019 0.03619827 0.001257927 0.04224452
11  Dec 2019 0.02905545 0.001323654 0.04982447
```

**Step 5: Calculate Excess Portfolio Return and Excess Market Return** We need to calculate two variables to implement Eq. (5.1). First, we need the excess portfolio return $R_i - R_f$ (**Ex.Ret**) and the excess market return or $R_m - R_f$ (**Ex.Market**).

```
1   > combo$Ex.Ret <- combo$Port.Ret - combo$Risk.Free
2   > combo$Ex.Market <- combo$Market - combo$Risk.Free
3   > head.tail(combo)
4                Market    Risk.Free      Port.Ret       Ex.Ret    Ex.Market
5   Jan 2015 -0.02962929 1.666514e-05  0.07306837  0.07305171 -0.02964595
6   Feb 2015  0.05620466 1.666514e-05  0.07258170  0.07256504  0.05618799
7   Mar 2015 -0.01570571 1.666514e-05 -0.02373268 -0.02374934 -0.01572237
8                Market    Risk.Free      Port.Ret     Ex.Ret   Ex.Market
9   Oct 2019 0.02210462 0.001504160 0.05596117 0.05445701 0.02060046
10  Nov 2019 0.03619827 0.001257927 0.04224452 0.04098659 0.03494034
11  Dec 2019 0.02905545 0.001323654 0.04982447 0.04850082 0.02773179
```

**Step 6: Run Regression of Excess Portfolio Return on Excess Market Return** Using `lm()`, we run an OLS regression of the excess portfolio return on the excess market return. We use `digits = 3` to show the first three non-zero digits of the output and then `scipen = 999` to increase the threshold before R converts the data into scientific notation. The latter is useful in particular for looking at the *p*-values that are very small.

```
1   > options(digits = 3, scipen = 999)
2   > capm <- lm(Ex.Ret ~ Ex.Market, data = combo)
3   > summary(capm)
4
5   Call:
6   lm(formula = Ex.Ret ~ Ex.Market, data = combo)
7
8   Residuals:
9       Min        1Q    Median       3Q       Max
10  -0.09318 -0.02069 -0.00259  0.02482  0.09868
11
12  Coefficients:
13               Estimate Std. Error t value      Pr(>|t|)
14  (Intercept)  0.01176    0.00516  2.28          0.026 *
15  Ex.Market    1.26135    0.14604  8.64 0.0000000000053 ***
16  ---
17  Signif. codes:  0 '***' 0.001 '**' 0.01 '*' 0.05 '.' 0.1 ' ' 1
18
19  Residual standard error: 0.0387 on 58 degrees of freedom
20  Multiple R-squared:  0.563, Adjusted R-squared:  0.555
21  F-statistic: 74.6 on 1 and 58 DF,  p-value: 0.0000000000053
22
23  > options(digits = 7, scipen = 0)
```

**CAPM Alpha** When making investments in a fund, investors often consider the contribution of the fund manager to the performance of the fund. As such, we need a way to measure whether a fund manager provided value that goes beyond simply investing in the index (i.e., is there a benefit to active management by the manager instead of simply using a passive investment strategy?). An alternative way to look at this problem is we want to know how well did the fund manager do compared to her benchmark.

To answer this question, we look at the **alpha** of the manager as a measure of this outperformance. The alpha of a portfolio can be calculated by using the excess return form of the CAPM. The regression controls for the sensitivity of the portfolio's return to its benchmark. Therefore, any return that is not captured by the market can be attributed to the manager. This attribution can be observed through the alpha or intercept term of the regression. If this alpha is positive and statistically significant, the manager is taken to have provided positive value. Conversely, a negative and statistically significant alpha is taken to mean that the manager has provided negative value (i.e., we would have been better off investing in the benchmark).

The alpha of the portfolio is the intercept term from the regression output above. The intercept is equal to 0.01176, which translates into a monthly return of 1.2%. This alpha is statistically significant at the 5% level (i.e., its *p*-value is less than or equal to 0.05), which is a conventional level of statistical significance used in practice. As such, assuming our hypothetical portfolio returns, the manager provided an incremental return of 1.2% per month beyond that of what is expected from its sensitivity to the benchmark.

**CAPM Beta**  Another important measure we can derive from the results of the CAPM regression is the portfolio's **beta**. The beta of a portfolio measures how sensitive the portfolio's return is to the movement of the overall market. Therefore, the beta of the portfolio measures what is called **systematic risk** or **market risk**. Systematic risk is the portion of a security's risk that cannot be diversified away and, as such, beta is commonly thought of as amount of compensation investors receive for taking on risk that cannot be diversified away.

From the regression results above, we can read off what the beta for the portfolio is. This is the coefficient of **Ex.Market**, which is 1.26. The $p$-value tells us the minimum significance level for which the beta is statistically significant. A $p$-value of 0.00 for the coefficient of **Ex.Market** means that, at any conventional level of significance (e.g., 1%, 5%, or 10%), the beta of our portfolio is statistically different from zero.

An alternative way to view the beta and the $p$-value is to call them directly from the regression summary results. Instead of simply stopping at summary (CAPM), we can include $coefficients[], where inside the square brackets we could place one number or a vector of numbers using c(...) to output. Looking at the regression summary output above, we see the coefficients have eight values. The order of the numbering works as follows. Number 1 is the estimate of the intercept, number 2 is the estimate of the excess market return (or the beta), number 3 goes back up as the Std. Err. of the intercept, then number 4 goes down as the Std. Err. of the beta, and so on. So, if we need to know what the beta and $p$-value of the beta are, we choose numbers 2 and 8.

```
1  > (beta<-summary(capm)$coefficients[2])
2  [1] 1.261345
3  >
4  > options(scipen = 999)
5  > (beta.pval<-summary(capm)$coefficients[8])
6  [1] 0.000000000005297137
7  > options(scipen = 0)
```

We could also have combined the output into one by using coefficients[c(2, 8)].

```
1  > options(scipen = 999)
2  > summary(capm)$coefficients[c(2, 8)]
3  [1] 1.261345272239198900 0.000000000005297137
4  > options(scipen = 0)
```

The results of the CAPM regression show that the CAPM beta is 1.26. This beta of 1.26 can then be used in the CAPM to calculate, say, the cost of equity for the company. This means that if the market goes up by 1%, we expect our portfolio to go up by 1.26%. However, if the market goes down by 1%, we expect our portfolio to go down by 1.26%.

**Using Calculated Beta to Estimate Cost of Equity in the CAPM**  To calculate the cost of equity using the CAPM, we need to find an equity risk premium (ERP) that is compatible. For our example, we should use an ERP that is calculated using a 3-Month Treasury security. One such source for the historical ERP is Professor Damodaran's website (http://pages.stern.nyu.edu/~adamodar/New_Home_Page/datafile/histretSP.html). Professor Damodaran calculates an ERP from 1928 to 2019 of 8.18% based on the difference between the arithmetic average of the S&P 500 return and the 3-month Treasury return. On December 31, 2019, the 3-month Treasury was 1.55%. Using Eq. (5.1) and based on these inputs for the ERP and risk-free rate, our beta of 1.26 yields an estimate of a CAPM cost of equity for this portfolio is 11.9%.

```
1   > ERP <- 0.0818
2   > Rf <- 0.0155
3   > (capm.coe <- Rf + beta * ERP)
4   [1] 0.118678
```

**Calculate Adjusted Beta**  There have been studies that show betas that are above the market beta of one tend to go down in the long term, while betas that are below the market beta of one tend to go up in the long term. Since valuations take a long-term view, some believe betas used in calculating the cost of equity need to be adjusted to reflect this reversion to the market beta. To the extent that a particular firm's beta does exhibit a regression toward one, one common adjustment is to apply 2/3 weight to the raw beta, which is the beta we calculated above, and 1/3 weight to the market beta of one. This adjustment is often attributed to [8, Blume (1975)]. Since our raw beta of 1.26 is above the market beta of 1.0, we would expect the adjusted beta to be lower than our raw beta. As the output below shows, the adjusted beta is 1.17. Instead of the CAPM beta, had we instead used the adjusted beta of 1.17 to estimate the CAPM cost of equity, we would have calculated a cost of equity of 11.2%.

```
1   > (adj.beta <- (2 / 3) * beta + (1 / 3) * 1)
2   [1] 1.17423
3   > (capm.adjbeta <- Rf + adj.beta * ERP)
4   [1] 0.111552
```

## 5.2  Market Model

The CAPM requires that we use expected returns and the true market portfolio. A more common way to calculate beta in practice is to use the **market model** because the market model uses a market proxy without the requirement that this market proxy be the true market portfolio. In addition, the market model does not require the use of a risk-free rate and, therefore, there is no need to calculate the excess returns of the firm and the market. That is,

$$R_i = \alpha + \beta_i R_m, \tag{5.3}$$

where all variables are defined the same way as in the CAPM regression. Since we already have all the inputs when we calculated the CAPM above, we only need to run the regression of the form in Eq. (5.3) to see our results.

```
1  > options(digits = 3, scipen = 999)
2  > mkt.model <- lm(Port.Ret ~ Market, data = combo)
3  > summary(mkt.model)
4
5  Call:
6  lm(formula = Port.Ret ~ Market, data = combo)
7
8  Residuals:
9       Min       1Q   Median       3Q      Max
10  −0.09291 −0.02079 −0.00259  0.02479  0.09874
11
12  Coefficients:
13              Estimate Std. Error t value      Pr(>|t|)
14  (Intercept)  0.01157    0.00519    2.23          0.03 *
15  Market       1.25714    0.14608    8.61 0.000000000006 ***
16  ——
17  Signif. codes:  0 '***' 0.001 '**' 0.01 '*' 0.05 '.' 0.1 ' ' 1
18
19  Residual standard error: 0.0387 on 58 degrees of freedom
20  Multiple R-squared:  0.561, Adjusted R-squared:  0.553
21  F-statistic: 74.1 on 1 and 58 DF,  p-value: 0.00000000000596
22
23  > options(digits = 7, scipen = 0)
```

The market model beta is 1.257, which is very close to the beta calculated using the CAPM of 1.261. The small difference is primarily driven by the low risk-free rate during the estimation period, but regardless since we are subtracting pretty much the same number on both sides of the equation the difference should still be a relatively small number. The small difference in the beta between the CAPM and market model is the typical result in practice, which is in part the reason why many practitioners use the market model when estimating beta.

Any inferences we made with the CAPM beta can also be applied to the market model beta. For example, the interpretation of a beta of 1.26 is that for a 1% increase (decrease) in the S&P 500 Index return, we expect that our portfolio would go up (down) by 1.26%. We can also implement the adjusted beta calculation using the beta calculated from the market model. Given that the two beta calculations are so close, there is virtually no difference from a practical perspective in the resulting betas from these two approaches. Our estimate cost of equity using a market model beta is 11.83%, which is very close to the 11.87% cost of equity estimated using the CAPM.

```
1  > (mm.beta <- summary(mkt.model)$coefficients[2])
2  [1] 1.257142
3  > (mm.coe <- Rf + mm.beta * ERP)
4  [1] 0.1183342
```

A practical benefit of using the market model gives us the flexibility to choose whatever equity risk premium (ERP) we would like to use. In the CAPM, the ERP must be based on a risk-free rate that has the same tenor as the risk-free rate used in the CAPM. For example, if we use a 3-month risk-free rate, the ERP in the CAPM has to be based on the difference between the market return and the return on a 3-month risk-free security. On the other hand, if we use a 20-year risk-free rate, the ERP in the CAPM has to be based on the difference between the market return and the return on a 20-year risk-free security. Those two ERPs will be different numbers. With a normal upward sloping yield curve, the ERP based on a 3-month risk-free security will be larger than the ERP based on a 20-year risk-free security. Since the market model does not specify a risk-free rate, we can apply an ERP based on any tenor for the risk-free security.

## 5.3  Rolling Window Regressions

The estimates of alpha and beta are sensitive to the assumptions we use when we estimate those variables. One of those key assumptions is the time period over which we estimate alpha and beta. In this section, we show how to calculate the alpha and beta over a long window on a rolling basis. In particular, we will use a 252 trading day (i.e., one year) rolling window and calculate the AMZN alpha and beta on a daily basis from 2016 to 2019.

**Step 1: Calculate AMZN and SPY Returns**  We use `load.data()` to import AMZN and SPY data into R. We use `Delt()` to calculate the total return for AMZN and SPY. Alternative, we could have calculated log returns using `diff(log(x))`, where $x$ is the adjusted close price. As a practical matter, there is typically a very small difference in the beta whether we calculate betas using arithmetic or logarithmic returns.

```
 1  > data.amzn <- load.data("AMZN Yahoo.csv", "AMZN")
 2  > data.spy <- load.data("SPY Yahoo.csv", "SPY")
 3  >
 4  > rets <- Delt(data.amzn$AMZN.Adjusted)
 5  > rets$SPY <- Delt(data.spy$SPY.Adjusted)
 6  > names(rets)[1] <- "AMZN"
 7  > rets <- rets[-1, ]
 8  > head.tail(rets)
 9                    AMZN             SPY
10  2015-01-02 -0.005896623 -0.0005352814
11  2015-01-05 -0.020517267 -0.0180595459
12  2015-01-06 -0.022833293 -0.0094190191
13                    AMZN             SPY
14  2019-12-27  0.0005511802 -0.0002477767
15  2019-12-30 -0.0122526652 -0.0055132196
16  2019-12-31  0.0005143517  0.0024292950
```

**Step 2: Run a Rolling Regression** We will use `rollapply()` in the   zoo package to run our rolling regression. The `width = 252` argument tells R we want to use 252 observations at a time. In our case, this means we want 252 trading days of data in our regression. Inside the braces is the code to run the regression using `lm()`, and we ask R to store the output of each regression in *coeffs* dataset. Since we need 252 days of data, the first observation from our dataset that would have alpha and beta values is December 31, 2015.

```
1   > require(zoo)
2   > coeffs <- rollapply(rets,
3   +     width = 252,
4   +     FUN = function(X)
5   +     {
6   +     roll.reg = lm(AMZN ~ SPY,
7   +         data = as.data.frame(X))
8   +         return(roll.reg$coef)
9   +     },
10  +     by.column = FALSE)
11  >
12  > coeffs[249:255, ]
13               X.Intercept.        SPY
14  2015-12-28            NA         NA
15  2015-12-29            NA         NA
16  2015-12-30            NA         NA
17  2015-12-31   0.003200066 1.129821
18  2016-01-04   0.003054242 1.155975
19  2016-01-05   0.003025456 1.153230
20  2016-01-06   0.003124768 1.141963
```

**Step 3: Clean Up Dataset** We first remove all the NAs from the dataset using `na.omit()`. As we can see, the first observation with data is December 31, 2015. Since we want to show the rolling alphas and betas beginning in 2016, we delete the first observation (Line 12). We then rename the variables to **Alpha** and **Beta** in Line 13.

```
1   > coeffs <- na.omit(coeffs)
2   > head.tail(coeffs)
3               X.Intercept.        SPY
4   2015-12-31   0.003200066 1.129821
5   2016-01-04   0.003054242 1.155975
6   2016-01-05   0.003025456 1.153230
7               X.Intercept.        SPY
8   2019-12-27  -0.0004424709 1.323418
9   2019-12-30  -0.0005179113 1.327407
10  2019-12-31  -0.0005438502 1.324721
11  >
12  > coeffs <- coeffs[-1, ]
13  > names(coeffs) <- c("Alpha", "Beta")
14  > options(digits = 3)
15  > head.tail(coeffs)
16                  Alpha Beta
17  2016-01-04 0.00305 1.16
```

```
18   2016–01–05 0.00303 1.15
19   2016–01–06 0.00312 1.14
20                 Alpha Beta
21   2019–12–27 –0.000442 1.32
22   2019–12–30 –0.000518 1.33
23   2019–12–31 –0.000544 1.32
24   > options(digits=7)
```

**Step 4: Plot the Data**  We create two charts: one for the alphas and for the betas. However, we put these two charts into one chart using `mfrow()` in Line 2 below. The first number in `c(2, 1)` tells R that we want the chart stacked one on top of the other, while the second number tells R that we only have one column. If we had four charts, for example, and we wanted to show them in a 2×2 chart, we would enter `c(2, 2)` instead. The output of this code is shown in Fig. 5.1.

```
1   > par(mfrow = c(2, 1))
2   > plot(y = coeffs$Alpha,
3   +       x = index(coeffs),
4   +       xlab = "",
5   +       ylab = "Alpha",
```

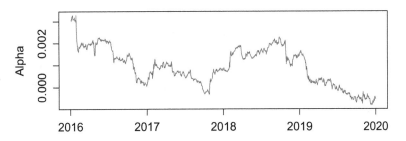

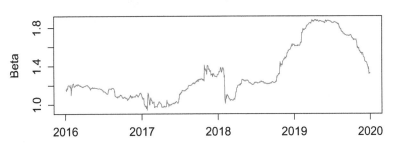

**Fig. 5.1**  AMZN Alpha and Beta estimated using a 252-day rolling window, 2016–2019. Data source: Price data reproduced with permission of CSI ©2020. www.csidata.com

```
6   +     type = "l",
7   +     col = "blue",
8   +     main = "AMZN Alpha, 252-Day Rolling Window ")
9   > plot(y = coeffs$Beta,
10  +     x = index(coeffs),
11  +     xlab = "",
12  +     ylab = "Beta",
13  +     type = "l",
14  +     col = "red",
15  +     main = "AMZN Beta, 252-Day Rolling Window")
16  > par(mfrow = c(1, 1))
```

## 5.4   Betas on Different Reference Days

Betas could also be sensitive to the day of the week or day of the month the returns are calculated. When weekly returns are calculated, it is typical for the calculation to be based on the change in price from the last trading day this week to the last trading day last week. During most weeks, these would be Friday-to-Friday returns. However, weekly betas calculated using any other day of the week could be different from the Friday-to-Friday returns. In theory, there should be no reason to prefer a return calculated using Friday-to-Friday returns with returns calculated using any other day of the week unless there is something peculiar about the beta calculated using a different reference day. Similarly, monthly returns are often calculated as the change in price from the last trading day this month to last trading day last month. The beta calculated at the end of the month could be different from the beta calculated at the middle of the month (or some other period during the month). In this section, we calculate AMZN betas as of the end of 2019 based on two years of weekly returns using different days of the week.

**Step 1: Enter Inputs and Assumptions**  We will use the zoo and dplyr packages for this calculation. We also enter the start and end dates of our calculation here. Since we are using two years of weekly returns, we would like 104 weekly returns observations (i.e., 52 weeks×2 years). Thus, we need a *start.date* and an *end.date* that give us a little bit of allowance (i.e., more than a two-year window) to make sure we can calculate at least 104 weekly returns. In the later steps, we will then only use the returns for the last 104 full weeks. This is the rationale behind the start date of December 1, 2017.

```
1   > library(zoo)
2   > library(dplyr)
3   > start.date <- "2017-12-01"
4   > end.date <- "2019-12-31"
```

**Step 2: Create Dataset of AMZN and SPY Prices**  We import the AMZN and SPY data using load.data(). We then combine the adjusted close prices for AMZN and SPY using cbind(). We then create an additional *date* vector that we will append to the *new.prices* dataset together with the AMZN and SPY adjusted close prices.

```
1  > data <- load.data("AMZN Yahoo.csv", "AMZN")
2  > mkt <- load.data("SPY Yahoo.csv", "SPY")
3  >
4  > prices <- cbind(data$AMZN.Adjusted, mkt$SPY.Adjusted)
5  > date <- as.Date(index(prices), format = "%Y-%m-%d")
6  > prices <- data.frame(prices)
7  > new.prices <- cbind(data.frame(date), prices)
8  > head.tail(new.prices)
9             date AMZN.Adjusted SPY.Adjusted
10 2014-12-31 2014-12-31       310.35     186.2590
11 2015-01-02 2015-01-02       308.52     186.1593
12 2015-01-05 2015-01-05       302.19     182.7974
13             date AMZN.Adjusted SPY.Adjusted
14 2019-12-27 2019-12-27      1869.80       322.86
15 2019-12-30 2019-12-30      1846.89       321.08
16 2019-12-31 2019-12-31      1847.84       321.86
```

**Step 3: Create a Vector of All Calendar Days** The dates in the *new.prices* dataset are only trading days. However, sometimes there are days when there is no trading (e.g., holidays and weekends). This may become an issue if there is a particular day of the week that does not have data because it is a non-trading day. To mitigate this problem, we will later copy over the last traded price prior to the trading holiday and use that as the price for that particular day. In this step, we first create the *all.dates* dataset, which houses all calendar days between the *start.date* and *end.date* we selected above. We do this using seq(), and for the increments, we use by = "day". As we can see in Line 5, *all.dates* has 761 observations, which is about what we would expect given that the date range we set in Step 1 is for two years (i.e., 365 days per year×2 years or 730 days) and one 31-day month of data.

```
1  > all.dates <- seq(as.Date(start.date), as.Date(end.date), by = "day")
2  > all.dates <- data.frame(all.dates)
3  > names(all.dates) <- "date"
4  > dim(all.dates)
5  [1] 761    1
6  > head.tail(all.dates)
7            date
8  1 2017-12-01
9  2 2017-12-02
10 3 2017-12-03
11           date
12 759 2019-12-29
13 760 2019-12-30
14 761 2019-12-31
```

**Step 4: Merge *all.dates* and *new.prices* Datasets** We use merge() but start with *all.dates* as the base. We then tell R to merge the *new.prices* into *all.dates* using date. The all = TRUE effectively keeps all the dates in *all.dates* regardless of whether that date is available in *new.prices* or not. As the output below shows, there is no AMZN and SPY data on December 28 and 29, 2019, which is a weekend. Using na.locf(), we retain the value of the last available data point to fill in the NAs.

```
 1  > combo <- merge(all.dates, new.prices, by = "date", all = TRUE)
 2  > head(combo)
 3          date AMZN.Adjusted SPY.Adjusted
 4  1 2014-12-31        310.35      186.2590
 5  2 2015-01-02        308.52      186.1593
 6  3 2015-01-05        302.19      182.7974
 7  4 2015-01-06        295.29      181.0756
 8  5 2015-01-07        298.42      183.3320
 9  6 2015-01-08        300.46      186.5852
10  > tail(combo)
11          date AMZN.Adjusted SPY.Adjusted
12  1492 2019-12-26       1868.77       322.94
13  1493 2019-12-27       1869.80       322.86
14  1494 2019-12-28            NA           NA
15  1495 2019-12-29            NA           NA
16  1496 2019-12-30       1846.89       321.08
17  1497 2019-12-31       1847.84       321.86
18  >
19  > combo <- na.locf(combo)
20  > tail(combo)
21          date AMZN.Adjusted SPY.Adjusted
22  1492 2019-12-26       1868.77       322.94
23  1493 2019-12-27       1869.80       322.86
24  1494 2019-12-28       1869.80       322.86
25  1495 2019-12-29       1869.80       322.86
26  1496 2019-12-30       1846.89       321.08
27  1497 2019-12-31       1847.84       321.86
```

**Step 5: Add a Day of the Week Variable**  We need to be able to identify the day of the week, so we can subset a specific day of the week during the estimation period in a later step. Thus, we create the variable **wkday** using weekdays().

```
 1  > combo$wkday <- weekdays(combo$date)
 2  > head.tail(combo)
 3          date AMZN.Adjusted SPY.Adjusted      wkday
 4  1 2014-12-31        310.35      186.2590 Wednesday
 5  2 2015-01-02        308.52      186.1593    Friday
 6  3 2015-01-05        302.19      182.7974    Monday
 7          date AMZN.Adjusted SPY.Adjusted    wkday
 8  1495 2019-12-29       1869.80       322.86  Sunday
 9  1496 2019-12-30       1846.89       321.08  Monday
10  1497 2019-12-31       1847.84       321.86 Tuesday
```

**Step 6: Create Objects Needed for Calculation**  We create two objects. First, we create *beta*, which is a data frame that currently has a temporary value of 999 for each of the five days of the week. We can actually enter any value, but 999 ensures that if there is an error in the program and the values were not overwritten, we can easily see the error. In the last line, we create a vector called *day.week*, which has the five days of the typical work week. We reference this vector when we subset the data based on the different days of the week.

```
1   > beta <- data.frame(rep(999, 5))
2   > colnames(beta) <- "beta"
3   > rownames(beta) <- c("Monday", "Tuesday", "Wednesday",
4   +       "Thursday", "Friday")
5   > head(beta)
6                 beta
7   Monday        999
8   Tuesday       999
9   Wednesday     999
10  Thursday      999
11  Friday        999
12  > day.week <- c("Monday", "Tuesday", "Wednesday", "Thursday", "Friday")
```

**Step 7: Calculate Betas for Every Day of the Week** Since we are repeating the same steps when we calculate beta for each day of the week, we use a for-loop to do the calculations more efficiently. Line 2 identifies from the *combo* dataset the particular day of the week that we are interested in. We then calculate the return on the stock and market in Lines 3 and 4. Next, we combine the returns data in *combo.ret*. In other places in this book, we were concerned with cleaning up the data. Here, since we will subset the last 104 observations in Line 9, we can opt not to clean up the dataset. In addition, since this is an intermediate calculation, in practice, it is not necessary to clean up the data. Finally, we replace the beta values in the *beta* dataset with the coefficient from the regression in Lines 10 and 11.

```
1   > for (j in 1:5) {
2   +     new.data <- subset(combo, wkday == paste(day.week[j]))
3   +     stock.ret <- new.data[, 2] / Lag(new.data[, 2], k = 1) - 1
4   +     mkt.ret <- new.data[, 3] / Lag(new.data[, 3], k = 1) - 1
5   +
6   +     combo.ret <- cbind(mkt.ret, stock.ret)
7   +     combo.ret <- data.frame(combo.ret)
8   +     colnames(combo.ret) <- c("mkt.ret", "stock.ret")
9   +     combo.ret <- combo.ret[ (nrow(combo.ret) - 104 + 1):nrow(combo.ret), ]
10  +     beta[j, ] <- summary(lm(stock.ret ~ mkt.ret,
11  +         data = combo.ret))$coeff[2]
12  + }
13  > beta
14                beta
15  Monday       1.480706
16  Tuesday      1.529798
17  Wednesday    1.563849
18  Thursday     1.278707
19  Friday       1.368291
20  > mean(beta$beta)
21  [1] 1.44427
22  > median(beta$beta)
23  [1] 1.480706
```

**Step 8: Plot the Betas**  From the above, we can already see that Thursday has the lowest beta and Wednesday has the highest beta for AMZN. However, it may be easier to visually see the different betas lined up. Line 1 in the code below rounds off the betas to two digits. We will use this as the data labels in the chart. Then, we use barplot() to create the chart. Note that for the *y*-axis range (ylim), we are adding 0.5 to the highest beta in order to give enough space when we add the beta values to the chart. In the last line, we use text() to add text to the chart. The pos = 3 allows us to add the values on top of the bars.

```
1   > vals <- round(beta$beta, digits = 2)
2   > mp <- barplot(vals,
3   +     names.arg=c("Monday", "Tuesday", "Wednesday", "Thursday", "Friday"),
4   +     ylim = c(0, max(beta) + .5),
5   +     ylab = "Beta",
6   +     border = 0,
7   +     col = c("gray40", "red", "blue", "darkgreen", "black"),
8   +     main = "AMZN Betas Based on Different Reference Days")
9   > text(mp, vals, labels = vals, pos = 3)
```

Figure 5.2 shows the Friday-to-Friday return is the second to the lowest beta among the betas calculated using different reference days. As shown in Step 7, the Friday-to-Friday return is lower than the median and average of the five betas. A priori, there is no reason to believe one beta calculation is better than the other, especially given that these are based on the weekly returns over the same time period. Thus, in some instances, we may want to consider taking the average or median of all five weekly beta calculations in our analysis. However, one thing we may want to consider is that these betas are estimates and that whether the other betas are within the 95% confidence interval of, say, the Friday-to-Friday beta. In other words, although the numbers appear different, are the numbers statistically different from the Friday-to-Friday beta?

I also did a larger scale analysis of reference betas using firms in the S&P 500 Index. I start off with the components of the S&P 500 Index as of May 18, 2020. I then looked at how many of those firms were first added to the index on or before January 1, 2018, so I can get two years of data. I end up with 356 firms. For each of those firms, I calculated the betas for each day of the week based on two years of weekly returns. Based on the above, I find that the Friday-to-Friday beta is the highest beta 22% of the time, 2nd to the highest beta 22% of the time, and is the median beta 21% of the time. In aggregate, the average rank of the Friday-to-Friday beta is 2.86 with 1 being the highest rank. In other words, the Friday-to-Friday beta appears to be slightly biased upwards relative to betas calculated during other days of the week. The magnitude of the deviations could also make a non-trivial difference in the cost of equity estimate. The range of deviations from the median beta in the above sample is −0.36 to 0.39. Assuming a 6% equity risk premium that could be a difference of −2.2% to 2.3% in the CAPM cost of equity. Therefore, to the extent that one believes there is no reason for betas calculated using the different days of the week should be different, the above indicates that there is a non-trivial chance that the Friday-to-Friday beta would be higher or lower than the median beta

## AMZN Betas Based on Different Reference Days

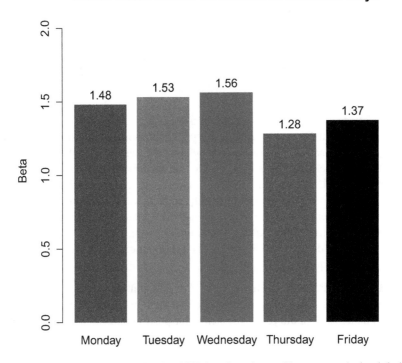

**Fig. 5.2** AMZN Beta as of December 31, 2019 based on the weekly returns calculated during a different day of the week. Data source: Price data reproduced with permission of CSI ©2020. www.csidata.com

and the deviation from the median beta could have a non-trivial impact on the cost of equity. Unfortunately, there is no way to know without performing the calculations what the position is of the Friday-to-Friday beta relative to betas calculated based on other days of the week.

## 5.5 Fama–French Three Factor Model

The popularity of the CAPM is due to its simplicity, and it is grounded in finance theory. However, the CAPM has not performed well in empirical testing. This suggests other factors may need to be added to help explain the remaining variation in asset returns unexplained by the market. One such model that has gained popularity, especially in the academic community, is the Fama–French three factor model (see [15, Fama and French (1993)]) (FF Model). For example, [10, Cochrane (2011)] points out that the FF Model has taken the place of the CAPM for routine risk adjustment in empirical work.

In the FF Model, we have

$$R_i = R_f + \beta_i(R_m - R_F) + h_i HML + s_i SMB, \qquad (5.4)$$

where $HML$ is the difference in the returns of portfolios with high B/M ratios and low B/M ratios, and $SMB$ is the difference in returns of portfolios of small company stocks and big company stocks. The rest of the variables are defined in the same way as in Eq. (5.1).

The data needed to implement the Fama–French model can be downloaded from Professor Kenneth French's Data Library (https://mba.tuck.dartmouth.edu/pages/faculty/ken.french/data_library.html). The data we will use is the Fama–French three factor model data. Note that the files on Professor French's data library are regularly updated, so downloading data at a later date results in retrieving more data albeit the start of the data should remain the same as what we report below.

The Fama–French data can be downloaded in separate ZIP files on a monthly, weekly, or daily frequency. After unpacking the ZIP file, be sure to put the text file in the R working directory so R can access it. For our example, we will use the data with monthly frequency.

**Step 1: Import Portfolio Returns Data**  We will continue to use the portfolio we constructed at the beginning of this chapter. As such, we should check to make sure the *port* data is still in the R environment. Our portfolio has 60 monthly returns.

```
1   > data.port <- read.csv("Hypothetical Portfolio (Monthly).csv")
2   > data.port <- data.port[, -1]
3   > head.tail(data.port)
4           Date     Port.Ret
5   1 Jan 2015  0.07306837
6   2 Feb 2015  0.07258170
7   3 Mar 2015 -0.02373268
8           Date     Port.Ret
9   58 Oct 2019 0.05596117
10  59 Nov 2019 0.04224452
11  60 Dec 2019 0.04982447
```

**Step 2: Import Fama–French Data**  Since we are using our portfolio of monthly returns above, we use the Fama–French monthly data series. We have the option to download the TXT or CSV version of the file. We download the CSV version of the file, so we can use `read.csv()` to import. When we open the CSV, we will notice that the headers start in Row 4, so use `skip = 3` to ignore the first three rows of the CSV file. Then, `header = TRUE` tells R that Row 4 should be used as the header. Also, by opening up the CSV file, we notice that there are annual returns at the bottom, so we would want R to only read the CSV file up to the last row of the monthly returns data. The last row of monthly returns data is Row 1126 in the CSV file, but since we skipped the first three rows and used one row as the header (i.e., total of four rows), we use `nrows = 1122` to tell R to only read the subsequent 1122 rows in after the header.

```
1   > data.ff <- read.csv(file = "F-F_Research_Data_Factors.csv",
```

```
 2  +     skip = 3,
 3  +     header = TRUE,
 4  +     nrows = 1122)
 5  > head.tail(data.ff)
 6           X Mkt.RF   SMB   HML   RF
 7  1 192607   2.96 −2.30 −2.87 0.22
 8  2 192608   2.64 −1.40  4.19 0.25
 9  3 192609   0.36 −1.32  0.01 0.23
10           X Mkt.RF  SMB   HML    RF
11  1120 201910   2.06 0.25 −2.07 0.15
12  1121 201911   3.87 0.87 −1.86 0.12
13  1122 201912   2.77 0.69  1.82 0.14
```

**Step 3: Create a Date Variable** Using `str()`, we see that the year–month looking variable is read in by R as an integer. To create a date variable, we use the `lubridate` package. In Line 9, we first add a 01 to the end of every value in X. This allows us to then read in the data as year–month–date easily using `ymd()`.

```
 1  > str(data.ff)
 2  'data.frame': 1122 obs. of  5 variables:
 3   $ X      : int  192607 192608 192609 192610 192611 192612 192701 192702 ...
 4   $ Mkt.RF: num  2.96 2.64 0.36 −3.24 2.53 2.62 −0.06 4.18 0.13 0.46 ...
 5   $ SMB    : num  −2.3 −1.4 −1.32 0.04 −0.2 −0.04 −0.56 −0.1 −1.6 0.43 ...
 6   $ HML    : num  −2.87 4.19 0.01 0.51 −0.35 −0.02 4.83 3.17 −2.67 0.6 ...
 7   $ RF     : num  0.22 0.25 0.23 0.32 0.31 0.28 0.25 0.26 0.3 0.25 ...
 8  > library(lubridate)
 9  > data.ff$dt <− paste(data.ff$X, "01", sep = "")
10  > data.ff$Date <− ymd(data.ff$dt)
11  > head.tail(data.ff)
12           X Mkt.RF   SMB   HML   RF        dt         Date
13  1 192607   2.96 −2.30 −2.87 0.22 19260701 1926−07−01
14  2 192608   2.64 −1.40  4.19 0.25 19260801 1926−08−01
15  3 192609   0.36 −1.32  0.01 0.23 19260901 1926−09−01
16           X Mkt.RF  SMB   HML    RF        dt       Date
17  1120 201910   2.06 0.25 −2.07 0.15 20191001 2019−10−01
18  1121 201911   3.87 0.87 −1.86 0.12 20191101 2019−11−01
19  1122 201912   2.77 0.69  1.82 0.14 20191201 2019−12−01
20  > class(data.ff$Date)
21  [1] "Date"
22  > data.ff <− data.ff[, −c(1, 6)]
```

In the last line, we clean up the *data.ff* dataset by removing the columns X (Column 1) and dt (Column 6).

**Step 4: Convert Percentage Returns into Decimals** Because the returns in *data.port* are in decimals, we would want to convert the values in the first four columns of *data.ff* into decimals as well. We do so by dividing the first four columns by 100.

```
 1  > data.ff[, 1:4] <− data.ff[, 1:4] / 100
 2  > head.tail(data.ff)
 3    Mkt.RF     SMB    HML    RF       Date
 4  1 0.0296 −0.0230 −0.0287 0.0022 1926−07−01
```

```
5    2 0.0264 −0.0140  0.0419 0.0025 1926−08−01
6    3 0.0036 −0.0132  0.0001 0.0023 1926−09−01
7        Mkt.RF    SMB      HML     RF        Date
8    1120 0.0206 0.0025 −0.0207 0.0015 2019−10−01
9    1121 0.0387 0.0087 −0.0186 0.0012 2019−11−01
10   1122 0.0277 0.0069  0.0182 0.0014 2019−12−01
```

**Step 5: Subset Fama–French Data to 2015–2019** Using subset (), we limit the
FF data to 2015–2019. We then replace the row labels with observation numbers.
We confirm that we have 60 observations, which is what you would expect from five
years of monthly returns.

```
1    > ff.sub <− subset(data.ff,
2    +     Date >= "2015−01−01" &
3    +     Date <= "2019−12−31")
4    > rownames(ff.sub) <− seq(1, nrow(ff.sub), 1)
5    > head.tail(ff.sub)
6        Mkt.RF     SMB      HML RF        Date
7    1 −0.0311 −0.0056 −0.0348   0 2015−01−01
8    2  0.0613  0.0049 −0.0181   0 2015−02−01
9    3 −0.0112  0.0303 −0.0046   0 2015−03−01
10       Mkt.RF    SMB      HML     RF        Date
11   58 0.0206 0.0025 −0.0207 0.0015 2019−10−01
12   59 0.0387 0.0087 −0.0186 0.0012 2019−11−01
13   60 0.0277 0.0069  0.0182 0.0014 2019−12−01
```

**Step 6: Combine Portfolio Returns Data and Fama–French Data** We use
cbind() to combine the Fama–French data and the portfolio returns data. We
then rename the portfolio return variable to Port.Ret.

```
1    > ff.sub <− cbind(ff.sub, data.port[, 2])
2    > names(ff.sub)[6] <− "Port.Ret"
3    > head.tail(ff.sub)
4        Mkt.RF     SMB      HML RF        Date     Port.Ret
5    1 −0.0311 −0.0056 −0.0348   0 2015−01−01  0.07306837
6    2  0.0613  0.0049 −0.0181   0 2015−02−01  0.07258170
7    3 −0.0112  0.0303 −0.0046   0 2015−03−01 −0.02373268
8        Mkt.RF    SMB     HML      RF        Date     Port.Ret
9    58 0.0206 0.0025 −0.0207 0.0015 2019−10−01  0.05596117
10   59 0.0387 0.0087 −0.0186 0.0012 2019−11−01  0.04224452
11   60 0.0277 0.0069  0.0182 0.0014 2019−12−01  0.04982447
```

**Step 7: Calculate Excess Portfolio Return** We subtract the risk-free rate RF from
the portfolio return  Port.Ret to arrive at the portfolio return in excess of the risk-
free rate Port.RF.

```
1    > ff.sub$Port.RF <− ff.sub$Port.Ret − ff.sub$RF
2    > head.tail(ff.sub)
3        Mkt.RF      SMB      HML RF        Date     Port.Ret        Port.RF
4    1 −0.0311 −0.0056 −0.0348   0 2015−01−01  0.07306837  0.07306837
5    2  0.0613  0.0049 −0.0181   0 2015−02−01  0.07258170  0.07258170
6    3 −0.0112  0.0303 −0.0046   0 2015−03−01 −0.02373268 −0.02373268
7        Mkt.RF    SMB      HML      RF        Date     Port.Ret        Port.RF
```

```
8   58 0.0206 0.0025 −0.0207 0.0015 2019−10−01 0.05596117 0.05446117
9   59 0.0387 0.0087 −0.0186 0.0012 2019−11−01 0.04224452 0.04104452
10  60 0.0277 0.0069  0.0182 0.0014 2019−12−01 0.04982447 0.04842447
```

**Step 8: Regress Port.RF on Fama–French Factors**  The regression output shows that all three Fama–French factors are statistically significant at conventional levels. Based on the Fama–French model, the alpha is insignificant. In other words, the portfolio manager did not seem to add value.

```
1   > ff.reg <− lm(Port.RF ~ Mkt.RF + SMB + HML,
2   +     data = ff.sub)
3   > summary(ff.reg)
4
5   Call:
6   lm(formula = Port.RF ~ Mkt.RF + SMB + HML, data = ff.sub)
7
8   Residuals:
9        Min        1Q     Median        3Q        Max
10  −0.069818 −0.022926  0.000997  0.023078  0.077066
11
12  Coefficients:
13               Estimate Std. Error t value Pr(>|t|)
14  (Intercept)  0.008449   0.004523   1.868 0.066993 .
15  Mkt.RF       1.269921   0.126658  10.026 4.16e−14 ***
16  SMB         −0.523079   0.191646  −2.729 0.008464 **
17  HML         −0.693077   0.167725  −4.132 0.000121 ***
18  −−−
19  Signif. codes:  0 '***' 0.001 '**' 0.01 '*' 0.05 '.' 0.1 ' ' 1
20
21  Residual standard error: 0.03346 on 56 degrees of freedom
22  Multiple R−squared:  0.6836, Adjusted R−squared:  0.6667
23  F−statistic: 40.34 on 3 and 56 DF,  p−value: 5.106e−14
```

**Step 9: Compare Fama–French Results with CAPM Results**  Using the same data, we regress the excess portfolio return on the excess market return. By contrast to the Fama–French model, the CAPM regression shows that the alpha is positive and significant. This can be taken by some as the portfolio manager adding value. Thus, whether the manager adds value or not depends on what factors we use to control the manager's performance.

```
1   > capm.reg <− lm(Port.RF ~ Mkt.RF,
2   +     data = ff.sub)
3   > summary(capm.reg)
4
5   Call:
6   lm(formula = Port.RF ~ Mkt.RF, data = ff.sub)
7
8   Residuals:
9        Min        1Q     Median        3Q        Max
10  −0.104655 −0.021540 −0.004027  0.021938  0.104073
11
12  Coefficients:
```

```
13                 Estimate Std. Error t value Pr(>|t|)
14   (Intercept) 0.012583    0.005344    2.355    0.0219 *
15   Mkt.RF      1.171758    0.145527    8.052 5.02e-11 ***
16   ——
17   Signif. codes:   0 '***' 0.001 '**' 0.01 '*' 0.05 '.' 0.1 ' ' 1
18
19   Residual standard error: 0.04016 on 58 degrees of freedom
20   Multiple R-squared:  0.5278, Adjusted R-squared: 0.5197
21   F-statistic: 64.83 on 1 and 58 DF,  p-value: 5.015e-11
```

## 5.6   Testing for Heteroskedasticity

Heteroskedasticity causes the standard errors of the ordinary least squares estimate to be positively or negatively biased. This could lead to incorrect inferences. For example, a parameter estimate could be statistically significant, but the result could be overturned when the standard error is corrected for heteroskedasticity. Thus, it is worthwhile testing for heteroskedasticity and knowing how, if we find heteroskedasticity, to correct it. First, we test for heteroskedasticity using the Breusch–Pagan test. In R, we do this by using bptest() in the lmtest package. We use the studentize = FALSE in Line 4 to use the original Breusch–Pagan test, which assumes the regression errors are normally distributed. The null hypothesis is that the residuals have constant variance (homoskedastic), and the high $p$-value below shows that we cannot reject the null hypothesis. In other words, we cannot reject the null hypothesis that the residuals are homoskedastic.

```
1   > library(lmtest, sandwich)
2   > bptest(formula(ff.reg),
3   +         data = ff.sub,
4   +         studentize = FALSE)
5
6     Breusch—Pagan test
7
8   data:  formula(ff.reg)
9   BP = 2.622, df = 3, p-value = 0.4537
```

R also allows us to run the Breusch–Pagan test while dropping the normality assumption. We do this by setting studentize = TRUE. The results show a similar result. That is, we cannot reject the null hypothesis that the residuals are homoskedastic.

```
1   > bptest(formula(ff.reg),
2   +       data = ff.sub,
3   +       studentize = TRUE)
4
5     studentized Breusch—Pagan test
6
7   data:  formula(ff.reg)
8   BP = 3.581, df = 3, p-value = 0.3104
```

We did not find an issue with heteroskedasticity in the residuals in our example, but if we did we can correct the standard errors in R using the White correction. To do the **White correction**, we use `coeftest()` in the `lmtest` package. The type equals HC0 uses the original White correction.

```
1  > library(lmtest)
2  > coeftest(ff.reg,
3  +    vcov. = vcovHC(ff.reg, type = "HC0"))
4
5  t-test of coefficients:
6
7                 Estimate Std. Error t value  Pr(>|t|)
8  (Intercept)  0.0084487  0.0043460  1.9440  0.056923 .
9  Mkt.RF       1.2699207  0.1189252 10.6783 4.017e-15 ***
10 SMB         -0.5230789  0.1706776 -3.0647  0.003349 **
11 HML         -0.6930773  0.1629276 -4.2539 8.060e-05 ***
12 ──
13 Signif. codes:  0 '***' 0.001 '**' 0.01 '*' 0.05 '.' 0.1 ' ' 1
```

We can compare the output above with the uncorrected results below. As we can see, the parameter estimates are identical in both, but the standard errors of the estimates are different. The standard errors above are called **White standard errors**.

```
1  > summary(ff.reg)
2
3  Call:
4  lm(formula = Port.RF ~ Mkt.RF + SMB + HML, data = ff.sub)
5
6  Residuals:
7        Min        1Q    Median        3Q       Max
8  -0.069818 -0.022926  0.000997  0.023078  0.077066
9
10 Coefficients:
11               Estimate Std. Error t value Pr(>|t|)
12 (Intercept)  0.008449   0.004523   1.868 0.066993 .
13 Mkt.RF       1.269921   0.126658  10.026 4.16e-14 ***
14 SMB         -0.523079   0.191646  -2.729 0.008464 **
15 HML         -0.693077   0.167725  -4.132 0.000121 ***
16 ──
17 Signif. codes:  0 '***' 0.001 '**' 0.01 '*' 0.05 '.' 0.1 ' ' 1
18
19 Residual standard error: 0.03346 on 56 degrees of freedom
20 Multiple R-squared:  0.6836, Adjusted R-squared:  0.6667
21 F-statistic: 40.34 on 3 and 56 DF,  p-value: 5.106e-14
```

Using type equals HC1 in the `coeftest()` above adjusts the degrees of freedom by a factor of $n/(n - k)$, which changes the standard errors by a little bit.

```
1  > coeftest(ff.reg,
2  +    vcov. = vcovHC(ff.reg, type = "HC1"))
3
4  t-test of coefficients:
5
6                 Estimate Std. Error t value  Pr(>|t|)
```

```
 7   (Intercept)   0.0084487   0.0044985  1.8781 0.0655774 .
 8   Mkt.RF        1.2699207   0.1230993 10.3162 1.462e-14 ***
 9   SMB          -0.5230789   0.1766681 -2.9608 0.0044934 **
10   HML          -0.6930773   0.1686461 -4.1097 0.0001306 ***
11   ──
12   Signif. codes:  0 '***' 0.001 '**' 0.01 '*' 0.05 '.' 0.1 ' ' 1
```

## 5.7   Testing for Non-Normality

Linear regression models generally assume that the residuals are normally distributed. If that is not the case, we have to correct the issue by transforming the variables such that the distribution of the errors approximates a normal distribution. Thus, we may want to test for non-normality, so we know we are making correct inferences with the data. We test for non-normality by analyzing the third moment (skewness) and the fourth moment (kurtosis). If the residuals are normally distributed, skewness should equal 0 and kurtosis should equal 3. We can calculate the skewness and kurtosis using the skewness() and kurtosis() in the moments package. As the output shows, there is some positive skew to the residuals and the kurtosis is slightly below 3.

```
1   > library(moments)
2   > skewness(ff.reg$residuals)
3   [1] 0.02888204
4   > kurtosis(ff.reg$residuals)
5   [1] 2.46439
```

We can use more formal tests to determine whether the above differences from what we would expect from normality are significant or not. We can separately test skewness and kurtosis. For skewness, we can use the **D'Agostino test**, which we invoke through agostino.test(). As the high $p$-value indicates, the D'Agostino skewness test does not reject the null hypothesis that the residuals are not skewed (i.e., same as a normal distribution).

```
1   > agostino.test(ff.reg$residuals)
2
3     D'Agostino skewness test
4
5   data:  ff.reg$residuals
6   skew = 0.028882, z = 0.100378, p-value = 0.92
7   alternative hypothesis: data have a skewness
```

We can then use the **Anscombe–Glynn test** to test kurtosis, which we invoke through anscombe.test(). As the high $p$-value indicates, the Anscombe–Glynn kurtosis test does not reject the null hypothesis that the residuals have a kurtosis of 3 (i.e., same as a normal distribution).

```
1   > anscombe.test(ff.reg$residuals)
2
```

```
3    Anscombe—Glynn kurtosis test
4
5    data:  ff.reg$residuals
6    kurt = 2.46439, z = −0.81139, p−value = 0.4171
7    alternative hypothesis: kurtosis is not equal to 3
```

A common test that combines both the skewness and kurtosis tests is the **Jarque–Bera test**. In R, we run the Jarque–Bera test using `jarque.test()`. As the high *p*-value from the output below indicates, we cannot reject that the residuals are normally distributed.

```
1    > jarque.test(ff.reg$residuals)
2
3    Jarque—Bera Normality Test
4
5    data:  ff.reg$residuals
6    JB = 0.72554, p−value = 0.6957
7    alternative hypothesis: greater
```

## 5.8  Testing for Autocorrelation

We may also want to analyze whether the residuals are serially correlated (i.e., do they follow a pattern). This issue is known as **autocorrelation** and is a violation of the assumptions of the classical linear regression model. A common test for autocorrelation is the Durbin–Watson (DW) test. To perform this test in R, we use `dwtest` in the `lmtest` package. As the output below shows, the DW statistic is 1.98 with a *p*-value of 0.49. This indicates that the residuals are not autocorrelated.

```
1    > library(lmtest)
2    > dwtest(ff.reg)
3
4    Durbin—Watson test
5
6    data:  ff.reg
7    DW = 1.9837, p−value = 0.4938
8    alternative hypothesis: true autocorrelation is greater than 0
```

A rule of thumb is a DW statistic of 2 indicates that the residuals are not autocorrelated. However, given that our output includes the *p*-value, we do not need to rely on such rule of thumb to make that determination.

## 5.9  Event Studies

An **event study** tests the effect of new information on a firm's stock price. Thus, event studies presume that markets are semi-strong form efficient. The **semi-strong** form of the EMH indicates that public information is quickly incorporated into stock prices. So, how quickly does information get incorporated into stock prices? Busse and Green [9] studied the stock price reaction to CNBC reports, and they find that stocks respond within seconds of initial mention. Therefore, if an event has an effect on the stock price, we should observe the stock price to react quickly.

Most of what we know in corporate finance has been attributed to event studies. However, the popularity of event studies also extends to other areas, such as the field of securities litigation in which experts are almost always expected to support their opinions on the effect of disclosed information on a company's stock price using an event study.

The foundation of the event study is based on the Capital Asset Pricing Model (CAPM). An event study estimates the effect of a specific event on a company or a group of companies by looking at the firm's stock price performance. The event study uses the historical relationship between the firm's stock return and one or more independent variables. The independent variables are often chosen to be a proxy for the market and/or the firm's industry. Based on this historical relationship, an event study predicts a return for the company on the event date. An abnormal return is then calculated as the difference between the firm's actual return on the event date and the predicted return on the event date.

As the name suggests, an event study requires an event to be identified. This event could be something that affects a particular firm (e.g., corporate earnings announcements) or something that affects an entire industry (e.g., passing of new legislation). Moreover, some events are not reported in isolation (i.e., events are announced together with other information that could also affect the firm's stock price). These complicate our ability to isolate the impact of the event with other news that affected the stock price that same day. Therefore, the hypothesis we test in an event study depends on the specifics of the event, and a more involved analysis may be necessary to determine the effect of only the event we intend to study. Put differently, we cannot simply apply a cookie-cutter approach to performing event studies, as the underlying economics may not match a purely mechanical application of event studies. This is more true for single-firm event studies because we cannot benefit from averaging our results across multiple firms.

To conduct an event study, we need to identify two time periods. The first time period is the day or days around which the effects of an event have occurred. We call this the **event window**. The length of the event window depends on the type of event being studied. In the USA, strong empirical evidence exists to show that stock prices react very quickly, usually within seconds or minutes of material news announcements. Therefore, a one day event window is likely sufficient when the timing of the event can be clearly identified. In some instances, news may be leaked days or weeks prior to the official announcement. This can be the case, for example, in some mergers in which we observe the target's stock price already begins to increase days before the official merger announcement. In other instances, the information may be much more complicated, and subsequent information (e.g., commentary by analysts) may shed light on the previously disclosed event. In these cases, the event window can be extended to longer periods to fully capture the effect of the event. Thus, typical event windows are one day (i.e., event day only), two days (i.e., either day before plus event day or event day plus day after), or three days (i.e., day before, event day, and day after).

Once the event window is defined, we can then select the appropriate **estimation period** from which we estimate the parameters under normal conditions unaffected by the event. The estimation procedure uses a market model of the following form:

$$R_{i,t} = \alpha + \beta R_{m,t} + \epsilon_{i,t}, \tag{5.5}$$

where $R_{i,t}$ is the return on day $t$ of the subject firm $i$ and $R_{m,t}$ is the return on day $t$ of the market proxy $m$. In addition, we assume $\text{cov}(R_{m,t}, \epsilon_{i,t}) = 0$, $\text{var}(\epsilon_{i,t}) = \sigma_i^2$ (i.e., constant variance), and $\text{cov}(\epsilon_{i,t}, \epsilon_{i,t-n}) = 0$ for all $n \neq 0$. It is common to see returns calculated using arithmetic returns as the above or in log returns, where in the latter we replace $R_{i,t}$ and $R_{m,t}$ with $r_{i,t} = \ln(1 + R_{i,t})$ and $r_{m,t} = \ln(1 + R_{m,t})$, respectively. A common choice for the market proxy is the S&P 500 Index or another broad-based index.

Note that what constitutes as a period unaffected by the event depends on the event being studied. For example, if we are studying earnings announcements, then placing the estimation period prior to the event window may be appropriate. However, if we are studying an event immediately after a large merger that changes the underlying economics of the firm, then placing the estimation period after the event window may be better. Further, one must also be cognizant of whether the volatility of the abnormal return is significantly different due to non-event factors. For example, we likely will find many significant days if we analyze stock price movements in March 2020 when we choose an estimation period that includes data prior to March 2020.

Aside from the location of the estimation period, we must also determine the length of the estimation period. Typically, when daily data is used, a six-month or one-year period is sufficient. Alternatives could be weekly or monthly data over a longer date range that gives us a sufficient number of observations. For example, one could use at least a two-year estimation period if using weekly returns and a five-year estimation period when using monthly returns. Note that the choice of the length of the estimation period results in a trade-off between including more data with the risk of including less relevant data to predict the performance of the stock during the event window.

We can then estimate the abnormal return as

$$AR_{i,\tau} = R_{i,\tau} - \alpha + \beta R_{m,\tau}, \tag{5.6}$$

where $\tau$ is the event window.

Since we are using statistical methods to estimate the abnormal return, not all non-zero abnormal returns should be interpreted as a reaction to the disclosed information. The reason is that we have to account for the normal volatility of the firm's stock return. For example, a highly volatile stock may move up and down 5–10% in any given day, which means that a 10% abnormal return for that stock cannot be distinguished from the normal volatility of the stock. Hence, we have to test whether the abnormal return is sufficiently large that it can be considered statistically significant. We use a $t$-test to perform this task.

If we assume that the estimation period length $L$ is sufficiently long, we can argue that $\text{var}(AR_{i,\tau}|R_{m,\tau})$ converges to $\sigma_i^2$ such that

$$\frac{AR_{i,\tau}}{\sigma_i} \sim \mathbb{N}(0, 1) \tag{5.7}$$

with $L - 2$ degrees of freedom. That is, the $t$-statistic is defined as

$$t_{i,\tau} = \frac{AR_{i,\tau}}{\sigma_i}, \tag{5.8}$$

where $\sigma_i$ is the root mean square error (RMSE) or standard error of the residual from the market model regression in Eq. (5.5). Whether $t_{i,\tau}$ is significant depends on whether the $p$-value at $L-2$ degrees of freedom is less than or equal to the threshold level of significance. The $p$-value will tell us the smallest possible significance level for which the $t$-statistic is considered statistically significant, where conventional levels of significance are 1%, 5%, and 10%.

### 5.9.1   Example: Drop in Tesla Stock After 1Q 2019 Earnings Release on April 24, 2019

During trading hours on April 24, 2019, Tesla (ticker TSLA) reported financial results for 1Q 2019. The company reported revenues of $4.54 billion, which missed investors' expectations of $5.19 billion. Tesla also reported a loss per share of $2.90, which was much larger than the $0.69 loss per share investors were expecting. All else equal, reporting revenues and earnings below investors' expectations should result in a stock price decline. Consistent with this expectation, Tesla stock fell approximately 2% by the close of trading. However, is this price decline in Tesla stock large enough that this could not be explained by the normal volatility of Tesla stock? We can use an event study to test this.

**Step 1: Identify the Estimation Period and Event Window**  The event date we are interested in is April 24, 2019. For our estimation period, we are using the one-year period closest to the event date, but not including the event date. We would not like to contaminate the estimation by the event. Thus, we subtract 365 calendar days from the event date to get the start of our estimation period, which is April 24, 2018. Thus, the event window in this example would be from April 24, 2018 to April 23, 2019.

```
1  > (event.dt <- as.Date("2019-04-24"))
2  [1] "2019-04-24"
3  > (estper.start <- event.dt - 365)
4  [1] "2018-04-24"
```

**Step 2: Import TSLA and SPY Data**  We use `load.data()` to load the price data for TSLA and SPY.

```
1   > data.firm <- load.data("TSLA Yahoo.csv", "TSLA")
2   > head.tail(data.firm)
3              TSLA.Open TSLA.High TSLA.Low TSLA.Close TSLA.Volume TSLA.Adjusted
4   2014-12-31    44.618    45.136   44.450     44.482    11487500        44.482
5   2015-01-02    44.574    44.650   42.652     43.862    23822000        43.862
6   2015-01-05    42.910    43.300   41.432     42.018    26842500        42.018
7              TSLA.Open TSLA.High TSLA.Low TSLA.Close TSLA.Volume TSLA.Adjusted
8   2019-12-27    87.000    87.062   85.222     86.076    49728500        86.076
9   2019-12-30    85.758    85.800   81.852     82.940    62932000        82.940
10  2019-12-31    81.000    84.258   80.416     83.666    51428500        83.666
11  >
12  > data.mkt <- load.data("SPY Yahoo.csv", "SPY")
13  > head.tail(data.mkt)
14            SPY.Open SPY.High SPY.Low SPY.Close SPY.Volume SPY.Adjusted
15  2014-12-31   207.99   208.19  205.39    205.54  130333800     186.2590
16  2015-01-02   206.38   206.88  204.18    205.43  121465900     186.1593
17  2015-01-05   204.17   204.37  201.35    201.72  169632600     182.7974
18            SPY.Open SPY.High SPY.Low SPY.Close SPY.Volume SPY.Adjusted
19  2019-12-27   323.74   323.80  322.28    322.86   42528800       322.86
20  2019-12-30   322.95   323.10  320.55    321.08   49729100       321.08
21  2019-12-31   320.53   322.13  320.15    321.86   57077300       321.86
```

**Step 3: Calculate Returns for TSLA and SPY**  In Lines 1 and 2 below, we calculate the log returns for TSLA and SPY. We then clean up the returns data to exclude the December 31, 2014 observation and rename the variables.

```
1   > rets <- cbind(diff(log(data.firm$TSLA.Adjusted)),
2   +      diff(log(data.mkt$SPY.Adjusted)))
3   > rets <- rets[-1, ]
4   > names(rets) <- c("Firm", "Market")
5   > head.tail(rets)
6                       Firm        Market
7   2015-01-02 -0.014036226 -0.0005354247
8   2015-01-05 -0.042950196 -0.0182246098
9   2015-01-06  0.005648234 -0.0094636586
10                      Firm        Market
11  2019-12-27 -0.001300423 -0.0002478074
12  2019-12-30 -0.037113102 -0.0055284735
13  2019-12-31  0.008715203  0.0024263490
```

**Step 4: Subset Returns Data to Estimation Period and Event Window**  We first create a variable called *window*, which is the date range that includes the estimation period and the event window. The format of the output is so that we can use it inside the brackets to subset an xts object (see Line 4).

```
1   > window <- paste(estper.start, "/", event.dt, sep = "")
2   > window
3   [1] "2018-04-24/2019-04-24"
4   > rets <- rets[window]
5   > head.tail(rets)
6                       Firm        Market
7   2018-04-24 0.0003175732 -0.013558751
```

```
8   2018–04–25 –0.0098201799  0.002468396
9   2018–04–26  0.0169211341   0.010114567
10                      Firm            Market
11  2019–04–22 –0.039220732   0.0008616182
12  2019–04–23  0.004367234   0.0089514669
13  2019–04–24 –0.020055805 –0.0022217592
```

**Step 5: Conduct the Event Study**  We present three ways to perform an event study as applied to our example. The first approach uses a single step in which dummy (or indicator) variables are used to denote the event date. The second approach uses two stages. The first stage estimates the market model parameters during the estimation period, and then the second stage takes those parameter estimates and applies them to the event date. The third approach uses a non-parametric, sample quantile approach to measure statistical significance. In all three methods, the **abnormal return** calculated is identical. The first two approaches differ in the standard error that is used to calculate the $t$-statistic and determine statistical significance. The third approach differs in that it uses a ranking of the abnormal returns during the estimation period to determine whether the abnormal return is large enough to be statistically significant.

### 5.9.2   Single Step Event Study

The single step event study considers the data in both the estimation period and the event date but adds a dummy variable on the event date. Generally, we modify the above equation to add a third term on the RHS. That is,

$$R_{i,t} = \alpha + \beta_i R_{m,t} + \sum_{n=1}^{N} \gamma_n D_n, \tag{5.9}$$

where $D_n$ is a dummy variable to let R know that April 24, 2019 is the event date (Line 1). Since we only have one event date in our example, the third term on the RHS is only $\gamma D$ and $\gamma$ is the abnormal return. $D$ takes on the value of 1 on April 24, 2019 and zero otherwise. The time $t$ in this single-stage event study goes through April 24, 2019 (i.e., includes the event date), and that is the only date in which the dummy variable has a value of 1. We then run the market model regression using $lm()$.

```
1  > rets$Dummy <- ifelse(index(rets) == "2019–04–24", 1, 0)
2  > head.tail(rets)
3                      Firm            Market Dummy
4  2018–04–24  0.0003175732 –0.013558751      0
5  2018–04–25 –0.0098201799  0.002468396      0
6  2018–04–26  0.0169211341   0.010114567      0
7                      Firm            Market Dummy
8  2019–04–22 –0.039220732   0.0008616182     0
```

```
 9  2019–04–23  0.004367234  0.0089514669      0
10  2019–04–24 −0.020055805 −0.0022217592      1
11  > event <- lm(Firm ~ Market + Dummy, data = rets)
12  > summary(event)
13
14  Call:
15  lm(formula = Firm ~ Market + Dummy, data = rets)
16
17  Residuals:
18        Min        1Q     Median        3Q       Max
19  −0.154874 −0.017476  0.001918  0.016711  0.156417
20
21  Coefficients:
22                Estimate Std. Error t value Pr(>|t|)
23  (Intercept) −0.0008574  0.0021979  −0.390    0.697
24  Market       1.2706843  0.2337810   5.435  1.3e–07 ***
25  Dummy       −0.0163753  0.0348554  −0.470    0.639
26  ——
27  Signif. codes:  0 '***' 0.001 '**' 0.01 '*' 0.05 '.' 0.1 ' ' 1
28
29  Residual standard error: 0.03478 on 249 degrees of freedom
30  Multiple R-squared:  0.1071, Adjusted R-squared:  0.09992
31  F-statistic: 14.93 on 2 and 249 DF,  p-value: 7.503e–07
```

In a one step event study, we can read off the abnormal return from the regression result. The coefficient of the dummy variable (i.e., −0.016) is the abnormal return estimate on the event date. That is, the model indicates that the abnormal return on April 24, 2019 for TSLA is −1.6%. However, as we look at the $p$-value, it is 0.470, which means that the abnormal return is *not* statistically significant.

The reason for this is because the normal volatility of the abnormal return for TSLA, which is reflected in the standard error of the dummy variable in the output, is quite large at 3.49%. This means that TSLA's price decline on April 24, 2019 was less than half that level, and so it cannot be distinguished from normal fluctuations in TSLA stock.

### 5.9.3  Two Stage Event Study

In contrast to the one stage event study above, a two stage event study splits the estimation period and event window apart. This is how the original event study was performed in [14, Fama, et al. (1969)].

**Step 1: Create Event Window and Estimation Period Dataset**  We first identify the event window (Line 1). We are using a one day event window in this example, which is on April 24, 2019. Since we are also using *rets*, we can drop the third column, which is the dummy variable used in the one-pass event study. In this step, we also construct our estimation period dataset (Line 4). Similarly, since we are using *rets*, we also drop the third column.

```
1   > (event.window <- rets["2019-04-24", -3])
2                   Firm        Market
3   2019-04-24 -0.02005581 -0.002221759
4   > est.period <- rets["2018-04-24/2019-04-23", -3]
5   > head.tail(est.period)
6                    Firm        Market
7   2018-04-24  0.0003175732 -0.013558751
8   2018-04-25 -0.0098201799  0.002468396
9   2018-04-26  0.0169211341  0.010114567
10                   Firm        Market
11  2019-04-18  0.007456590 0.0019672729
12  2019-04-22 -0.039220732 0.0008616182
13  2019-04-23  0.004367234 0.0089514669
```

**Step 2: Regress Stock Return on Market Return**  In the first of the two steps, we
run a regression of the TSLA return on the SPY return over the estimation period
data. We can compare the estimates and standard errors of the coefficients, and they
are identical to that of the regression in the one-pass event study.

```
1   > event2 <- lm(Firm ~ Market, data = est.period)
2   > summary(event2)
3
4   Call:
5   lm(formula = Firm ~ Market, data = est.period)
6
7   Residuals:
8        Min       1Q    Median        3Q       Max
9   -0.154874 -0.017547  0.002061  0.016810  0.156417
10
11  Coefficients:
12               Estimate Std. Error t value Pr(>|t|)
13  (Intercept) -0.0008574  0.0021979  -0.390    0.697
14  Market       1.2706843  0.2337810   5.435 1.3e-07 ***
15  ---
16  Signif. codes:  0 '***' 0.001 '**' 0.01 '*' 0.05 '.' 0.1 ' ' 1
17
18  Residual standard error: 0.03478 on 249 degrees of freedom
19  Multiple R-squared:  0.1061, Adjusted R-squared:  0.1025
20  F-statistic: 29.54 on 1 and 249 DF,  p-value: 1.303e-07
```

**Step 3: Identify the Regression Alpha, Beta, and RMSE**  To calculate the
predicted TSLA return, we need the alpha (intercept), beta, and root mean square
error (sigma) from the regression results. We will use these in the second step
to calculate the abnormal return on the event date and to determine the statistical
significance of the abnormal returns.

```
1   > (alpha <- summary(event2)$coeff[1])
2   [1] -0.0008573843
3   > (beta <- summary(event2)$coeff[2])
4   [1] 1.270684
5   > (rmse <- summary(event2)$sigma)
6   [1] 0.03478057
```

**Step 4: Use Alpha and Beta to Predict Stock Returns During Event Window**
In the second step of the regression, we use the alpha and beta to estimate the
predicted TSLA return on April 24, 2019 (Line 1). We then take the difference
between the actual return and predicted return to get to the abnormal return (Line
2). We calculate the $t$-statistic by dividing the abnormal return by the root mean
square error or RMSE (Line 3). We calculate the two-tailed $p$-value of the $t$-statistic
in Lines 4 and 5.

```
 1  > event.window$Pred <- alpha + beta * event.window$Market
 2  > event.window$Ab.Ret <- event.window$Firm - event.window$Pred
 3  > event.window$tStat <- event.window$Ab.Ret / rmse
 4  > event.window$pval <- 2 * (1 - pt(abs(event.window$tStat),
 5  +      df = nrow(rets) - (2 + 1)))
 6  > options(digits = 3)
 7  > event.window
 8                 Firm    Market     Pred  Ab.Ret  tStat  pval
 9  2019-04-24 -0.0201 -0.00222 -0.00368 -0.0164 -0.471 0.638
10  > options(digits = 7)
```

As the output in Line 9 shows, the abnormal return is $-0.0164$ or $-1.64\%$, which
is identical to the abnormal return calculated above in the one-pass regression.
The difference is that the $t$-statistic is $-0.471$ in the two-step versus $-0.470$ in
the one-pass event study. Since the abnormal return is the same, the difference in
$t$-statistics arises from the standard error used to calculate the $t$-statistic. In the two-
step event study, we used the RMSE of the regression over the estimation period
as our estimate of the standard error. In the one-pass event study, it uses a slightly
different standard error calculation because it includes some information related to
the event date (in addition to the estimation period data). The difference between the
two methods with respect to the $t$-statistics does not alter the qualitative conclusions
from our event study. That is, the TSLA abnormal (i.e., firm-specific) return on April
24, 2019 could be explained by the market factors.

The precision of the inputs and assumptions depends on the ultimate application
of the event study. In many academic studies, the final event study result is often an
average across a large cross-sectional sample of firms or events. This averaging has
the benefit of reducing the individual firm-specific effects on the abnormal returns,
and it may be less important to have very precise inputs and assumptions. However,
when dealing with single-firm event studies, we have to be more careful with the
inputs and assumptions that we use.

### 5.9.4   Sample Quantiles/Non-Parametric

In most academic articles that use event studies, the results rely on cross-sectional
aggregation of the abnormal returns and rely on the average effect of the event.
When inferences are made based on an event study for a single firm, we may also
consider an alternative approach that could be used as a complement to the standard
event study approach. In this subsection, we use **sample quantiles** of the abnormal

return during the estimation period to determine whether the abnormal return on the event date is large enough to be statistically significant. Because we do not make any assumption on the distribution of the abnormal returns during the estimation period, this approach is considered **non-parametric**.

To implement this approach, we can continue from our earlier example of a two-step event study. From that analysis, we found that the abnormal return for TSLA on April 24, 2019 is $-1.64\%$. Since we want to know whether the stock price decline on April 24, 2019 was statistically significant, we will perform a one-tailed test. Assuming a significance level or $\alpha$ of 5%, we will have to find the $\alpha * N$ observation, where $N$ is the number of abnormal return observations during the estimation period, which is 250.

**Step 1: Calculate Abnormal Returns During Estimation Period**   The abnormal return during the estimation period is equal to the actual return less the predicted return. We estimate the predicted return using the coefficients of the regression output.

```
 1  > ab.ret <- est.period$Firm - alpha + beta * est.period$Market
 2  > head.tail(ab.ret)
 3                      Firm
 4  2018-04-24 -0.016053935
 5  2018-04-25 -0.005826244
 6  2018-04-26  0.030630940
 7                      Firm
 8  2019-04-18  0.01081376
 9  2019-04-22 -0.03726850
10  2019-04-23  0.01659911
```

**Step 2: Sort the Abnormal Returns Data**   We first convert the object into a data frame using `data.frame()`. Then, using `order()`, we sort the abnormal returns from smallest (i.e., most negative) to largest.

```
 1  > sorted <- data.frame(ab.ret)
 2  > sorted <- sorted[order(sorted$Firm), ]
 3  > head.tail(sorted)
 4  [1] -0.1486900 -0.1212715 -0.1124748
 5  [1] 0.1578002 0.1623663 0.1652305
```

**Step 3: Calculate Observation Number for Cut-Off**   To get the cut-off observation number, we multiply the significance level we select by the number of abnormal return observations in the estimation period. The $\alpha$ we select is 5% in a one-tailed test. This is a one-tailed test because we would like to analyze whether the price of TSLA declined on April 24, 2019. We have $N = 251$ observations during the estimation period, which means that the cut-off would have been observation number 12.55. To avoid having fractions of an observation, we choose the next largest integer for the cut-off. We can find that by using `ceiling()`.

```
 1  > (alpha <- 0.05)
 2  [1] 0.05
 3  > (N <- nrow(est.period))
```

```
4   [1] 251
5   > (cut.off <- alpha * N)
6   [1] 12.55
7   > (cut.off <- ceiling(alpha * N))
8   [1] 13
```

**Step 4: Identify the Abnormal Return Cut-Off**  The *cut.off* variable equals 13, so we only need 13 observations. Below, we output the 13th most negative abnormal return, which is −6.67%. If our abnormal return is more negative than this number, we would consider it statistically significant. Since the abnormal return on April 24, 2019 is only −1.64%, using sample quantiles indicates that the TSLA price decline on that day is not statistically significant.

```
1   > sorted[cut.off]
2   [1] -0.066725
```

**Step 5: Alternative Way to Determine Cut-Off Return**  Instead of doing Steps 3 and 4, we could get to a similar result if we instead used quantile(). The first argument in quantile() is the dataset *sorted*, and the second argument tells R which quantiles we want to output. We show below the lower and upper tail at a 5% significance level. For our example, the relevant number is the lower quantile value of −6.48%. This number is different from our result in Step 4 and is equal to the midpoint of the 13th and 14th most negative abnormal returns.

```
1   > quantile(sorted, c(alpha, 1 - alpha))
2          5%           95%
3   -0.06476826  0.05678372
4   > mean(c(sorted[13], sorted[14]))
5   [1] -0.06476826
```

The difference is because **quantiles** can be calculated in many different ways. There are seven ways to calculate quantiles in R. As we can see, types 1, 2, and 3 give us the same answer as Step 4, while the other four types do not. For the differences between each type, visit: https://stat.ethz.ch/R-manual/R-devel/library/stats/html/quantile.html.

```
1    > quantile(sorted, c(alpha, 1 - alpha), type = 1)
2          5%           95%
3    -0.0667250   0.0581327
4    > quantile(sorted, c(alpha, 1 - alpha), type = 2)
5          5%           95%
6    -0.0667250   0.0581327
7    > quantile(sorted, c(alpha, 1 - alpha), type = 3)
8          5%           95%
9    -0.06672500  0.05543475
10   > quantile(sorted, c(alpha, 1 - alpha), type = 4)
11         5%           95%
12   -0.06748050  0.05664883
13   > quantile(sorted, c(alpha, 1 - alpha), type = 5)
14         5%           95%
15   -0.06652933  0.05799780
16   > quantile(sorted, c(alpha, 1 - alpha), type = 6)
```

```
17            5%           95%
18  −0.06739656  0.06038842
19  > quantile(sorted, c(alpha, 1 − alpha), type = 7)
20            5%           95%
21  −0.06476826  0.05678372
```

## 5.10   Selecting Best Regression Variables

The CAPM is a single factor model, which means there is only one factor that is used to explain the company returns. The Fama–French model has three or five factors, which means that there are multiple factors that are used to explain the company returns. There are plenty of different factor models used in practice. How then do we determine which set of factors to include in our analysis? There is no bright-line test for this. We can think about the factors individually and analyze whether each one should be included or not. We can run various regressions and then test to determine what is a better model. We can use a metric like adjusted $R$-squared to help us reach a conclusion in those situations.

Alternatively, we can use approaches that rely on the analysis of variance and the lowest (largest negative) **Akaike Information Criterion** (AIC) to select the best model. AIC deals with the trade-off between the goodness of fit of the model and the simplicity of the model and considers both the risk of overfitting and the risk of underfitting.

We can perform the model selection by either going forward or backward. Going forward means we start with the simplest model and then work our way to the full model. Going backward means we start with the full model and work our way to a reduced model. The steps can be automated in R, which is helpful if we have many factors that we are considering. For our illustration below, we will use the hypothetical monthly portfolio returns we constructed in Appendix C and the Fama–French 5 factor model.

### 5.10.1   Create Dataset of Returns

We first create a combined dataset of the portfolio returns and the Fama–French 5 factor model.

**Step 1: Import Portfolio Returns**   We import the portfolio returns we constructed using `read.csv()`. The output shows our portfolio has 60 monthly returns.

```
1  > data.port <- read.csv("Hypothetical Portfolio (Monthly).csv")
2  > data.port <- data.port[, −1]
3  > head.tail(data.port)
4         Date     Port.Ret
5  1 Jan 2015  0.07306837
```

```
6    2 Feb 2015  0.07258170
7    3 Mar 2015 −0.02373268
8           Date   Port.Ret
9   58 Oct 2019 0.05596117
10  59 Nov 2019 0.04224452
11  60 Dec 2019 0.04982447
```

**Step 2: Import Fama–French 5 Factor Data**  On Ken French's website, we can obtain a CSV, the Fama–French 5 factor model data with monthly returns. Opening the CSV file in Excel shows that the header starts in Row 4, so we want to skip the first three rows and tell R that the first row it reads in is a header. The FF 5 factor model data starts in 1963 and has 682 rows of monthly returns data. Below the monthly returns data are annual data we are not currently interested in. Thus, we tell R to only read in the next 678 rows following the header, where 678 is the total 682 rows less the four rows (i.e., three we skipped plus the header row).

```
1   > data.ff5 <- read.csv("F–F_Research_Data_5_Factors_2x3.csv",
2   +      skip = 3,
3   +      header = TRUE,
4   +      nrows = 678)
5   > head.tail(data.ff5)
6          X Mkt.RF   SMB   HML   RMW   CMA   RF
7   1 196307  −0.39 −0.47 −0.83  0.66 −1.15 0.27
8   2 196308   5.07 −0.79  1.67  0.39 −0.40 0.25
9   3 196309  −1.57 −0.48  0.18 −0.76  0.24 0.27
10         X Mkt.RF  SMB   HML   RMW   CMA   RF
11  676 201910   2.06 0.21 −2.07  0.43 −0.96 0.15
12  677 201911   3.87 0.50 −1.86 −1.50 −1.29 0.12
13  678 201912   2.77 0.97  1.82  0.23  1.29 0.14
```

We then use the `lubridate` package to create a date variable R reads in as a date. To make this process easy, we create a year–month–date looking variable dt by pasting in 01 to the end of the values in column X.

```
1   > data.ff5$dt <- paste(data.ff5$X, "01", sep = "")
2   > library(lubridate)
3   > data.ff5$Date <- ymd(data.ff5$dt)
4   > head.tail(data.ff5)
5          X Mkt.RF   SMB   HML   RMW   CMA   RF       dt        Date
6   1 196307  −0.39 −0.47 −0.83  0.66 −1.15 0.27 19630701 1963−07−01
7   2 196308   5.07 −0.79  1.67  0.39 −0.40 0.25 19630801 1963−08−01
8   3 196309  −1.57 −0.48  0.18 −0.76  0.24 0.27 19630901 1963−09−01
9          X Mkt.RF  SMB   HML   RMW   CMA   RF       dt        Date
10  676 201910   2.06 0.21 −2.07  0.43 −0.96 0.15 20191001 2019−10−01
11  677 201911   3.87 0.50 −1.86 −1.50 −1.29 0.12 20191101 2019−11−01
12  678 201912   2.77 0.97  1.82  0.23  1.29 0.14 20191201 2019−12−01
```

Next, we first do a little bit of cleaning by removing unnecessary columns from *data.ff5* and then subset the data to 2015 to 2019.

```
1   > data.ff5 <- data.ff5[, −c(1, 8)]
2   > ff5.sub <- subset(data.ff5,
3   +      Date >= "2015−01−01" &
```

```
4  +    Date <= "2019-12-31")
5  > rownames(ff5.sub) <- seq(1, nrow(ff5.sub), 1)
6  > head.tail(ff5.sub)
7    Mkt.RF   SMB   HML   RMW   CMA RF       Date
8  1  -3.11 -0.87 -3.48  1.67 -1.67   0 2015-01-01
9  2   6.13  0.21 -1.81 -1.07 -1.75   0 2015-02-01
10 3  -1.12  3.04 -0.46  0.10 -0.51   0 2015-03-01
11   Mkt.RF  SMB   HML   RMW   CMA RF        Date
12 58   2.06 0.21 -2.07  0.43 -0.96 0.15 2019-10-01
13 59   3.87 0.50 -1.86 -1.50 -1.29 0.12 2019-11-01
14 60   2.77 0.97  1.82  0.23  1.29 0.14 2019-12-01
```

Finally, we convert the returns into decimals instead of percentage points by dividing them by 100.

```
1  > ff5.sub[, 1:6] <- ff5.sub[, 1:6] / 100
2  > head.tail(ff5.sub)
3     Mkt.RF      SMB      HML      RMW      CMA RF       Date
4  1 -0.0311 -0.0087 -0.0348  0.0167 -0.0167   0 2015-01-01
5  2  0.0613  0.0021 -0.0181 -0.0107 -0.0175   0 2015-02-01
6  3 -0.0112  0.0304 -0.0046  0.0010 -0.0051   0 2015-03-01
7     Mkt.RF     SMB      HML      RMW      CMA      RF       Date
8  58 0.0206 0.0021 -0.0207  0.0043 -0.0096 0.0015 2019-10-01
9  59 0.0387 0.0050 -0.0186 -0.0150 -0.0129 0.0012 2019-11-01
10 60 0.0277 0.0097  0.0182  0.0023  0.0129 0.0014 2019-12-01
```

**Step 3: Combine the Portfolio and Fama–French Datasets**  Using cbind(), we combine the two datasets. Then, we calculate the excess portfolio return Port.RF by subtracting from portfolio return Port.Ret the risk-free rate RF.

```
1  > options(digits = 3)
2  > combine$Port.RF <- combine$Port.Ret - combine$RF
3  > head.tail(combine)
4    Port.Ret  Mkt.RF     SMB     HML     RMW     CMA RF       Date Port.RF
5  1   0.0731 -0.0311 -0.0087 -0.0348  0.0167 -0.0167  0 2015-01-01  0.0731
6  2   0.0726  0.0613  0.0021 -0.0181 -0.0107 -0.0175  0 2015-02-01  0.0726
7  3  -0.0237 -0.0112  0.0304 -0.0046  0.0010 -0.0051  0 2015-03-01 -0.0237
8    Port.Ret Mkt.RF    SMB     HML     RMW     CMA      RF       Date Port.RF
9  58   0.0560 0.0206 0.0021 -0.0207  0.0043 -0.0096 0.0015 2019-10-01
   0.0545
10 59   0.0422 0.0387 0.0050 -0.0186 -0.0150 -0.0129 0.0012 2019-11-01
   0.0410
11 60   0.0498 0.0277 0.0097  0.0182  0.0023  0.0129 0.0014 2019-12-01
   0.0484
12 > options(digits = 7)
```

## 5.10.2  Forward Step Approach

**Step 1: Identify Minimum Model**  In the forward step approach, we start with the
minimum model and work our way to the full model. The minimum model has a
vector of 1s.

```
1   > min.model <- lm(Port.RF ~ 1, data = combine)
2   > summary(min.model)
3
4   Call:
5   lm(formula = Port.RF ~ 1, data = combine)
6
7   Residuals:
8        Min        1Q     Median        3Q       Max
9   -0.135053 -0.044315  0.004036  0.038522  0.135138
10
11  Coefficients:
12               Estimate Std. Error t value Pr(>|t|)
13  (Intercept) 0.023002   0.007481   3.075  0.00319 **
14  ---
15  Signif. codes:  0 '***' 0.001 '**' 0.01 '*' 0.05 '.' 0.1 ' ' 1
16
17  Residual standard error: 0.05795 on 59 degrees of freedom
```

**Step 2: Run Steps**  We then work our way to the full model, which is the Fama–
French 5 factor model. As we can see, the starting AIC of the minimum model is
−341. The AIC of the model with only Mkt.RF is −384. The AIC of the model
with Mkt.RF and CMA is −412. The AIC of the model with Mkt.RF, CMA, and
SMB is −420. So the best model is the last one.

```
1   > fwd.model <- step(min.model,
2   +    direction = "forward",
3   +    scope = (~ Mkt.RF + SMB + HML + RMW + CMA))
4   Start:  AIC=-340.79
5   Port.RF ~ 1
6
7            Df Sum of Sq      RSS      AIC
8   + Mkt.RF  1  0.104582 0.093562 -383.81
9   + CMA     1  0.074604 0.123539 -367.13
10  + HML     1  0.022888 0.175256 -346.15
11  <none>                0.198144 -340.79
12  + SMB     1  0.000539 0.197605 -338.95
13  + RMW     1  0.000067 0.198077 -338.81
14
15  Step:  AIC=-383.81
16  Port.RF ~ Mkt.RF
17
18          Df Sum of Sq      RSS      AIC
19  + CMA    1  0.036632 0.056929 -411.62
20  + HML    1  0.022536 0.071026 -398.34
21  + SMB    1  0.016486 0.077076 -393.44
22  <none>              0.093562 -383.81
```

```
23   + RMW   1  0.000797 0.092764 −382.32
24
25   Step: AIC=−411.62
26   Port.RF ~ Mkt.RF + CMA
27
28           Df Sum of Sq     RSS     AIC
29   + SMB   1 0.0087903 0.048139 −419.68
30   + RMW   1 0.0019455 0.054984 −411.70
31   <none>              0.056929 −411.62
32   + HML   1 0.0009467 0.055983 −410.62
33
34   Step: AIC=−419.68
35   Port.RF ~ Mkt.RF + CMA + SMB
36
37           Df  Sum of Sq     RSS     AIC
38   <none>              0.048139 −419.68
39   + HML   1 1.9852e−04 0.047940 −417.93
40   + RMW   1 7.6512e−05 0.048062 −417.78
41   > summary(fwd.model)
42
43   Call:
44   lm(formula = Port.RF ~ Mkt.RF + CMA + SMB, data = combine)
45
46   Residuals:
47        Min       1Q    Median        3Q       Max
48   −0.064899 −0.016831 −0.004939  0.015285  0.069306
49
50   Coefficients:
51               Estimate Std. Error t value Pr(>|t|)
52   (Intercept)  0.009086   0.003949   2.301  0.02514 *
53   Mkt.RF       1.102979   0.117771   9.365 4.66e−13 ***
54   CMA         −1.459816   0.251608  −5.802 3.19e−07 ***
55   SMB         −0.528665   0.165323  −3.198  0.00228 **
56   ──
57   Signif. codes:  0 '***' 0.001 '**' 0.01 '*' 0.05 '.' 0.1 ' ' 1
58
59   Residual standard error: 0.02932 on 56 degrees of freedom
60   Multiple R−squared:  0.757, Adjusted R−squared:  0.744
61   F−statistic: 58.17 on 3 and 56 DF,  p−value: < 2.2e−16
```

### 5.10.3  Backward Step Approach

**Step 1: Start with Full Model**  In the backward step approach, we start with the full model and then work our way to the reduced model. The full model has all five factors included.

```
1   > full.model <− lm(Port.RF ~ Mkt.RF + SMB + HML + RMW + CMA,
2   +    data = combine)
```

**Step 2: Run the Steps** We start with the Fama–French 5 factor model (all independent variables) and then work our way to the reduced models.

```
1   > reduced.model <- step(full.model,
2   +     direction = "backward")
3   Start: AIC=-416.04
4   Port.RF ~ Mkt.RF + SMB + HML + RMW + CMA
5
6              Df Sum of Sq      RSS     AIC
7   - RMW    1   0.000090 0.047940 -417.93
8   - HML    1   0.000212 0.048062 -417.78
9   <none>                0.047850 -416.04
10  - SMB    1   0.006245 0.054095 -410.68
11  - CMA    1   0.014571 0.062422 -402.09
12  - Mkt.RF 1   0.074224 0.122075 -361.85
13
14  Step: AIC=-417.93
15  Port.RF ~ Mkt.RF + SMB + HML + CMA
16
17             Df Sum of Sq      RSS     AIC
18  - HML    1   0.000199 0.048139 -419.68
19  <none>                0.047940 -417.93
20  - SMB    1   0.008042 0.055983 -410.62
21  - CMA    1   0.014484 0.062425 -404.09
22  - Mkt.RF 1   0.074634 0.122575 -363.60
23
24  Step: AIC=-419.68
25  Port.RF ~ Mkt.RF + SMB + CMA
26
27             Df Sum of Sq      RSS     AIC
28  <none>                0.048139 -419.68
29  - SMB    1   0.008790 0.056929 -411.62
30  - CMA    1   0.028937 0.077076 -393.44
31  - Mkt.RF 1   0.075399 0.123538 -365.13
```

**Step 3: Show Summary of Best Model Regression** The regression results of the best model are saved under the *reduced.model* object. Thus, we can use summary() to output the regression results. As we can see, whether we go forward or backward, we end up in the same place. That is, the model that includes Mkt.RF, SMB, and CMA as the factors.

```
1   > summary(reduced.model)
2
3   Call:
4   lm(formula = Port.RF ~ Mkt.RF + SMB + CMA, data = combine)
5
6   Residuals:
7        Min        1Q    Median        3Q       Max
8   -0.064899 -0.016831 -0.004939  0.015285  0.069306
9
10  Coefficients:
11             Estimate Std. Error t value Pr(>|t|)
12  (Intercept)  0.009086   0.003949   2.301  0.02514 *
```

```
13   Mkt.RF        1.102979    0.117771    9.365 4.66e-13 ***
14   SMB          -0.528665    0.165323   -3.198  0.00228 **
15   CMA          -1.459816    0.251608   -5.802 3.19e-07 ***
16   ——
17   Signif. codes:  0 '***' 0.001 '**' 0.01 '*' 0.05 '.' 0.1 ' ' 1
18
19   Residual standard error: 0.02932 on 56 degrees of freedom
20   Multiple R-squared:  0.757, Adjusted R-squared:  0.744
21   F-statistic: 58.17 on 3 and 56 DF,  p-value: < 2.2e-16
```

## *5.10.4   Suppressing Steps in Output*

Once we are comfortable with using the model, we can suppress the individual steps that get to the result. We do this by adding `trace = 0` inside `step()`. The output below is identical to the output of the backward approach above. The only difference is we did not print out the output of all the different steps R took to get to the best model in Step 2 above.

```
 1   > reduced.model2 <- step(full.model,
 2   +      direction = "backward",
 3   +      trace = 0)
 4   > summary(reduced.model2)
 5
 6   Call:
 7   lm(formula = Port.RF ~ Mkt.RF + SMB + CMA, data = combine)
 8
 9   Residuals:
10        Min        1Q     Median        3Q        Max
11   -0.064899 -0.016831 -0.004939  0.015285  0.069306
12
13   Coefficients:
14               Estimate Std. Error t value Pr(>|t|)
15   (Intercept)  0.009086   0.003949    2.301  0.02514 *
16   Mkt.RF       1.102979   0.117771    9.365 4.66e-13 ***
17   SMB         -0.528665   0.165323   -3.198  0.00228 **
18   CMA         -1.459816   0.251608   -5.802 3.19e-07 ***
19   ——
20   Signif. codes:  0 '***' 0.001 '**' 0.01 '*' 0.05 '.' 0.1 ' ' 1
21
22   Residual standard error: 0.02932 on 56 degrees of freedom
23   Multiple R-squared:  0.757, Adjusted R-squared:  0.744
24   F-statistic: 58.17 on 3 and 56 DF,  p-value: < 2.2e-16
```

## 5.11  Further Reading

The original paper on the CAPM is [25, Sharpe (1964)]. In the above example, we calculated the beta using 60 months of returns data. For example, this is the return frequency and estimation period length used by Yahoo Finance. As noted above, the return frequencies and length of the estimation can vary. For example, Morningstar uses three years of monthly returns data, while the default on the Bloomberg terminal is two years of weekly returns. Alternative return frequencies and estimation period lengths are also used. However, we have to be aware of the following two trade-offs. The first trade-off is that higher frequency (i.e., daily returns data) is more susceptible to asynchronous trading issues than lower frequency (i.e., monthly data). The second trade-off is that shorter estimation periods may be more relevant to the current beta but may be more affected by extreme events than longer estimation periods, but longer estimation periods may include older and less relevant data.

We also used a 3-month risk-free rate to estimate the CAPM. Depending on the applications, we may want to use a longer-term risk-free rate. For example, when calculating the cost of equity in a valuation context, we may want to use a 10-year or 20-year US Treasury security.

The original paper on the FF Model is [15, Fama and French (1993)]. In [16, Fama and French (2015)], the authors develop a five factor model. A lot of data related to the FF Models can be found on Ken French's website (https://mba.tuck. dartmouth.edu/pages/faculty/ken.french/data_library.html). An excellent treatment of factor models can be found in [13, Damodaran (2012)] and [20, Koller, et al. (2010)].

For a discussion of factor investing related to asset management, see [1, Ang (2014)]. A detailed discussion of some common factors, like value, momentum, and liquidity, from an empirical perspective can be found in [5, Bali, et al. (2016)]. There are a number of studies that question the reliability of papers that find factors. In [2, Ang (2017)], [3, Ang (2018)], and [4, Ang (2020)], I discuss how size is no longer a factor. Other studies discuss issues with a wider array of factors. Examples of such studies include [11, Cornell (2020)], [19, Hou, Xue, and Zhang (2020)], and [18, Harvey, Liu, and Zhu (2016)].

The seminal paper on event studies is [14, Fama, et al. (1969)]. A good background on event studies can be found in [21, Kothari and Warner (2005)]. The authors report over 575 published articles in five top finance journals made use of event studies from 1974 to 2000. A good discussion of the approach we use is in [22, MacKinlay (1997)] and [7, Binder (1985)]. For more specific applications of event studies to securities litigation, see [12, Crew, et al. (2007)].

# References

1. Ang, A. (2014). *Asset management: a systematic approach to factor investing*. Oxford: Oxford University Press.
2. Ang, C. (2017). Why we shouldn't add a size premium to the CAPM cost of equity. NACVA QuickRead. http://quickreadbuzz.com/2017/02/15/shouldnt-add-size-premium-capm-cost-equity/.
3. Ang, C. (2018). The absence of a size effect relevant to the cost of equity. *Business Valuation Review, 37*, 87–92.
4. Ang, C. (2020). It's time for valuation experts to let go of the size premium. Law360. https://www.law360.com/articles/1283192.
5. Bali, T., Engle, R., & Murray, S. (2016). *Empirical asset pricing: the cross section of stock returns*. Wiley.
6. Berk, J, & van Binsbergen, J. (2017). How do investors compute the discount rate? They use the CAPM. *Financial Analysts Journal, 73*, 25–32.
7. Binder, J. (1985) On the use of the multivariate regression model in event studies. *Journal of Accounting Research, 23*, 370–383.
8. Blume, S. (1975). Betas and their regression tendencies. *Journal of Finance, 30*, 785–795.
9. Busse, J., & Green, T. C. (2002). Market efficiency in real time. *Journal of Financial Economics, 65*, 415–437.
10. Cochrane, J. (2011). Presidential address: Discount rates. *Journal of Finance, 66*, 1047–1108.
11. Cornell, B. (2020). Stock characteristics and stock returns: A skeptic's look at the cross section of expected returns. *Journal of Portfolio Management, 46*, 131–142.
12. Crew, N., Gold, K., & Moore, M. (2007). Federal securities acts and areas of expert analysis, In R. Weil, et al. (Eds.) *Litigation services handbook* (4th edn.). Wiley.
13. Damodaran, A. (2012). *Investment valuation: Tools and techniques for determining the value of any asset* (3rd edn.). Wiley.
14. Fama, E., Fischer, L., Jensen, M., & Roll, R. (1969). The speed of adjustment of stock prices to new information. *International Economic Review, 10*, 1–21.
15. Fama, E., & French, K. (1993). Common risk factors in the returns on stocks and bonds. *Journal of Financial Economics, 33*, 3–56.
16. Fama, E., & French, K. (2015). A five-factor asset pricing model. *Journal of Financial Economics, 116*, 1–22.
17. Graham, J., & Harvey, C. (2002). How do CFOs make capital budgeting and capital structure decisions? *Journal of Applied Corporate Finance, 15*, 8–23.
18. Harvey, C., Liu, Y., & Zhu, H. (2016). . . . and the cross-section of expected returns. *Review of Financial Studies, 29*, 5–68.
19. Hou, K., Xue, C., & Zhang, L. (2020). Replicating anomalies. *Review of Financial Studies, 33*, 2019–2133.
20. Koller, T., Goedhart, M., & Wessels, D. (2010). *Valuation: Measuring and managing the value of companies* (5th edn.). Wiley.
21. Kothari, S. P., & Warner, J. B. (2005). Econometrics of event studies. In B. E. Eckbo (Ed.) *Handbook of corporate finance: empirical corporate finance*. Elsevier/North-Holland.
22. MacKinlay, A. C. (1997). Event studies in economics and finance. *Journal of Economic Literature, 35*, 13–39.
23. Pinto, J., Robinson, T., & Stowe, J. (2015). Equity valuation: a survey of professional practice. Working paper, accessed at the Social Science Research Network https://ssrn.com/abstract= 2657717.
24. Roll, R. (1977). A critique of the asset pricing theory's tests. *Journal of Financial Economics, 4*, 129–176.
25. Sharpe, W. (1964). Capital asset prices: A theory of market equilibrium under conditions of risk. *Journal of Finance, 3*, 425–442

# Chapter 6
# Risk-Adjusted Portfolio Performance Measures

Our discussion on factor models in Chap. 5 showed that we should only expect higher returns if we take on more risk. Therefore, one way to measure performance is by putting it in the context of the level of risk taken to achieve that performance. For example, the return of a manager who invests in stocks should not be compared with the return of a manager who invests in bonds, as we would expect the latter to return less because of the lower level of risk undertaken. A more appropriate comparison would be to somehow normalize the returns from the different investments by their respective risk levels. The complicating issue is that there is no single definition of returns as well as no single definition of risk.

In this chapter, we implement several risk-adjusted performance measures that differ in their definition of returns and/or risks. In particular, we will look at the Sharpe Ratio in Sect. 6.2, Roy's Safety First Ratio in Sect. 6.3, Treynor Ratio in Sect. 6.4, Sortino Ratio in Sect. 6.5, and Information Ratio in Sect. 6.6. We then compare the results of these different measures in Sect. 6.7. We can use these different risk-adjusted performance measures to help us rank portfolios. However, because of the differences in the definition of return and risk, these metrics may rank investments differently.

## 6.1 Portfolio and Benchmark Data

Prior to implementing these different risk-adjusted performance measures, we first need to have portfolios whose performance we wish to analyze. In practice, we would have portfolios of our own, and we would be analyzing those portfolios. However, for our simplified analysis, we will compare the performance of two portfolios: an Actual Portfolio, which is a proxy for our investment portfolio, and an Alternative Portfolio, which is a proxy for an investment alternative. In addition, we need to set a benchmark portfolio for use with some metrics.

© The Author(s), under exclusive license to Springer Nature Switzerland AG 2021
C.S. Ang, *Analyzing Financial Data and Implementing Financial Models Using R*,
Springer Texts in Business and Economics,
https://doi.org/10.1007/978-3-030-64155-9_6

**Step 1: Import Portfolio Returns** In practice, if we were measuring the risk-adjusted performance of our Actual Portfolio, we would be importing returns data of that portfolio. However, for illustrative purposes, we import the returns of our hypothetical portfolio from January 2015 to December 2016. The file we import is labeled Hypothetical Portfolio (Monthly).csv. The details of how this portfolio is calculated are described in Appendix C. We convert the portfolio data into a numerical vector to make it easier to combine with the data of the other portfolios in Step 4.

```
1   > hypo <- read.csv("Hypothetical Portfolio (Monthly).csv", header = TRUE)
2   > dim(hypo)
3   [1] 60   3
4   > head.tail(hypo)
5      X     Date     Port.Ret
6   1 1 Jan 2015  0.07306837
7   2 2 Feb 2015  0.07258170
8   3 3 Mar 2015 −0.02373268
9      X      Date    Port.Ret
10  58 58 Oct 2019 0.05596117
11  59 59 Nov 2019 0.04224452
12  60 60 Dec 2019 0.04982447
13  >
14  > Port <- hypo[, 3]
15  > head.tail(Port)
16  [1]  0.07306837  0.07258170 −0.02373268
17  [1] 0.05596117 0.04224452 0.04982447
```

**Step 2: Import Alternative Portfolio Returns Data** For our Alternative Portfolio, we consider an investment in the S&P 500 Index Growth ETF (SPYG). SPYG is comprised of large capitalization growth stocks.

```
1   > SPYG <- load.data("SPYG Yahoo.csv", "SPYG")
2   > data.spyg <- to.monthly(SPYG)
3   > altport <- Delt(data.spyg$SPYG.Adjusted)
4   > altport <- altport[−1, ]
5   > names(altport) <- paste("Alt.Port")
6   > dim(altport)
7   [1] 60   1
8   > head.tail(altport)
9               Alt.Port
10  Jan 2015 −0.01910356
11  Feb 2015  0.05969036
12  Mar 2015 −0.01735287
13              Alt.Port
14  Oct 2019 0.01748521
15  Nov 2019 0.03436953
16  Dec 2019 0.02786833
```

**Step 3: Import Benchmark Portfolio Returns Data** Since the Actual Portfolio and Alternative Portfolio are comprised of large capitalization stocks, we use SPY as our benchmark portfolio.

```
1   > SPY <- load.data("SPY Yahoo.csv", "SPY")
```

```
2   > spy <- to.monthly(SPY)
3   > benchmark <- Delt(spy$SPY.Adjusted)
4   > benchmark <- benchmark[-1, ]
5   > names(benchmark) <- paste("Benchmark")
6   > dim(benchmark)
7   [1] 60  1
8   > head.tail(benchmark)
9                Benchmark
10  Jan 2015 -0.02962929
11  Feb 2015  0.05620466
12  Mar 2015 -0.01570571
13               Benchmark
14  Oct 2019 0.02210462
15  Nov 2019 0.03619827
16  Dec 2019 0.02905545
```

**Step 4: Combine Data** We use dim() above to show that each of the three datasets have 60 observations. We can then use cbind() to bring all three returns data into one dataset. As we can see, *port.rets* is an xts object. It took the properties of SPY and SPYV. Because the hypothetical portfolio data Port was a numeric vector, we saved some steps in trying to clean up the data. The data in *port.rets* will then be used in subsequent sections of this chapter.

```
1   > port.rets <- cbind(benchmark, altport, Port)
2   > head.tail(port.rets)
3                Benchmark    Alt.Port       Port
4   Jan 2015 -0.02962929 -0.01910356  0.07306837
5   Feb 2015  0.05620466  0.05969036  0.07258170
6   Mar 2015 -0.01570571 -0.01735287 -0.02373268
7                Benchmark    Alt.Port       Port
8   Oct 2019 0.02210462 0.01748521 0.05596117
9   Nov 2019 0.03619827 0.03436953 0.04224452
10  Dec 2019 0.02905545 0.02786833 0.04982447
11  > str(port.rets)
12  An ``xts'' object on Jan 2015/Dec 2019 containing:
13    Data: num [1:60, 1:3] -0.02963 0.0562 -0.01571 0.00983 0.01286 ...
14    - attr(*, "dimnames")=List of 2
15    ..$ : NULL
16    ..$ : chr [1:3] "Benchmark" "Alt.Port" "Port"
17    Indexed by objects of class: [yearmon] TZ: UTC
18    xts Attributes:
19    NULL
```

## 6.2   Sharpe Ratio

One of the most common performance metrics is the **Sharpe Ratio**. This was named after Bill Sharpe, who wrote the seminal paper of what has become known as the Capital Asset Pricing Model [26]. The Sharpe Ratio compares the excess return of a portfolio relative to the risk-free rate with the portfolio's standard deviation.

That is,

$$Sharpe_p = \frac{\mathbb{E}(R_p) - R_f}{\sigma_p},\tag{6.1}$$

where $\mathbb{E}(R_p)$ is the expected annualized average return of portfolio $p$, $R_f$ is the annual risk-free rate, and $\sigma_p$ is the annualized standard deviation of portfolio $p$. The higher the Sharpe Ratio, the higher the risk-adjusted performance of the portfolio. Below, we show the steps in calculating the Sharpe Ratio for the Actual Portfolio and the Alternative Portfolio.

**Step 1: Identify the Risk-Free Rate**  One of the components in Eq. (6.1) is the risk-free rate. For our purposes, we will use the 3-month Treasury Bill (DGS3MO) with an annualized yield as of December 31, 2019, of 1.55%. We obtained this data from the Federal Reserve Electronic Database (FRED).

```
1   rf <- 0.0155
```

**Step 2: Calculate Annualized Portfolio Returns and Standard Deviation**  In this exercise, we would like to measure the performance of our portfolio as of the end of 2019. However, we first have to determine what frequency of returns and the length of the estimation period we will use to estimate the returns and standard deviation. For illustration purposes, we will use 60 monthly returns. We use mean() to calculate the average return and sd() to calculate the standard deviation over the 60-month period. The resulting return and standard deviation from using monthly data would be a monthly return and a monthly standard deviation. To annualize these metrics, we multiply the average monthly portfolio return by 12, and we multiply the monthly standard deviation by $\sqrt{12}$. The output shows that the annualized portfolio return is 28.6%, and the annualized standard deviation is 20.1%.

```
1   > (ret.p <- mean(port.rets$Port) * 12)
2   [1] 0.2858659
3   > (sd.p <- sd(port.rets$Port) * sqrt(12))
4   [1] 0.2005606
```

**Step 4: Calculate the Sharpe Ratio**  Using Eq. (6.1), we find the Sharpe Ratio for our portfolio is 1.35.

```
1   > (sharpe.port <- (ret.p - rf) / sd.p)
2   [1] 1.348051
```

**Step 5: Repeat Steps 1 to 4 for the Alternative Portfolio**  Following Steps 1 through 4 above, we find the Sharpe Ratio for the Alternative Portfolio is lower at 0.95. This indicates that our portfolio has higher risk-adjusted returns than the Alternative Portfolio.

```
1   > (ret.alt <- mean(port.rets$Alt.Port) * 12)
2   [1] 0.1330132
3   > (sd.alt <- sd(port.rets$Alt.Port) * sqrt(12))
```

```
4   [1] 0.1237228
5   > (sharpe.alt <- (ret.alt - rf) / sd.alt)
6   [1] 0.9498102
```

## 6.3  Roy's Safety First Ratio

**Roy's Safety First** (SF) Ratio makes a slight modification to the Sharpe Ratio. Specifically, instead of using the risk-free rate, Roy's SF Ratio instead uses a target or minimum acceptable return to calculate the excess return. That is,

$$RoySF_p = \frac{\mathbb{E}(R_p) - MAR}{\sigma_p}, \tag{6.2}$$

where $MAR$ is the **minimum acceptable return** and all other variables are defined similarly to Eq. (6.1). Again, the higher the Roy's Safety First Ratio, the better the risk-adjusted performance of the portfolio.

**Step 1: Determine the MAR**  To implement Roy's SF Ratio, we need to come up with a MAR. The MAR can potentially depend on many factors, such as investors' return objectives, risk preferences, etc. For our example, we use a simple MAR that assumes at least a real return of zero. In that case, the MAR should match inflation expectations. Based on the December 2019 Livingston Survey, the long-term outlook for inflation is 2.2%. We use this inflation estimate as our proxy for MAR when calculating Roy's SF Ratio for our portfolio and Alternative Portfolio.

```
1   > mar <- 0.022
```

**Step 2: Calculate Annualized Portfolio Returns and Standard Deviation**  We can use the values from Step 2 of the Sharpe Ratio calculation for our portfolio's annualized return and standard deviation.

```
1   > ret.p
2   [1] 0.2858659
3   > sd.p
4   [1] 0.2005606
```

**Step 3: Calculate the Roy's Safety First Ratio for the Actual Portfolio**  Using Eq. (6.2), we find the Roy's Safety First Ratio for our portfolio of 1.32.

```
1   > (roysf.port <- (ret.p - mar) / sd.p)
2   [1] 1.315642
```

**Step 4: Repeat Steps 1 to 3 for the Alternative Portfolio**  Following Steps 1 through 3 above, we find the Roy's Safety First Ratio for the Alternative Portfolio is lower at 0.90. This indicates that our portfolio has higher risk-adjusted returns than the Alternative Portfolio.

```
1   > ret.alt
2   [1] 0.1330132
3   > sd.alt
4   [1] 0.1237228
5   > (roysf.alt <- (ret.alt - mar) / sd.alt)
6   [1] 0.8972735
```

## 6.4   Treynor Ratio

The **Treynor Ratio** modifies the denominator in the Sharpe Ratio to use beta instead of standard deviation. Using beta instead of standard deviation implies that the Treynor Ratio only considers systematic risk, while the Sharpe Ratio considers total risk (i.e., systematic and firm-specific risk). Therefore, the ratios could differ if the ratio of systematic risk to total risk is different in the two portfolios we are analyzing. That is,

$$Treynor_p = \frac{\mathbb{E}(R_p) - R_f}{\beta_p},\tag{6.3}$$

where $\beta_p$ is the beta of the portfolio, and the rest of the terms are defined similarly to Eq. (6.1). The higher the Treynor Ratio, the better the risk-adjusted performance of the portfolio.

**Step 1: Estimate Portfolio Beta**   We run a market model regression to estimate the beta of our portfolio. To calculate beta, we use a 5-year estimation period with 60 monthly returns.

```
1   > reg <- lm(Port ~ Benchmark, data = port.rets)
2   > (beta.p <- summary(reg)$coefficients[2])
3   [1] 1.257142
```

**Step 2: Calculate Annualized Portfolio Return and Risk-Free Rate**   Since we already calculated the annualized portfolio return and obtained the risk-free rate earlier, we only need to make sure that the data is still in the R environment.

```
1   > ret.p
2   [1] 0.2858659
3   > rf
4   [1] 0.0155
```

**Step 3: Calculate the Treynor Ratio**   Using Eq. (6.3), we find the Treynor Ratio of our portfolio is 0.22.

```
1   > (treynor.port <- (ret.p - rf) / beta.p)
2   [1] 0.215064
```

**Step 6: Repeat Steps 1 to 3 for the Alternative Portfolio**   Following Steps 1 through 3 above, we find the Treynor Ratio for the Alternative Portfolio is 0.12.

This indicates that our portfolio has higher risk-adjusted returns than the Alternative Portfolio.

```
1  > reg.alt <- lm(Alt.Port ~ Benchmark, data = port.rets)
2  > (beta.alt <- summary(reg.alt)$coefficients[2])
3  [1] 1.001638
4  > ret.alt
5  [1] 0.1330132
6  > rf
7  [1] 0.0155
8  > (treynor.alt<-(ret.alt - rf) / beta.alt)
9  [1] 0.117321
```

## 6.5  Sortino Ratio

The **Sortino Ratio** has the same numerator as Roy's SF Ratio in Eq. (6.2). However, the denominator measures only the deviation of values lower than the MAR, which is referred to as **downside deviation** (DD). That is,

$$Sortino_p = \frac{\mathbb{E}(R_p) - MAR}{DD_p}, \tag{6.4}$$

where $\mathbb{E}(R_p)$ is the average return of portfolio $p$, $MAR$ is the minimum acceptable return, and $DD_p$ is the downside deviation. The higher the Sortino Ratio, the better the risk-adjusted performance of the portfolio.

**Step 1: Determine the Period MAR**  In the Sortino Ratio, we have to compare monthly MARs to the monthly portfolio returns. From the Roy's SF Ratio calculation, we used an annual MAR of 2.2%. We convert the annual MAR into a monthly MAR by dividing the MAR by 12.

```
1  > mar
2  [1] 0.022
3  > (period.mar <- mar / 12)
4  [1] 0.001833333
```

**Step 2: Identify Monthly Portfolio Returns that Fall Below the Monthly MAR**
We use subset(), to export to a new dataset *down.port* observations in which the monthly portfolio return falls below the monthly MAR. In our sample, there are 22 months in which this has happened.

```
1  > down.port <- subset(port.rets$Port, Port < period.mar)
2  > dim(down.port)
3  [1] 22  1
4  > head.tail(down.port)
5                  Port
6  Mar 2015 -0.02373268
7  Jun 2015 -0.01590165
```

```
 8    Aug 2015 −0.04044800
 9                        Port
10    Feb 2019  0.0006835328
11    May 2019 −0.0914066269
12    Aug 2019 −0.0294750940
```

**Step 3: Calculate Downside Deviation** The downside deviation can be calculated as the standard deviation of the returns in *down.port*. In other words, it is the standard deviation of monthly returns that fell below the monthly MAR. We then annualize the monthly downside deviation by multiplying it by $\sqrt{12}$.

```
1    > (dd.port <- sd(down.port) * sqrt(12))
2    [1] 0.104987
```

**Step 4: Calculate the Sortino Ratio** Using Eq. (6.4), we find the Sortino Ratio of our portfolio is 2.51.

```
1    > (sortino.port <- (ret.p − mar) / dd.port)
2    [1] 2.51332
```

**Step 6: Repeat Steps 1 to 4 for the Alternative Portfolio** Following Steps 1 through 4 above, we find the Sortino Ratio for the Alternative Portfolio is 1.25. This indicates that our portfolio has higher risk-adjusted returns than the Alternative Portfolio.

```
 1   > down.alt <- subset(port.rets$Alt.Port, Alt.Port < period.mar)
 2   > dim(down.alt)
 3   [1] 20  1
 4   > head.tail(down.alt)
 5              Alt.Port
 6   Jan 2015 −0.01910356
 7   Mar 2015 −0.01735287
 8   Jun 2015 −0.01832590
 9              Alt.Port
10   Dec 2018 −0.084887450
11   May 2019 −0.052740465
12   Aug 2019 −0.006892938
13   > (dd.alt <- sd(down.alt) * sqrt(12))
14   [1] 0.08868624
15   > (sortino.alt <- (ret.alt − mar) / dd.alt)
16   [1] 1.251752
```

## 6.6  Information Ratio

While the four ratios above are close variants of one another, the **Information Ratio** (IR) is a risk-adjusted return measure that requires more modification and its interpretation is not as simple. The IR is the ratio of the portfolio alpha over the tracking error. The **alpha** of the portfolio is the annualized **active return**, which is the difference between the portfolio's return and the benchmark return. The **tracking error** is the annualized standard deviation of the active return. As such, if we use monthly returns, the alpha is equal to the monthly active return multiplied by 12 and the tracking error is equal to the standard deviation of the monthly returns multiplied by the square root of 12.

**Step 1: Calculate Portfolio Active Return** The Active Return is equal to the portfolio return less the benchmark return.

```
1  > active.p  <- port.rets$Port - port.rets$Benchmark
2  > head.tail(active.p)
3                  Port
4   Jan 2015  0.10269766
5   Feb 2015  0.01637705
6   Mar 2015 -0.00802697
7                  Port
8   Oct 2019 0.033856551
9   Nov 2019 0.006046251
10  Dec 2019 0.020769022
```

**Step 2: Calculate Annualized Portfolio Alpha** The alpha of the our portfolio is the annualized average monthly active return. We calculate the average monthly active return using mean(). To annualize, we multiply the monthly average active return by 12. Our portfolio's annualized alpha is 16.9%.

```
1  > (alpha.p <- mean(active.p) * 12)
2  [1] 0.1689296
```

**Step 3: Calculate Tracking Error** The tracking error is the annualized standard deviation of the monthly active returns. We calculate the standard deviation of the monthly active returns using sd(). To annualize, we multiply the monthly standard deviation by the square root of 12.

```
1  > (tracking.p <- sd(active.p) * sqrt(12))
2  [1] 0.1364182
```

**Step 4: Calculate the Information Ratio** The Information Ratio is equal to the alpha divided by the tracking error. We find the Information Ratio of the Actual Portfolio equals 1.24.

```
1  > (ir.port <- alpha.p / tracking.p)
2  [1] 1.238322
```

**Step 5: Repeat Steps 1 through 4 for the Alternative Portfolio** Following Steps 1 through 4 above, we find the Information Ratio for the Alternative Portfolio is 0.51. This indicates that our portfolio has higher risk-adjusted returns than the Alternative Portfolio.

```
1  > active.alt <- port.rets$Alt.Port - port.rets$Benchmark
2  > head.tail(active.alt)
3                  Alt.Port
```

```
 4  Jan 2015  0.010525729
 5  Feb 2015  0.003485704
 6  Mar 2015 -0.001647165
 7                  Alt.Port
 8  Oct 2019 -0.004619412
 9  Nov 2019 -0.001828735
10  Dec 2019 -0.001187120
11  > (alpha.alt <- mean(active.alt) * 12)
12  [1] 0.01607695
13  > (tracking.alt <- sd(active.alt) * sqrt(12))
14  [1] 0.03141241
15  > (ir.alt <- alpha.alt / tracking.alt)
16  [1] 0.5118026
```

## 6.7   Combining Results

We can summarize the different risk measures by first combining the different measures for our portfolio and the Alternative Portfolio in two different vectors. We then use rbind() to stack the two vectors. We clean up the output by renaming the columns and rows. To make the output easier to read, we use digits = 3. With all the different measures in one table, it is now easier to see that, under all five metrics, our portfolio performs better than the Alternative Portfolio.

```
 1  > port <- c(sharpe.port, roysf.port, treynor.port, sortino.port, ir.port)
 2  > alt.port <- c(sharpe.alt, roysf.alt, treynor.alt, sortino.alt, ir.alt)
 3  > risk.adj.ret <- rbind(port, alt.port)
 4  > colnames(risk.adj.ret) <- paste(c("Sharpe", "Roy SF",
 5  +      "Treynor", "Sortino","Info Ratio"))
 6  > rownames(risk.adj.ret) <- paste(c("Portfolio", "Alt Portfolio"))
 7  > options(digits = 3)
 8  > risk.adj.ret
 9                 Sharpe Roy SF Treynor Sortino Info Ratio
10  Portfolio        1.35  1.316   0.215    2.51      1.238
11  Alt Portfolio    0.95  0.897   0.117    1.25      0.512
12  > options(digits = 7)
```

## 6.8   Further Reading

For an excellent treatment of portfolio theory, see [1]. Maginn et al. [3] contain a detailed discussion of the portfolio performance measures we discussed in this chapter. Applications of risk-adjusted performance measures for hedge funds can be found in [4] and [2].

# References

1. Elton, E., Gruber, M., Brown, S., & Goetzmann, W. (2010). *Modern portfolio theory and investment analysis* (8th ed.). Hoboken: Wiley.
2. Gregoriou, G., Hubner, G., Papageorgiou, N., & Rouah, F. (2005). *Hedge funds: Insights in performance measurement, risk analysis, and portfolio allocation.* Hoboken: Wiley.
3. Maginn, J., Tuttle, D., Pinto, J., & McLeavey, D. (2007). *Managing investment portfolios: A dynamic process* (3rd ed.). Hoboken: Wiley.
4. Tran, V. (2006). *Evaluating hedge fund performance.* Hoboken: Wiley.

# Chapter 7
# Markowitz Mean–Variance Optimization

In [3], Harry Markowitz put forth a theory of how to use mean and variance to determine portfolios that offer the highest return for a given level of risk. This approach showed us that what is important is how the assets in the portfolio moved together (i.e., **covariance**) and not the risk of the individual securities by themselves. The implication is that by properly diversifying your portfolio, you can reduce the risk of the entire portfolio from that of the weighted average risk of the assets that make up the portfolio. Subsequent research has shown that most of the benefits of diversification can be attained by holding 12 to 18 securities in our portfolio.

In this chapter, we show how to construct the **Mean–Variance (MV) efficient frontier**, which is the set of all portfolios generated by various combinations of the securities in the portfolio that yield the highest return for a given level of risk. In Sect. 7.1, we begin by manually constructing the MV efficient frontier using two assets, so we can gain the intuition of what we are trying to accomplish. Then, in Sect. 7.2, we convert this manual approach for two assets into an approach that uses **quadratic programming**. Since many investors (e.g., some pension funds) do not or cannot short sell securities for various reasons, we implement the two-asset case not allowing short selling.

We then extend the quadratic programming approach to show how we find the **MV efficient portfolios** when we have multiple securities. In the multiple asset scenario, we first continue with the no short sale restriction in Sect. 7.3 and then demonstrate the effect on the MV efficient frontier of allowing short selling in Sect. 7.4. We will see that allowing short selling extends outward the MV efficient frontier such that we are able to achieve higher returns for a given level of risk.

© The Author(s), under exclusive license to Springer Nature Switzerland AG 2021
C.S. Ang, *Analyzing Financial Data and Implementing Financial Models Using R*,
Springer Texts in Business and Economics,
https://doi.org/10.1007/978-3-030-64155-9_7

## 7.1 Two Assets the Long Way

In this section, we first demonstrate the intuition of how to construct the MV efficient frontier using the case of a two-asset portfolio and manually laying out each step of the calculation. This approach is termed the "Long Way" because we lay out each step without the benefit of any simplifying techniques.

Let us consider the scenario in which we can either put money in US large capitalization stocks (our proxy for this is the S&P 500 Index ETF with symbol SPY) or US investment grade bonds (our proxy for this is the iShares Long-Term Corporate Bond ETF with symbol IGLB). In determining how much money to invest in each of the two securities, we may want to know two types of portfolios: minimum variance portfolio (i.e., the combination of these two securities that yields the smallest portfolio risk) and tangency portfolio (i.e., the combination of these two securities that yields the highest Sharpe Ratio). We demonstrate below how to identify these two portfolios. In addition, not all portfolios generated by a combination of SPY and IGLB yield the highest return for a given level of risk. We identify the set of portfolios that meet those criteria, and these portfolios are called MV efficient portfolios.

**Step 1: Calculate Monthly Returns for SPY and IGLB**  We import the SPY and IGLB data using `load.data()`. We then convert the daily data into monthly data using `to.monthly()`. Then, we calculate the monthly returns by applying `Delt()` on the monthly adjusted prices of SPY and IGLB. We then clean up the returns dataset by deleting the first observation (December 2014) and rename the return variables.

```
 1  > SPY <- load.data("SPY Yahoo.csv", "SPY")
 2  > SPY <- to.monthly(SPY)
 3  >
 4  > IGLB <- load.data("IGLB Yahoo.csv", "IGLB")
 5  > IGLB <- to.monthly(IGLB)
 6  >
 7  > rets <- cbind(Delt(SPY$SPY.Adjusted),
 8  +     Delt(IGLB$IGLB.Adjusted))
 9  > rets <- rets[-1, ]
10  > names(rets) <- c("SPY", "IGLB")
11  > head.tail(rets)
12                  SPY           IGLB
13  Jan 2015 -0.02962929  5.315343e-02
14  Feb 2015  0.05620466 -2.737789e-02
15  Mar 2015 -0.01570571 -4.791363e-05
16                  SPY          IGLB
17  Oct 2019 0.02210462 0.001911647
18  Nov 2019 0.03619827 0.006902584
19  Dec 2019 0.02905545 0.003505375
```

**Step 2: Calculate Mean and Standard Deviation of SPY and IGLB Returns**  As the name suggests, the Mean–Variance efficient frontier requires us to find the mean and variance (technically, standard deviation) of the portfolio returns. As such, we would need to find the mean and standard deviation of the returns of the individual securities. The mean return is calculated using mean(), while the standard deviation is calculated using sd().

```
1  > (avg.SPY <- mean(rets$SPY))
2  [1] 0.009744689
3  > (avg.IGLB <- mean(rets$IGLB))
4  [1] 0.004984509
5  >
6  > (sd.SPY <- sd(rets$SPY))
7  [1] 0.03448893
8  > (sd.IGLB <- sd(rets$IGLB))
9  [1] 0.02363155
```

**Step 3: Calculate the Covariance of SPY and IGLB Returns**  We discussed in Chap. 4 that, in the context of portfolios, what is important is how the assets in the portfolio move together. The metric we use to measure how assets move together is the **covariance**, which we calculate using cov(). We apply as.numeric() to change the output to a number without any headers/labels that may end up causing more confusion.

```
1  > (covar <- as.numeric(cov(rets$SPY, rets$IGLB)))
2  [1] 0.0001280108
```

**Step 4: Create a Vector of Assumed Weights for SPY and IGLB**  When we calculated portfolio risk in Chap. 4, we assumed the weights of the two securities. To find out a whole spectrum of the risk-return combination of SPY and IGLB, we need to create a vector of possible weight combinations. Since many individual investors and institutional investors do not or are not allowed to short sell stock, we can restrict our weights from 100% SPY/0% IGLB to 0% SPY/100% IGLB. Since the weights must sum to one, we can create a vector of weights for SPY between 0% to 100% by 1% increments. The IGLB weight is one minus the SPY weight. Note that the number of increments is a trade-off between speed and accuracy. For our current purpose, using 1% increments would be sufficiently accurate and would not affect the speed of running the program substantially.

```
1  > w.SPY <- seq(0, 1, by = 0.01)
2  > w.IGLB <- 1 - w.SPY
3  > port <- data.frame(cbind(w.SPY, w.IGLB))
4  > head.tail(port)
5     w.SPY w.IGLB
6  1   0.00   1.00
7  2   0.01   0.99
8  3   0.02   0.98
9     w.SPY w.IGLB
10 99   0.98   0.02
11 100  0.99   0.01
12 101  1.00   0.00
```

**Step 5: Calculate Portfolio Return for Each Pair of Weights**  Since we have 101 pairs of SPY and IGLB weights, we can calculate portfolio returns given each of those pairs. The portfolio return is the weighted average of the returns of the securities in the portfolio.

```
1  > port$Port.Ret <- port$w.SPY * avg.SPY +
2  +     port$w.IGLB * avg.IGLB
3  > head.tail(port)
4     w.SPY w.IGLB     Port.Ret
5  1  0.00   1.00 0.004984509
6  2  0.01   0.99 0.005032111
7  3  0.02   0.98 0.005079713
8     w.SPY w.IGLB     Port.Ret
9  99   0.98   0.02 0.009649486
10 100  0.99   0.01 0.009697088
11 101  1.00   0.00 0.009744689
```

**Step 6: Calculate Portfolio Risk for Each Pair of Weights** Similarly, we calculate portfolio risks for each pair of SPY and IGLB weights. Chapter 4 describes the calculation in more detail.

```
1  > port$Port.Risk <- sqrt((port$w.SPY^2 * sd.SPY^2) +
2  +     (port$w.IGLB^2 * sd.IGLB^2) +
3  +     (2 * covar * port$w.SPY * port$w.IGLB))
4  > head.tail(port)
5     w.SPY w.IGLB     Port.Ret  Port.Risk
6  1  0.00   1.00 0.004984509 0.02363155
7  2  0.01   0.99 0.005032111 0.02345188
8  3  0.02   0.98 0.005079713 0.02327723
9     w.SPY w.IGLB     Port.Ret  Port.Risk
10 99   0.98   0.02 0.009649486 0.03387660
11 100  0.99   0.01 0.009697088 0.03418195
12 101  1.00   0.00 0.009744689 0.03448893
```

**Step 7: Identify Minimum–Variance Portfolio** The **minimum variance portfolio** is the portfolio that has the smallest portfolio risk. To identify this portfolio, we use a combination of subset() and min(). The output below shows that the minimum variance portfolio *minvar.port* is attained if we invest 29% of our portfolio in SPY and 71% in IGLB. This portfolio gives us a monthly portfolio return of 0.64% and a monthly portfolio standard deviation of 2.08%.

```
1  > (minvar.port <- subset(port, Port.Risk == min(Port.Risk)))
2     w.SPY w.IGLB     Port.Ret  Port.Risk    Sharpe
3  30  0.29   0.71 0.006364961 0.02083904 0.2434515
```

**Step 8: Identify the Tangency Portfolio** The **tangency portfolio** is the portfolio that yields the highest Sharpe Ratio. The **Sharpe Ratio** requires a risk-free rate, and we use the 3-Month Constant Maturity Treasury as a proxy for that rate. Since the yield as of December 31, 2019, of 1.55% is an annual yield, we convert this into a monthly yield by dividing the rate by 12.

```
1  > rf <- 0.0155 / 12
2  > port$Sharpe <- (port$Port.Ret - rf) / port$Port.Risk
3  > head.tail(port)
```

```
 4     w.SPY w.IGLB    Port.Ret  Port.Risk     Sharpe
 5   1  0.00   1.00 0.004984509 0.02363155 0.1562675
 6   2  0.01   0.99 0.005032111 0.02345188 0.1594944
 7   3  0.02   0.98 0.005079713 0.02327723 0.1627361
 8       w.SPY w.IGLB    Port.Ret  Port.Risk     Sharpe
 9  99   0.98   0.02 0.009649486 0.03387660 0.2467137
10 100   0.99   0.01 0.009697088 0.03418195 0.2459023
11 101   1.00   0.00 0.009744689 0.03448893 0.2450938
12 >
13 > (tangency.port <- subset(port, Sharpe == max(Sharpe)))
14     w.SPY w.IGLB   Port.Ret  Port.Risk     Sharpe
15  57  0.56   0.44 0.00765021 0.02332858 0.2725646
```

Then, using `subset()` and `max()`, we identify the portfolio with the highest Sharpe Ratio. This portfolio turns out to be the portfolio in which we put 56% weight on SPY and 44% weight on IGLB. This portfolio yields a Sharpe Ratio of 0.27.

**Step 9: Identify Efficient Portfolios** Not all combinations of SPY and IGLB can be classified as efficient portfolios. **Efficient portfolios** are portfolios that yield the highest return for a given level of risk. We extract these efficient portfolios by using `subset()`, so we keep portfolios with returns greater than the return of the minimum variance portfolio. As the output below shows, there are 72 such portfolios.

```
 1 > eff.frontier <- subset(port, Port.Ret >= minvar.port$Port.Ret)
 2 > dim(eff.frontier)
 3 [1] 72  5
 4 > head.tail(eff.frontier)
 5      w.SPY w.IGLB    Port.Ret  Port.Risk     Sharpe
 6  30   0.29   0.71 0.006364961 0.02083904 0.2434515
 7  31   0.30   0.70 0.006412563 0.02084368 0.2456810
 8  32   0.31   0.69 0.006460165 0.02085547 0.2478245
 9      w.SPY w.IGLB    Port.Ret  Port.Risk     Sharpe
10  99   0.98   0.02 0.009649486 0.03387660 0.2467137
11 100   0.99   0.01 0.009697088 0.03418195 0.2459023
12 101   1.00   0.00 0.009744689 0.03448893 0.2450938
```

**Step 10: Plot the Mean–Variance Efficient Frontier** We can chart the mean–variance efficient frontier as follows. First, we plot in blue hollow circles all the portfolios we formed from all combinations of SPY and IGLB. In Lines 9 to 11, we plot the efficient frontier as solid blue circles using `pch = 16`. In Lines 14 to 16, we then add a red square marker using `points()` and `pch = 17` to mark the minimum variance portfolio. We use `cex = 2.5` to increase the size of the square marker. In Lines 18 to 21, we add a green triangle marker using `pch = 19` and `cex = 2.5` to mark the tangency portfolio.

```
 1 > plot(x = port$Port.Risk,
 2 +     xlab = "Portfolio Risk",
 3 +     y = port$Port.Ret,
 4 +     ylab = "Portfolio Return",
 5 +     col = "blue",
```

```
 6  +      main = "Mean—Variance Efficient Frontier of Two Assets
 7  +           (Based on the Long Way)")
 8  >
 9  > points(x = eff.frontier$Port.Risk,
10  +          col = "blue",
11  +          y = eff.frontier$Port.Ret, pch = 16)
12  >
13  > points(x = minvar.port$Port.Risk,
14  +        y = minvar.port$Port.Ret,
15  +        col = "red",
16  +        pch = 15, cex = 2.5)
17  >
18  > points(x = tangency.port$Port.Risk,
19  +        y = tangency.port$Port.Ret,
20  +        col = "darkgreen",
21  +        pch = 17, cex = 2.5)
```

As shown in Fig. 7.1, the blue hollow circles below the efficient frontier have the same level of risk as some of the portfolios on the frontier but have lower returns.

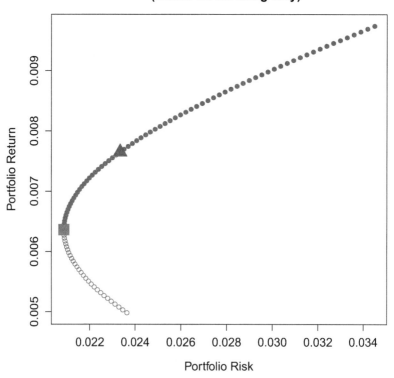

**Fig. 7.1** Mean–variance efficient frontier of portfolios consisting of SPY and IGLB calculated using the Long Way

This means that these portfolios below the frontier are not efficient because they provide lower returns for the same level of risk. This is why when identifying points on the mean–variance efficient frontier, we only looked at returns of portfolios above the minimum variance point.

## 7.2   Two Assets Using Quadratic Programming

The Long Way is tractable in the two-asset case. However, most portfolios have more than two assets, and using the Long Way becomes extremely challenging very quickly. To see why, note that the more securities we have in our portfolio, the number of variance–covariance terms grows quickly. In fact, given $N$ securities in our portfolio, the **variance–covariance matrix** would have $N(N + 1)/2$ elements, which consists of $N$ variance terms and $N(N − 1)/2$ covariance terms. Given two assets, we would have three $[= 2(2 + 1)/2]$ terms in total, consisting of two variance terms and one $[= 2(2 - 1)/2]$ covariance term. When we have four assets, as in our later example, we would have a total of 10 $[= 4(4 + 1)/2]$ terms, consisting of four variance terms and six $[= 4(4 - 1)/2]$ covariance terms. If we had 15 securities, we would end up with a total of 120 $[= 15(15 + 1)/2]$ total terms, consisting of 15 variance terms and 105 $[= 15(15 - 1)/2]$ covariance terms. As such, using the Long Way is going to be difficult in practice.

The Long Way is useful to gain an intuition of what we are doing when we apply **quadratic programming**. The technique is called quadratic programming because we are performing an optimization on a quadratic function. A quadratic function is a function that has a variable (in this case weight) that is squared. In the two-asset case, we are minimizing Eq. (4.2). The weights in that formula are squared. For our purpose, quadratic programming finds the combination of securities that yields the minimum risk for a specific level of return.

**Step 1: Convert Returns Dataset into a Matrix**   Recall in Chap. 4, we used matrix algebra in calculating portfolio risk. We apply what we have learned there when we calculate the MV efficient frontier using quadratic programming. Note that we have already calculated the returns of SPY and IGLB above in the *rets* dataset, so we can apply `matrix()` to convert the *rets* dataset into a matrix. The second argument in `matrix()` is the number of rows of the matrix, which in this case is 60. We then rename the column headers. Otherwise, we would see a generic [, 1] and [, 2] as the headers.

```
1  > mat.ret <- matrix(rets, nrow(rets))
2  > colnames(mat.ret) <- c("SPY", "IGLB")
3  > head.tail(mat.ret)
4               SPY          IGLB
5  [1,] −0.02962929   5.315343e−02
6  [2,]   0.05620466 −2.737789e−02
7  [3,] −0.01570571 −4.791363e−05
8               SPY          IGLB
```

```
 9   [58,] 0.02210462 0.001911647
10   [59,] 0.03619827 0.006902584
11   [60,] 0.02905545 0.003505375
```

**Step 2: Calculate Variance–Covariance Matrix**  Applying cov() on *mat.ret*
calculates the variance–covariance matrix. The terms on the main diagonal (i.e.,
top-left to bottom-right) are the variance terms. In particular, the variance of SPY
returns is 0.001189 and the variance of IGLB returns is 0.000558. The 0.000128
in the off-diagonal are the covariance terms. Note that the two numbers in the off-
diagonal are the same because the covariance of SPY v. IGLB and covariance of
IGLB v. SPY are equivalent.

```
1   > (vcov <- cov(mat.ret))
2                SPY          IGLB
3   SPY  0.0011894861 0.0001280108
4   IGLB 0.0001280108 0.0005584502
```

**Step 3: Construct the Target Portfolio Return Vector**  Instead of creating a series
of weights for SPY and IGLB, in the quadratic programming approach we construct
a series of target portfolio returns. The term vector means a row or column of
numbers. To create this series of potential portfolio returns, we have to come up
with upper and lower bounds. When short sales are not allowed, as in the case here,
the natural bounds for the returns would be the SPY return on one end and the
IGLB return on the other end. This is because the two extreme combinations of
returns would be 100% SPY/0% IGLB, which yields a portfolio return equal to the
SPY return, and 0% SPY/100% IGLB, which yields a portfolio return equal to the
IGLB return. Thus, the first step is to find the average return for SPY and IGLB,
respectively.

```
1   > avg.ret <- matrix(apply(mat.ret, 2, mean))
2   > colnames(avg.ret) <- paste("Avg.Ret")
3   > rownames(avg.ret) <- paste(c("SPY","IGLB"))
4   > avg.ret
5           Avg.Ret
6   SPY  0.009744689
7   IGLB 0.004984509
```

Then, we set the smaller of the average returns as the minimum return *min.ret* and
the larger of the average returns as the maximum return *max.ret*.

```
1   > (min.ret <- min(avg.ret))
2   [1] 0.004984509
3   > (max.ret <- max(avg.ret))
4   [1] 0.009744689
```

Next, we create a sequence that begins with *min.ret* and ends with *max.ret* with
100 increments in between. Again, the number of increments is a trade-off between
accuracy and speed. Having 100 increments is a reasonable choice for what we are
doing. The *tgt.ret* is the target return portfolio vector.

```
1   > increments <- 100
```

```
2  > tgt.ret <- seq(min.ret, max.ret, length = increments)
3  > head.tail(tgt.ret)
4  [1] 0.004984509 0.005032592 0.005080674
5  [1] 0.009648524 0.009696607 0.009744689
```

**Step 4: Construct Temporary Portfolio Standard Deviation Vector** An output
of our optimization program are the portfolio standard deviations. However, the
output needs to overwrite the values of an existing vector after each iteration. Since
the number of iterations equals the number of increments, we create a series of 100
zeroes for our temporary portfolio standard deviation vector. To do this, we use
rep(), which repeats the value of zero, with length() equal to the number of
increments.

```
1  > (tgt.sd <- rep(0, length = increments))
2    [1] 0 0 0 0 0 0 0 0 0 0 0 0 0 0 0 0 0 0 0 0 0 0 0 0 0 0 0 0 0 0 0 0 0 0 0 0
3   [37] 0 0 0 0 0 0 0 0 0 0 0 0 0 0 0 0 0 0 0 0 0 0 0 0 0 0 0 0 0 0 0 0 0 0 0 0
4   [73] 0 0 0 0 0 0 0 0 0 0 0 0 0 0 0 0 0 0 0 0 0 0 0 0 0 0 0 0
```

**Step 5: Construct Temporary Portfolio Weights** Similarly, an output of the
optimization program is the portfolio weights. However, the output needs to
overwrite the values of an existing vector of weights after each iteration. Since the
number of iterations equals the number of increments, we create a series of 100
zeroes for our temporary portfolio weights vector. To do this, we use rep(), which
repeats the value of zero, with length() equal to the number of increments.

```
1  > wgt <- matrix(0, nrow = increments, ncol = length(avg.ret))
2  > head.tail(wgt)
3          [,1] [,2]
4  [1,]     0    0
5  [2,]     0    0
6  [3,]     0    0
7          [,1] [,2]
8  [98,]    0    0
9  [99,]    0    0
10 [100,]   0    0
```

**Step 6: Run the quadprog Optimizer** To run the optimizer, we use
solve.QP(), which is part of the quadprog package. The optimizer loops
through the calculations 100 times (i.e., the number of *increments* we set). During
each iteration, we want the optimizer to minimize portfolio risk as in Eq. (4.5)
subject to three constraints: (1) the weights in the portfolio have to equal one, (2)
the portfolio return has to equal a target return, which we have specified in *tgt.ret*,
and (3) the weights for each security have to be greater than or equal to zero (i.e.,
short selling is not allowed). What makes the code below appear to be complicated
is because much of the code is required to modify the built-in setup of solveQP()
to conform to the above objective function and constraints.

```
1  > library(quadprog)
2  > for (i in 1:increments){
3  +     Dmat <- 2 * vcov
```

```
 4   +      dvec <- c(rep(0, length(avg.ret)))
 5   +      Amat <- cbind(rep(1, length(avg.ret)), avg.ret,
 6   +         diag(1, nrow = ncol(mat.ret)))
 7   +      bvec <- c(1, tgt.ret[i], rep(0, ncol(mat.ret)))
 8   +      soln <- solve.QP(Dmat, dvec, Amat, bvec, meq = 2)
 9   +      tgt.sd[i] <- sqrt(soln$value)
10   +      wgt[i, ] <- soln$solution
11   + }
```

In Line 8, `solve.QP` attempts to solve quadratic problem of the form

$$\min(-d^\mathsf{T} b + 0.5 b^\mathsf{T} D b) \tag{7.1}$$

subject to

$$A^\mathsf{T} b \geq b_0. \tag{7.2}$$

These are reflected in the above code as follows. **Dmat** $(D)$ is the variance–covariance matrix. **dvec** $(d)$ is the target return. Then, **Amat** $(A)$ is the matrix defining the constraints. This has three constraints. The third constraint in Line 6 constraints the weights to be non-negative (i.e., no short sales are allowed). **bvec** $(b)$ corresponds to each of the three constraints in **Amat**. We set **meq** equal to 2, which means the first two constraints are set to equality. Specifically, the weights have to sum to 1 (constraint 1) and the return equals the target return (constraint 2).

From the output of the optimizer, we use two lists. First, we report **value**, which is the target standard deviation that is the solution to the quadratic programming problem. Second, we also use the **solution**, which gives us the weights that let us achieve the target standard deviation. Below, we report these outputs of the optimizer for the portfolio standard deviation, which corresponds to each target portfolio return level we constructed.

```
1   > head.tail(tgt.sd)
2   [1] 0.02363155 0.02345009 0.02327375
3   [1] 0.03387045 0.03417886 0.03448893
```

The output of **wgt** below shows the weights of SPY and IGLB for each target return level.

```
1   > head.tail(wgt)
2                 [,1]       [,2]
3   [1,] 0.00000000 1.000000
4   [2,] 0.01010101 0.989899
5   [3,] 0.02020202 0.979798
6                 [,1]         [,2]
7   [98,] 0.979798 0.02020202
8   [99,] 0.989899 0.01010101
9   [100,] 1.000000 0.00000000
```

**Step 7: Combine the Returns, Standard Deviation, and Weights of the Portfolio**
Using `cbind()`, we combine *tgt.ret*, *tgt.sd*, and *wgt* into *tgt.port*.

```
1   > tgt.port <- data.frame(cbind(tgt.ret, tgt.sd, wgt))
2   > names(tgt.port)[c(3, 4)] <- c("w.SPY", "w.IGLB")
3   > head.tail(tgt.port)
4           tgt.ret       tgt.sd      w.SPY     w.IGLB
5   1 0.004984509 0.02363155 0.00000000 1.000000
6   2 0.005032592 0.02345009 0.01010101 0.989899
7   3 0.005080674 0.02327375 0.02020202 0.979798
8           tgt.ret       tgt.sd      w.SPY     w.IGLB
9   98  0.009648524 0.03387045 0.979798 0.02020202
10  99  0.009696607 0.03417886 0.989899 0.01010101
11  100 0.009744689 0.03448893 1.000000 0.00000000
```

**Step 8: Identify Minimum Variance Portfolio**  We identify the minimum variance portfolio by finding the pair of weights that yields the smallest portfolio standard deviation (i.e., `min(tgt.sd)`). To extract that portfolio, we use `subset()`.

```
1   > (minvar.port <- subset(tgt.port, tgt.sd == min(tgt.sd)))
2           tgt.ret       tgt.sd      w.SPY     w.IGLB
3   30 0.006378905 0.02083965 0.2929293 0.7070707
```

Consistent with what we found in the Long Way, the minimum variance portfolio is given by the portfolio that has a weight of 29% for SPY and 71% for IGLB. However, notice that the weight allocation, portfolio return, and portfolio standard deviation of the minimum variance portfolio are slightly different when calculated using the quadratic programming approach compared to the Long Way. This difference is a result of creating target weights as the starting point in the Long Way and using those to find the rest of the results, while in the quadratic programming case we create the target returns as the starting point and then use those to find the rest of the results.

**Step 9: Identify the Tangency Portfolio**  Recall that the **tangency portfolio** is the portfolio with the highest **Sharpe Ratio**. Note that we still have `rf` of 1.55% in the R environment. We calculate the Sharpe Ratio of each pair of weights in Line 1. In Line 12, we identify the tangency portfolio using `subset()` and finding `max(Sharpe)`.

```
1   > tgt.port$Sharpe <- (tgt.port$tgt.ret - rf) / tgt.port$tgt.sd
2   > head.tail(tgt.port)
3           tgt.ret       tgt.sd      w.SPY     w.IGLB     Sharpe
4   1 0.004984509 0.02363155 0.00000000 1.000000 0.1562675
5   2 0.005032592 0.02345009 0.01010101 0.989899 0.1595271
6   3 0.005080674 0.02327375 0.02020202 0.979798 0.1628017
7           tgt.ret       tgt.sd      w.SPY     w.IGLB     Sharpe
8   98  0.009648524 0.03387045 0.979798 0.02020202 0.2467301
9   99  0.009696607 0.03417886 0.989899 0.01010101 0.2459105
10  100 0.009744689 0.03448893 1.000000 0.00000000 0.2450938
11  >
12  > (tangency.port <- subset(tgt.port, Sharpe == max(Sharpe)))
13          tgt.ret       tgt.sd      w.SPY     w.IGLB     Sharpe
14  57 0.007677136 0.0234276 0.5656566 0.4343434 0.2725618
```

The output above shows the tangency portfolio is the one with a 57% weight in SPY and 43% in IGLB. In the Long Way, the tangency portfolio has a weight of 56%

in SPY and 44% in IGLB. The difference is because in the quadratic programming approach, we start with target returns and then find the target weights. Thus, the target weights are not necessarily in 1% increments like the approach we took in the Long Way.

**Step 10: Identify Efficient Portfolios**  We identify the set of efficient portfolios by finding portfolios that have returns higher than the return of the minimum variance portfolio. The output below shows we have 71 such portfolios.

```
 1  > eff.frontier <- subset(tgt.port, tgt.ret >= minvar.port$tgt.ret)
 2  > dim(eff.frontier)
 3  [1] 71  5
 4  > head.tail(eff.frontier)
 5         tgt.ret     tgt.sd     w.SPY    w.IGLB     Sharpe
 6  30 0.006378905 0.02083965 0.2929293 0.7070707 0.2441134
 7  31 0.006426988 0.02084650 0.3030303 0.6969697 0.2463398
 8  32 0.006475071 0.02086064 0.3131313 0.6868687 0.2484777
 9         tgt.ret     tgt.sd     w.SPY    w.IGLB     Sharpe
10  98  0.009648524 0.03387045 0.979798 0.02020202 0.2467301
11  99  0.009696607 0.03417886 0.989899 0.01010101 0.2459105
12  100 0.009744689 0.03448893 1.000000 0.00000000 0.2450938
```

**Step 11: Plot the Mean–Variance Efficient Frontier**  We follow the same steps in Step 10 of the Long Way to chart the mean–variance efficient frontier. Figure 7.2 shows the chart generated by the code below.

```
 1  > plot(x = tgt.sd,
 2  +      xlab = "Portfolio Risk",
 3  +      y = tgt.ret,
 4  +      ylab = "Portfolio Return",
 5  +      col = "blue",
 6  +      main = "Mean—Variance Efficient Frontier of Two Assets
 7  +      (Based on the Quadratic Programming Approach)")
 8  >
 9  > points(x = eff.frontier$tgt.sd,
10  +      col = "blue",
11  +      y = eff.frontier$tgt.ret, pch = 16)
12  >
13  > points(x = minvar.port$tgt.sd,
14  +      y = minvar.port$tgt.ret,
15  +      col = "red",
16  +      pch = 15, cex = 2.5)
17  >
18  > points(x = tangency.port$tgt.sd,
19  +      y = tangency.port$tgt.ret,
20  +      col = "darkgreen",
21  +      pch = 17, cex = 2.5)
```

S

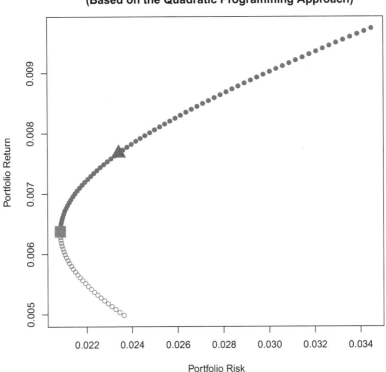

**Fig. 7.2** Mean–variance efficient frontier of portfolios consisting of SPY and IGLB calculated using quadratic programming

**Compare Long Way and Quadratic Programming**

We noted that the results of the quadratic programming approach seem to be slightly different from the results of the Long Way for the minimum variance portfolio and tangency portfolio. We explained that the difference is due to the different method with which the calculations are undertaken. Specifically, the Long Way fits the calculations to a set combination of portfolio weights, while the quadratic programming approach fits the calculations to a set portfolio return level. Regardless of this issue, we show below that a plot of the efficient frontiers calculated using the Long Way and quadratic programming approaches is identical for all practical purposes. Figure 7.3 shows the results of the code below. As the chart demonstrates, there is virtually no difference in the results of the two approaches.

(continued)

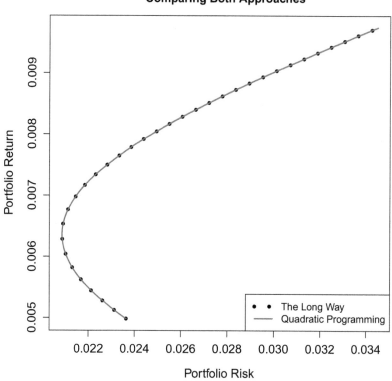

**Fig. 7.3** Comparison of mean–variance efficient frontiers calculated using the Long Way and quadratic programming

```
1  > plot(x = port$Port.Risk,
2  +      xlab = "Portfolio Risk",
3  +      y = port$Port.Ret,
4  +      ylab = "Portfolio Return",
5  +      type = "l",
6  +      lwd = 6,
7  +      lty = 3,
8  +      main = "Mean—Variance Efficient Frontier of Two Assets
9  +      Comparing Both Approaches")
10 > lines(x = tgt.sd, y = tgt.ret, col = "red", type = "l", lwd = 2)
11 > legend("bottomright",
12 +      c("The Long Way", "Quadratic Programming"),
13 +      col = c("black", "red"),
14 +      lty = c(3, 1),
15 +      lwd = c(6, 2))
```

# 7.3  Multiple Assets Using Quadratic Programming

Our portfolio above is comprised of only two assets: SPY and IGLB. Although these assets provide us with exposure to many securities, our exposure was limited to only large capitalization US stocks and US investment grade bonds. We can split the large cap exposure of SPY into value and growth stocks by using SPYV and SPYG, respectively. We can also add exposure to global equities using CWI. Thus, in our example below, we show how to perform mean–variance optimization with four assets.

Before we start, this maybe a good time to clear the R environment using `rm(list = ls()`. Make sure that before we run this code that everything temporary has been saved. We should also run the pre-loaded code in Appendix B so we have the `load.data()` and `head.tail()` functions.

**Step 1: Calculate Monthly Returns for SPYV, SPYG, CWI, and IGLB**  Using `load.data()`, we import the price data for SPYV, SPYG, CWI, and IGLB. We then convert the daily data into monthly data using `to.monthly()` and then calculate the monthly returns by applying `Delt()` to each security's adjusted close price.

```
1  > SPYV <- load.data("SPYV Yahoo.csv", "SPYV")
2  > spyv <- to.monthly(SPYV)
3  >
4  > SPYG <- load.data("SPYG Yahoo.csv", "SPYG")
5  > spyg <- to.monthly(SPYG)
6  >
7  > CWI <- load.data("CWI Yahoo.csv", "CWI")
8  > cwi <- to.monthly(CWI)
9  >
10 > IGLB <- load.data("IGLB Yahoo.csv", "IGLB")
11 > iglb <- to.monthly(IGLB)
12 >
13 > rets <- cbind(Delt(spyv$SPYV.Adjusted), Delt(spyg$SPYG.Adjusted),
14 +    Delt(cwi$CWI.Adjusted), Delt(iglb$IGLB.Adjusted))
15 > rets <- rets[-1, ]
16 > names(rets) <- c("SPYV", "SPYG", "CWI", "IGLB")
17 > head.tail(rets)
18                 SPYV          SPYG          CWI          IGLB
19 Jan 2015 -0.04470711 -0.01910356 -0.005678171  5.315343e-02
20 Feb 2015  0.05648936  0.05969036  0.058010028 -2.737789e-02
21 Mar 2015 -0.01486472 -0.01735287 -0.015909157 -4.791363e-05
22                 SPYV          SPYG          CWI          IGLB
23 Oct 2019 0.02690863 0.01748521 0.03410982 0.001911647
24 Nov 2019 0.03869599 0.03436953 0.01246980 0.006902584
25 Dec 2019 0.03028203 0.02786833 0.03953882 0.003505375
```

**Step 2: Convert Returns Dataset into a Matrix**  Using `matrix()`, convert *rets* into a matrix labeled *mat.ret*.

```
1  > mat.ret <- matrix(rets, nrow(rets))
2  > colnames(mat.ret) <- c("SPYV", "SPYG", "CWI", "IGLB")
```

```
3   > head.tail(mat.ret)
4               SPYV         SPYG            CWI          IGLB
5   [1,] -0.04470711 -0.01910356  -0.005678171  5.315343e-02
6   [2,]  0.05648936  0.05969036   0.058010028 -2.737789e-02
7   [3,] -0.01486472 -0.01735287  -0.015909157 -4.791363e-05
8               SPYV         SPYG            CWI          IGLB
9   [58,] 0.02690863 0.01748521 0.03410982 0.001911647
10  [59,] 0.03869599 0.03436953 0.01246980 0.006902584
11  [60,] 0.03028203 0.02786833 0.03953882 0.003505375
```

**Step 3: Calculate Variance–Covariance Matrix of Returns** Using cov(), we can calculate the **variance–covariance matrix** of the securities in *mat.ret*. The elements in the diagonal are the variances of the securities, while the elements in the off-diagonal are covariances between the securities.

```
1   > (vcov <- cov(mat.ret))
2               SPYV         SPYG          CWI          IGLB
3   SPYV 0.0013417926 0.0011281879 0.0010668378 0.0000839524
4   SPYG 0.0011281879 0.0012756119 0.0010461777 0.0001651636
5   CWI  0.0010668378 0.0010461777 0.0013046755 0.0002553844
6   IGLB 0.0000839524 0.0001651636 0.0002553844 0.0005584502
```

**Step 4: Construct the Target Portfolio Return Vector** Similar to the two-asset case, we now create a vector of target portfolio returns. We first have to calculate the average return of each security (Line 1). Then, we identify the minimum and maximum returns among the four securities (Lines 11 to 14). Then, in Lines 16–17, we create a vector of target returns that go from the minimum return of 0.5% to the maximum of 1.1% with 100 increments in between. Using length(), we can see that this gives us a total of 100 target returns (Line 18).

```
1   > avg.ret <- matrix(apply(mat.ret, 2, mean))
2   > rownames(avg.ret) <- c("SPYV", "SPYG",  "CWI", "IGLB")
3   > colnames(avg.ret) <- c("Avg.Ret")
4   > avg.ret
5             Avg.Ret
6   SPYV 0.008180608
7   SPYG 0.011084436
8   CWI  0.005402004
9   IGLB 0.004984509
10  >
11  > (min.ret <- min(avg.ret))
12  [1] 0.004984509
13  > (max.ret <- max(avg.ret))
14  [1] 0.01108444
15  >
16  > increments <- 100
17  > tgt.ret <- seq(min.ret, max.ret, length = increments)
18  > length(tgt.ret)
19  [1] 100
20  > head.tail(tgt.ret)
21  [1] 0.004984509 0.005046125 0.005107740
22  [1] 0.01096120 0.01102282 0.01108444
```

**Step 5: Construct Temporary Standard Deviation Vector** The optimizer will
output a portfolio standard deviation for each of the 100 iterations it will perform.
Therefore, we need to create a placeholder so that the optimizer can overwrite the
elements of that temporary vector with the standard deviation values it generates
through each loop.

```
1  > (tgt.sd <- rep(0, length = increments))
2   [1] 0 0 0 0 0 0 0 0 0 0 0 0 0 0 0 0 0 0 0 0 0 0 0 0 0 0 0 0 0 0 0 0 0 0 0 0
3   [37] 0 0 0 0 0 0 0 0 0 0 0 0 0 0 0 0 0 0 0 0 0 0 0 0 0 0 0 0 0 0 0 0 0 0 0 0
4   [73] 0 0 0 0 0 0 0 0 0 0 0 0 0 0 0 0 0 0 0 0 0 0 0 0 0 0 0 0
```

**Step 6: Construct Temporary Portfolio Weights Vector** Similarly, the optimizer
also outputs portfolio weights and needs to overwrite a placeholder for those
weights. Since our portfolio has four assets, we want four columns of 100 rows
each.

```
1  > wgt <- matrix(0, nrow = increments, ncol = length(avg.ret))
2  > colnames(wgt) <- c("SPYV", "SPYG", "CWI", "IGLB")
3  > head.tail(wgt)
4        SPYV SPYG CWI IGLB
5  [1,]   0    0   0    0
6  [2,]   0    0   0    0
7  [3,]   0    0   0    0
8        SPYV SPYG CWI IGLB
9  [98,]   0    0   0    0
10  [99,]   0    0   0    0
11  [100,]   0    0   0    0
```

**Step 7: Run the `quadprog` Optimizer** The code below is similar to the code we
ran in the two-asset case.

```
1  > library(quadprog)
2  > for (i in 1:increments){
3  +     Dmat <- 2*vcov
4  +     dvec <- c(rep(0, length(avg.ret)))
5  +     Amat <- cbind(rep(1, length(avg.ret)), avg.ret,
6  +         diag(1, nrow = ncol(mat.ret)))
7  +     bvec <- c(1, tgt.ret[i], rep(0, ncol(mat.ret)))
8  +     soln <- solve.QP(Dmat, dvec, Amat, bvec, meq = 2)
9  +     tgt.sd[i] <- sqrt(soln$value)
10  +     wgt[i, ] <- soln$solution
11  + }
12  > head.tail(tgt.sd)
13  [1] 0.02363155 0.02232601 0.02204782
14  [1] 0.03509073 0.03540243 0.03571571
15  >
16  > colnames(wgt) <- c("w.SPYV", "w.SPYG", "w.CWI", "w.IGLB")
17  > head.tail(wgt)
18           w.SPYV          w.SPYG          w.CWI      w.IGLB
19  [1,] 0.00000000 -9.029089e-18 2.359224e-15 1.0000000
20  [2,] 0.00000000  5.149084e-18 1.475837e-01 0.8524163
21  [3,] 0.01355094  0.000000e+00 1.914292e-01 0.7950199
22       w.SPYV  w.SPYG      w.CWI       w.IGLB
```

```
23   [98,]    0 0.979798 6.823144e–17  2.020202e–02
24   [99,]    0 0.989899 6.938190e–17  1.010101e–02
25  [100,]    0 1.000000 7.053236e–17 –8.881784e–16
```

### Check to See Weights Still Sum to One

When hard coding entries to override results, it is good practice to check that the results still make sense. In the above example, we assert that the values should be zero because they are so small and we have already assigned full weight of 100% to one security. We can easily check by summing all the weights in each row using the rowSums() command and output the results. As we can see, our check results in 100 observations with 1 as the sum of the weights for each observation.

```
1  > (CHECK.wgt <- rowSums(wgt))
2    [1] 1 1 1 1 1 1 1 1 1 1 1 1 1 1 1 1 1 1 1 1 1 1 1 1 1 1 1 1 1 1 1
3   [32] 1 1 1 1 1 1 1 1 1 1 1 1 1 1 1 1 1 1 1 1 1 1 1 1 1 1 1 1 1 1 1
4   [63] 1 1 1 1 1 1 1 1 1 1 1 1 1 1 1 1 1 1 1 1 1 1 1 1 1 1 1 1 1 1 1
5   [94] 1 1 1 1 1 1 1
```

Depending on the size of your screen, the numbers to the left (i.e., what observation number the first element of the row is) may change. The point here is that the correct answer should result in a hundred 1s.

### Step 8: Combine the Returns, Standard Deviation, and Weights of the Portfolio

We now combine the different data objects we created, so we can see what combination of the four securities generates a specific portfolio return/portfolio standard deviation mix. We use cbind() to do this.

```
1  > tgt.port <- data.frame(cbind(tgt.ret, tgt.sd, wgt))
2  > options(digits = 3)
3  > head.tail(tgt.port)
4    tgt.ret tgt.sd w.SPYV   w.SPYG     w.CWI w.IGLB
5  1 0.00498 0.0236 0.0000 –9.03e–18 2.36e–15  1.000
6  2 0.00505 0.0223 0.0000  5.15e–18 1.48e–01  0.852
7  3 0.00511 0.0220 0.0136  0.00e+00 1.91e–01  0.795
8      tgt.ret tgt.sd w.SPYV w.SPYG    w.CWI    w.IGLB
9  98   0.0110 0.0351      0   0.98 6.82e–17  2.02e–02
10 99   0.0110 0.0354      0   0.99 6.94e–17  1.01e–02
11 100  0.0111 0.0357      0   1.00 7.05e–17 –8.88e–16
12 > options(digits = 7)
13 >
14 > no.short.tgt.port <- tgt.port
```

Note that we added the last line of code above so that we can store the results for later use. In a later section, we will compare these results when short selling is not allowed with the results when short selling is allowed.

**Step 9: Identify the Minimum Variance Portfolio**  The minimum variance portfolio is the portfolio with the lowest standard deviation. The output below shows
this is the portfolio that has a weight of 27% in SPYV, 73% in IGLB, and virtually
no weight in SPYG and CWI.

```
1  > options(digits = 3)
2  > (minvar.port <- subset(tgt.port, tgt.sd == min(tgt.sd)))
3      tgt.ret tgt.sd w.SPYV    w.SPYG   w.CWI w.IGLB
4  15 0.00585 0.0207   0.27 -3.35e-18 2.2e-18   0.73
5  > options(digits = 7)
```

**Step 10: Identify the Tangency Portfolio**  The tangency portfolio is the portfolio
with the highest Sharpe Ratio. Thus, we first calculate the Sharpe Ratio for each
combination of portfolio weights. Note that the risk-free rate has to be converted
into a monthly rate. We do this by dividing the annual risk-free rate by 12. As the
below output shows, the tangency portfolio has a portfolio return of 0.87% with a
standard deviation of 2.52%. This is achieved with 61% weight in SPYG and 39%
in IGLB with virtually no weight to SPYV and CWI.

```
1  > rf <- 0.0155 / 12
2  > options(digits = 3)
3  > tgt.port$Sharpe <- (tgt.port$tgt.ret - rf) / tgt.port$tgt.sd
4  > head.tail(tgt.port)
5      tgt.ret tgt.sd w.SPYV    w.SPYG    w.CWI w.IGLB Sharpe
6  1 0.00498 0.0236 0.0000 -9.03e-18 2.36e-15  1.000  0.156
7  2 0.00505 0.0223 0.0000  5.15e-18 1.48e-01  0.852  0.168
8  3 0.00511 0.0220 0.0136  0.00e+00 1.91e-01  0.795  0.173
9      tgt.ret tgt.sd w.SPYV w.SPYG    w.CWI   w.IGLB Sharpe
10 98    0.0110 0.0351      0  0.98 6.82e-17  2.02e-02 0.276
11 99    0.0110 0.0354      0  0.99 6.94e-17  1.01e-02 0.275
12 100   0.0111 0.0357      0  1.00 7.05e-17 -8.88e-16 0.274
13 >
14 > (tangency.port <- subset(tgt.port, Sharpe == max(Sharpe)))
15     tgt.ret tgt.sd w.SPYV w.SPYG    w.CWI w.IGLB Sharpe
16 61 0.00868 0.0252      0  0.606 2.57e-17  0.394  0.293
17 > options(digits = 7)
```

**Step 11: Identify Efficient Portfolios**  The set of efficient portfolios are those
portfolios with returns that are higher than the return of the minimum variance
portfolio. Line 3 shows that there are 86 efficient portfolios.

```
1  > eff.frontier <- subset(tgt.port, tgt.ret >= minvar.port$tgt.ret)
2  > dim(eff.frontier)
3  [1] 86  7
4  > options(digits = 3)
5  > head.tail(eff.frontier)
6      tgt.ret tgt.sd w.SPYV    w.SPYG    w.CWI w.IGLB Sharpe
7  15 0.00585 0.0207  0.270 -3.35e-18 2.20e-18  0.730  0.220
8  16 0.00591 0.0207  0.266  1.21e-02 3.85e-18  0.722  0.223
9  17 0.00597 0.0207  0.256  2.75e-02 8.26e-18  0.717  0.226
10     tgt.ret tgt.sd w.SPYV w.SPYG    w.CWI   w.IGLB Sharpe
11 98    0.0110 0.0351      0  0.98 6.82e-17  2.02e-02 0.276
```

```
12   99    0.0110 0.0354     0    0.99 6.94e–17  1.01e–02  0.275
13   100   0.0111 0.0357     0    1.00 7.05e–17 –8.88e–16  0.274
14   > options(digits = 7)
```

**Step 12: Plot the Mean–Variance Efficient Frontier** We now plot the efficient frontier of this four-asset portfolio. Note that these portfolios are generated by the optimizer with a restriction on short selling. Figure 7.4 shows the mean–variance efficient frontier for the 4-asset portfolio.

```
1   > plot(x = tgt.sd,
2   +     xlab = "Portfolio Risk",
3   +     y = tgt.ret,
4   +     ylab = "Portfolio Return",
5   +     col = "blue",
6   +     main = "Mean—Variance Efficient Frontier of Four Assets
7   +     (Not Allowing Short Selling)")
8   >
9   > points(x = eff.frontier$tgt.sd,
```

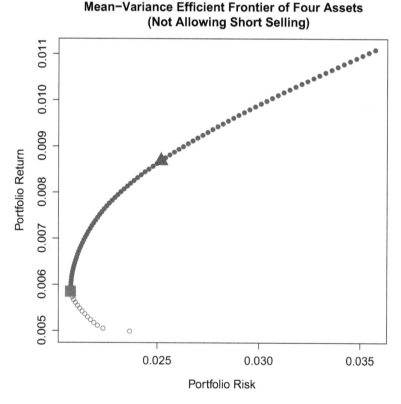

**Fig. 7.4** Mean–variance efficient frontier of portfolios consisting of SPYV, SPYG, CWI, and IGLB without allowing short selling

```
10   +      y = eff.frontier$tgt.ret,
11   +      col = "blue", pch = 16)
12   >
13   > points(x = minvar.port$tgt.sd,
14   +      y = minvar.port$tgt.ret,
15   +      col = "red", pch = 15, cex = 2.5)
16   >
17   > points(x = tangency.port$tgt.sd,
18   +      y = tangency.port$tgt.ret,
19   +      col = "darkgreen", pch = 17, cex = 2.5)
```

## 7.4  Effect of Allowing Short Selling

Before we proceed with running some code, let us take some time to think about what we should expect to see as the effect of allowing short selling. Note that by not allowing short selling we are adding a constraint that restricts the optimizer from putting negative weights on securities that could enhance return if we were allowed to take on short positions. As such, we should expect two things. First, we cannot be worse off when we do not impose a short selling restriction. Second, short selling should enhance returns in situations where taking a short position in a security can benefit us. Indeed, these results hold as we will see at the end of this section.

Now, let us turn to the portions of the code that need to be modified when short selling is allowed. No modification is necessary for Steps 1 to 3. We will begin the modifications with Step 4 of the last section.

**Step 4ss: Construct the Target Portfolio Return Vector (Modifying to Allow for Short Selling)** The portion of Step 4 that is different has to do with how we construct the bounds of the portfolio return vector. Recall that when short selling was not allowed, we can set limits on the portfolio returns by the maximum and minimum average returns from the four securities. However, when short selling is allowed, we can in theory sell the other securities and invest the proceeds from the sale of those securities in the security with the largest return. For the upper bound, we will construct it to be twice the maximum return of the four securities in our portfolio. This number is arbitrary but gives us a good enough range of portfolio returns for our purpose.

```
1   > tgt.ret <- seq(min.ret, max.ret*2, length = increments)
2   > head.tail(tgt.ret)
3   [1] 0.004984509 0.005158088 0.005331668
4   [1] 0.02182171 0.02199529 0.02216887
```

**Step 5ss: Construct Dummy Portfolio Standard Deviation Vector**  There is no change here.

**Step 6ss: Construct Dummy Portfolio Weights Vector**  There is no change here.

**Step 7ss: Run the `quadprog` Optimizer (Modifying to Allow for Short Selling)**
The two changes are in the `Amat` and `bvec` lines (Lines 5–6). What we took out are terms related to the third constraint placed on the optimizer, which is that the weights have to be greater or equal to 0.

```
1   > library(quadprog)
2   > for (i in 1:length(tgt.ret)){
3   +     Dmat <- 2 * vcov
4   +     dvec <- c(rep(0, length(avg.ret)))
5   +     Amat <- cbind(rep(1, length(avg.ret)), avg.ret)
6   +     bvec <- c(1, tgt.ret[i])
7   +     soln <- solve.QP(Dmat, dvec, Amat, bvec, meq = 2)
8   +     tgt.sd[i] <- sqrt(soln$value)
9   +     wgt[i, ] <- soln$solution
10  + }
11  > head.tail(tgt.sd)
12  [1] 0.02110113 0.02096358 0.02084466
13  [1] 0.06060900 0.06120956 0.06181088
14  > head.tail(wgt)
15          w.SPYV      w.SPYG        w.CWI     w.IGLB
16  [1,] 0.3954250 -0.2089324  0.025521142 0.7879863
17  [2,] 0.3874432 -0.1748190  0.003966194 0.7834096
18  [3,] 0.3794614 -0.1407056 -0.017588754 0.7788330
19          w.SPYV      w.SPYG        w.CWI     w.IGLB
20  [98,] -0.3788107 3.100066 -2.065309 0.3440534
21  [99,] -0.3867925 3.134179 -2.086864 0.3394768
22  [100,] -0.3947743 3.168293 -2.108419 0.3349002
```

As we can see, the negative weights placed on certain assets are very far from zero. For example, portfolio number 100 has 317% of the portfolio weight in SPYG, which is financed partly by short selling CWI by an amount equal to 211% of the value of the portfolio. To convince ourselves, we may want to check that all the weights still sum to one.

```
1   > (CHECK.wgt <- rowSums(wgt))
2     [1] 1 1 1 1 1 1 1 1 1 1 1 1 1 1 1 1 1 1 1 1 1 1 1 1 1 1 1 1 1 1 1 1 1 1 1 1
3    [37] 1 1 1 1 1 1 1 1 1 1 1 1 1 1 1 1 1 1 1 1 1 1 1 1 1 1 1 1 1 1 1 1 1 1 1 1
4    [73] 1 1 1 1 1 1 1 1 1 1 1 1 1 1 1 1 1 1 1 1 1 1 1 1 1 1 1 1
```

**Step 8ss: Combine the Returns, Standard Deviation, and Weights of the Portfolio**  There is no change in the code here.

```
1   > tgt.port <- data.frame(cbind(tgt.ret, tgt.sd, wgt))
2   > head.tail(tgt.port)
3        tgt.ret     tgt.sd     w.SPYV      w.SPYG        w.CWI     w.IGLB
4   1 0.004984509 0.02110113 0.3954250 -0.2089324  0.025521142 0.7879863
5   2 0.005158088 0.02096358 0.3874432 -0.1748190  0.003966194 0.7834096
6   3 0.005331668 0.02084466 0.3794614 -0.1407056 -0.017588754 0.7788330
7        tgt.ret     tgt.sd     w.SPYV      w.SPYG        w.CWI     w.IGLB
```

```
 8   98   0.02182171 0.06060900 −0.3788107 3.100066 −2.065309 0.3440534
 9   99   0.02199529 0.06120956 −0.3867925 3.134179 −2.086864 0.3394768
10  100  0.02216887 0.06181088 −0.3947743 3.168293 −2.108419 0.3349002
11   >
12   > with.short.tgt.port <− tgt.port
```

We added the last line to store the results, which we will later use to compare with
the earlier result when short selling was not allowed.

**Step 9ss: Identify the Minimum Variance Portfolio**  The output below shows the
minimum variance portfolio is the portfolio that has weights of 75% in IGLB, 33%
in SPYV, 6% in SPYG, and *negative* 15% in CWI. In other words, this portfolio
indicates short selling CWI.

```
1   > (minvar.port <− subset(tgt.port, tgt.sd == min(tgt.sd)))
2        tgt.ret      tgt.sd      w.SPYV      w.SPYG        w.CWI      w.IGLB
3    9 0.006373144 0.02053677 0.3315705 0.06397468 −0.1469184 0.7513732
```

**Step 10ss: Identify the Tangency Portfolio**  We now turn to calculating the
tangency portfolio, which is the portfolio with the highest Sharpe Ratio. As such,
we first calculate the Sharpe Ratios of all the portfolios we constructed. The last line
of the output shows that the tangency portfolio has a Sharpe Ratio of 0.366 and has
weights of 126% in SPYG, 59% in IGLB, 5% in SPYV, and *negative* 90% in CWI.

```
 1   > tgt.port$Sharpe <− (tgt.port$tgt.ret − rf) / tgt.port$tgt.sd
 2   > options(digits = 3)
 3   > head.tail(tgt.port)
 4     tgt.ret tgt.sd w.SPYV w.SPYG    w.CWI w.IGLB Sharpe
 5   1 0.00498 0.0211  0.395 −0.209  0.02552  0.788  0.175
 6   2 0.00516 0.0210  0.387 −0.175  0.00397  0.783  0.184
 7   3 0.00533 0.0208  0.379 −0.141 −0.01759  0.779  0.194
 8      tgt.ret tgt.sd w.SPYV w.SPYG w.CWI w.IGLB Sharpe
 9    98   0.0218 0.0606 −0.379   3.10 −2.07  0.344  0.339
10    99   0.0220 0.0612 −0.387   3.13 −2.09  0.339  0.338
11   100   0.0222 0.0618 −0.395   3.17 −2.11  0.335  0.338
12   >
13   > (tangency.port <− subset(tgt.port, Sharpe == max(Sharpe)))
14      tgt.ret tgt.sd w.SPYV w.SPYG  w.CWI w.IGLB Sharpe
15   44   0.0124 0.0305 0.0522   1.26 −0.901  0.591  0.366
16   > options(digits = 7)
```

**Step 11ss: Identify Efficient Portfolios**  We identify the efficient portfolios, which
are the portfolios with returns that are higher than the return of the minimum
variance portfolios.

```
1  > eff.frontier <- subset(tgt.port, tgt.ret >= minvar.port$tgt.ret)
2  > options(digits = 3)
3  > head.tail(eff.frontier)
4     tgt.ret tgt.sd w.SPYV w.SPYG  w.CWI w.IGLB Sharpe
5  9  0.00637 0.0205  0.332 0.0640 -0.147  0.751  0.247
6  10 0.00655 0.0206  0.324 0.0981 -0.168  0.747  0.256
7  11 0.00672 0.0206  0.316 0.1322 -0.190  0.742  0.264
8     tgt.ret tgt.sd w.SPYV w.SPYG w.CWI w.IGLB Sharpe
9  98    0.0218 0.0606 -0.379   3.10 -2.07  0.344  0.339
10 99    0.0220 0.0612 -0.387   3.13 -2.09  0.339  0.338
11 100   0.0222 0.0618 -0.395   3.17 -2.11  0.335  0.338
12 > options(digits = 7)
```

**Step 12ss: Plot the Mean–Variance Efficient Frontier**  We now plot the efficient frontier of this four-asset portfolio when short selling is not restricted. Figure 7.5

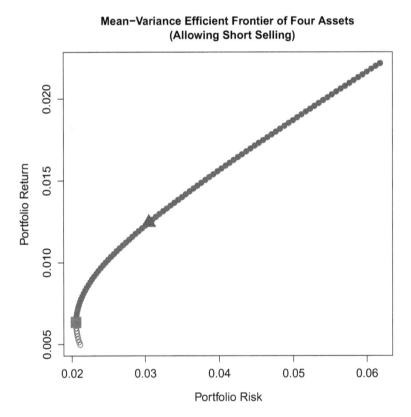

**Fig. 7.5** Mean–variance efficient frontier of portfolios consisting of SPYV, SPYG, CWI, and IGLB and allowing short selling

shows the MV efficient frontier of the four-asset portfolio when short selling
restrictions are removed.

```
1   > plot(x = tgt.sd,
2   +      xlab = "Portfolio Risk",
3   +      y = tgt.ret,
4   +      ylab = "Portfolio Return",
5   +      col = "blue",
6   +      main = "Mean—Variance Efficient Frontier of Four Assets
7   +      (Allowing Short Selling)")
8   >
9   > points(x = eff.frontier$tgt.sd,
10  +      y = eff.frontier$tgt.ret,
11  +      col = "blue", pch = 16)
12  >
13  > points(x = minvar.port$tgt.sd,
14  +      y = minvar.port$tgt.ret,
15  +      col = "red", pch = 15, cex = 2.5)
16  >
17  > points(x = tangency.port$tgt.sd,
18  +      y = tangency.port$tgt.ret,
19  +      col = "darkgreen", pch = 17, cex = 2.5)
```

**Comparing the Efficient Frontier with and Without Short Sales**

With two separate charts, it is hard to visualize the difference between the
mean–variance efficient frontier with and without short sales, so we plot the
two mean–variance efficient frontiers together.

```
1   > plot(x = no.short.tgt.port$tgt.sd,
2   +      y = no.short.tgt.port$tgt.ret,
3   +      xlab = "Portfolio Risk",
4   +      ylab = "Portfolio Return",
5   +      type = "l",
6   +      main = "MV Efficient Frontier for Four Assets
7   +      With and Without Short Selling")
8   > lines(x = with.short.tgt.port$tgt.sd,
9   +      y = with.short.tgt.port$tgt.ret,
10  +      col = "red",
11  +      type = "l")
12  > legend("bottomright",
13  +      c("Not Allow Short", "Allow Short"),
14  +      col = c("black", "red"),
15  +      lty = 1)
```

Figure 7.6 shows the output of the above code. The black line represents
the efficient frontier when short selling is not allowed, while the red line
represents the efficient frontier when short selling is allowed. The results show
that, if we allow short selling, we are able to attain higher levels of portfolio

(continued)

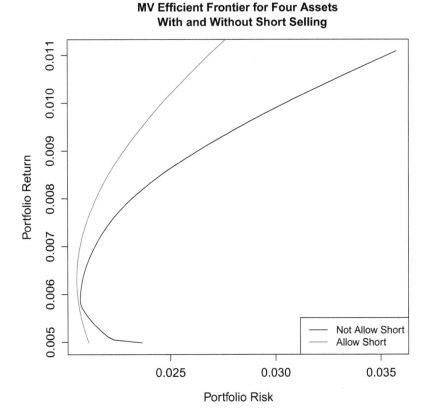

**Fig. 7.6** Comparison of mean–variance efficient frontiers of portfolios consisting of SPYV, SPYG, CWI, and IGLB with and without short selling

returns for a given level of risk or that we are able to get lower risk for a given level of portfolio return.

## 7.5  Further Reading

More details on the Sharpe Ratio are found in [4] and [5]. Markowitz [3] is the original paper on portfolio selection. A more comprehensive and modern treatment of asset allocation and portfolio management can be found in [2]. Additional techniques used to calculate the efficient frontier are found in [1].

# References

1. Elton, E., Gruber, M., Brown, S., & Goetzmann, W. (2010). *Modern portfolio theory and investment analysis* (8th ed.). Hoboken: Wiley.
2. Maginn, J., Tuttle, D., Pinto, J., & McLeavey, D. (2007). *Managing investment portfolios: A dynamic process* (3rd ed.). Hoboken: Wiley.
3. Markowitz, H. (1952). Portfolio selection. *Journal of Finance, 7*, 77–91.
4. Sharpe, W. (1975). Adjusting for risk in portfolio performance. *Journal of Portfolio Management, 1*, 29–34.
5. Sharpe, W. (1994). The Sharpe ratio. *Journal of Portfolio Management, 21*, 49–58.

# Chapter 8
# Equities

The **value** of an asset is the present value of expected free cash flows generated by the asset discounted at the appropriate risk-adjusted discount rate. Thus, one of the primary components of conducting a valuation is developing projections of free cash flows. Free cash flows are cash flows that the firm can distribute to capital providers without affecting the prospects of the company. To arrive at free cash flows, we often start with revenue projections and forecast most of the remaining items as some function of revenues. More details on forecasting cash flows can be found in [10]. In this chapter, we begin by discussing how to obtain a company's financial data in Sect. 8.1 by scraping the data off the company's SEC filings. The company's SEC filings are sources of historical financial information. However, equity values are based on forecasts of future performance. Thus, we ultimately have to work with projections of financial data, which we discuss in Sect. 8.2.

The other component in a valuation is the discount rate or the cost of capital. As discussed in Chap. 5, the CAPM is the most commonly used model to estimate the cost of equity. The CAPM requires three components. The first one is the risk-free rate, which we have previously discussed in Chap. 5. The second component is the equity risk premium (ERP). In Sect. 8.3, we discuss one way of estimating the ERP, which is to use historical data. The third component is the beta of the subject firm, which is the sensitivity of the stock's return on the market's return. We discussed how to calculate beta in Chap. 5, which is known as a levered beta because it incorporates the firm's financial leverage. All else being equal, firms with more debt are riskier. However, we often need to adjust the beta before it can be used in the discount rate by the difference in leverage between the average regression period leverage ratio and the target leverage ratio. We show how to do this in Sect. 8.4.

Valuations, such as discounted cash flow analysis, rely on several key inputs and assumptions. Although great care could have been used to select these inputs and assumptions, there may be a reasonable variation around some of the critical inputs and assumptions. Thus, it could be helpful to see what the effect of changing those assumptions is on value. In Sect. 8.5, we show how to perform a simple **sensitivity**

© The Author(s), under exclusive license to Springer Nature Switzerland AG 2021    225
C.S. Ang, *Analyzing Financial Data and Implementing Financial Models Using R*,
Springer Texts in Business and Economics,
https://doi.org/10.1007/978-3-030-64155-9_8

**analysis** when we change the discount rate and perpetuity growth rate assumptions in DCF valuation.

Another common equity valuation approach is to perform **relative valuation**. This approach uses multiples of comparable companies to the subject firm. A price multiple, for example, uses the market price of the security in the numerator and a financial metric, like earnings per share, in the denominator. This multiple is referred to as the Price-to-Earnings Ratio or P-E Ratio and is one of the most common multiples used in investments. To determine the subject firm's value, some multiple within the range of the comparable company's multiples is often used to infer the value of the subject firm. A common choice when the peer firms are fairly similar to the subject firm is the median or average of the comparable companies' multiples. The closer the comparable companies are to the subject firm, the better the median or average multiple will be in estimating the value of the subject firm. However, this decision requires a mix of qualitative and quantitative analysis. In Sect. 8.6, we show how to use regression analysis to quantitatively determine the subject firm's multiple based on the multiples of the comparable companies. This approach is more quantitative and could take into account the differences in attributes between comparable firms when estimating the appropriate multiple for the subject firm.

We then show how to perform tests on stock returns data in R. In Sect. 8.7.1, we show how to statistically test whether the average returns over two periods are the same using a t-test. Then, we show how to identify breakpoints in the data in Sect. 8.7.2 to show whether the relationship between the stock return and selected independent variables has changed over different periods. In Sect. 8.7.3, we statistically test the structural break in the data using a Chow Test. The Chow Test shows whether there have been changes in the intercept and beta. We can also test the changes in beta alone (i.e., not including any change in the intercept). We do this in Sect. 8.7.4.

Note that the discussion in this chapter is not meant to be comprehensive but is more focused on showing examples of quantitative techniques related to analyzing issues related to equities.

## 8.1   Company Financials

A common place to start when analyzing the equity of companies is its financial statements. Financial statements are standardized reports, which means that how they are constructed and reported is based on a set of rules that is generally understood by market participants. Specifically, a great majority of information contained in a company's financial statements is compliant with Generally Accepted Accounting Principles (GAAP) or International Accounting Standards (IAS). These also reflect certain positions of accounting and regulatory bodies (e.g., the U.S. Securities and Exchange Commission). Being standardized does not necessarily imply that we should take the values on its face, but what it means is that we have a general understanding of how those values are calculated and can make adjustments accordingly. For example, we know that net income contains certain one-time adjustments. Therefore, if we want an income figure that does not include these one-time adjustments, we understand that we cannot take net income on its face but have to make adjustments to it.

There are many sources of a company's financial statements. For firms that have to file with the U.S. Securities and Exchange Commission (SEC), we can use the SEC's electronic portal called EDGAR. Public firms generally file quarterly reports (Form 10-Q) or annual reports (Form 10-K). Foreign firms usually file their half-year and annual reports (Form 20-F). Sometimes, we can also obtain these periodic filings from the company's website or through other databases.

There are tools in R that allow us to scrape information from these SEC filings. Below, we show how to use R to obtain key financial metrics from the firm's SEC filings. We use as an example selected financial information for Amazon from 2015 to 2019 as reported in Item 6 "Selected Consolidated Financial Data" of Amazon's 2019 Form 10-K. To efficiently import the information we would like to use in our example requires a combination of using R packages and manual inspection of the data.

**Step 1: Obtain Financial Data from SEC Filings**  We start by scraping data from tables in the HTML version of Amazon's 2019 10-K. For this, we need to load tidyverse and rvest. In Line 4, we put the URL of the HTML version of Amazon's 2019 10-K. We do this as follows. First, we go to the SEC EDGAR website (https://www.sec.gov/edgar/searchedgar/companysearch.html), put "AMZN" in the search box, and type "10-K" in the filing type. We then see that the first filing is a 10-K that was filed on January 31, 2020. This is Amazon's 2019 10-K. We can click on that link, which opens the XBRL version of the 10-K. Next, the menu on the upper-left of the screen has an option to "Open as HTML," which we will click. This brings us to the HTML version of Amazon's 2019 10-K. We can copy and paste that URL in Line 4 below. We can then use this page to later compare the values that we obtain below to make sure that we are capturing the correct figures. The full URL of the Amazon 2019 10-K is https://www.sec.gov/Archives/edgar/data/1018724/000101872420000004/amzn-20191231x10k.htm.

```
 1  > library(tidyverse)
 2  > library(rvest)
 3  > temp <-
 4  +     "https://www.sec.gov/Archives/edgar/data/1018724/ . . . " %>%
 5  +     read_html() %>%
 6  +     html_table() %>%
 7  +     map_df(bind_cols) %>%
 8  +     as_tibble()
 9  > head.tail(temp)
10  # A tibble: 3 x 33
11    X1    X2    X3    X4    X5    X6    X7    X8    X9    X10   X11
12    <chr> <chr> <chr> <chr> <chr> <chr> <chr> <chr> <chr> <chr> <chr>
13  1 NA    NA    NA    NA    NA    NA    NA    NA    NA    NA    NA
14  2 NA    NA    NA    NA    NA    NA    NA    NA    NA    NA    NA
15  3 NA    NA    NA    NA    NA    NA    NA    NA    NA    NA    NA
```

```
16   # ... with 22 more variables: X12 <chr>, X13 <chr>, X14 <chr>,
17   #   X15 <chr>, X16 <chr>, X17 <chr>, X18 <chr>, X19 <chr>, X20 <chr>,
18   #   X21 <chr>, X22 <chr>, X23 <chr>, X24 <chr>, X25 <chr>, X26 <chr>,
19   #   X27 <chr>, X28 <chr>, X29 <chr>, X30 <lgl>, X31 <chr>, X32 <chr>,
20   #   X33 <chr>
21   # A tibble: 3 x 33
22     X1        X2    X3     X4    X5    X6    X7    X8    X9    X10   X11
23     <chr>     <chr> <chr>  <chr> <chr> <chr> <chr> <chr> <chr> <chr> <chr>
24   1 ""        NA    ""     NA    NA    NA    NA    NA    NA    NA    NA
25   2 /s/ Wend~ NA    ""     NA    NA    NA    NA    NA    NA    NA    NA
26   3 Wendell ~ NA    Dire~  NA    NA    NA    NA    NA    NA    NA    NA
27   # ... with 22 more variables: X12 <chr>, X13 <chr>, X14 <chr>,
28   #   X15 <chr>, X16 <chr>, X17 <chr>, X18 <chr>, X19 <chr>, X20 <chr>,
29   #   X21 <chr>, X22 <chr>, X23 <chr>, X24 <chr>, X25 <chr>, X26 <chr>,
30   #   X27 <chr>, X28 <chr>, X29 <chr>, X30 <lgl>, X31 <chr>, X32 <chr>,
31   #   X33 <chr>
```

**Step 2: Manually Identify Relevant Rows and Columns**  Although it is tempting to write code to identify the appropriate rows and columns. It is much more efficient and simpler to look at the underlying data. This can be done by viewing the *temp* dataset. Doing so, we will see that we are interested in data between rows 258 and 278, inclusive. Once we have subset those rows, we can then view this smaller dataset and see that there are extraneous rows and columns. We remove those in Line 2 below. Then, we rename the column headers to something that makes more sense than X1, X2, etc. (Line 3).

```
1   > finl.data <- temp[258:278, ]
2   > finl.data <- finl.data[-c(1:4, 10, 13, 15:19), c(1, 4, 8, 12, 16, 20)]
3   > names(finl.data) <- c("Metric", 2015, 2016, 2017, 2018, 2019)
4   > head(finl.data, 3)
5   # A tibble: 3 x 6
6     Metric            `2015` `2016`  `2017`  `2018`  `2019`
7     <chr>             <chr>  <chr>   <chr>   <chr>   <chr>
8   1 Net sales         107,006 135,987 177,866 232,887 280,522
9   2 Operating income  2,233   4,186   4,106   12,421  14,541
10  3 Net income (loss) 596     2,371   3,033   10,073  11,588
```

**Step 3: Clean Up Dataset Names and Values**  As the above shows, the Metric values are sometimes long, so we may want to shorten those so that they look nicer, but we have to be careful to make sure we maintain a sufficient amount of detail that we understand what the line items are. We do this in Lines 1 to 6. Next, we also notice that the values under the columns 2015 to 2019 are being read in by R as characters, so we need to change those to numbers. That way, we can work with them. We do this in Lines 7 to 11. We use gsub() to remove the comma from the character values and then convert those values to numeric using as.numeric().

```
1   > finl.data[4, 1] <- "Basic EPS"
2   > finl.data[5, 1] <- "Diluted EPS"
3   > finl.data[6, 1] <- "Wtd avg shares — Basic"
4   > finl.data[7, 1] <- "wtd avg shares — Diulted"
5   > finl.data[8, 1] <- "CF from operations"
```

```
 6  > finl.data[10, 1] <- "LT obligations"
 7  > finl.data$"2015" <- as.numeric(gsub(",", "", finl.data$"2015"))
 8  > finl.data$"2016" <- as.numeric(gsub(",", "", finl.data$"2016"))
 9  > finl.data$"2017" <- as.numeric(gsub(",", "", finl.data$"2017"))
10  > finl.data$"2018" <- as.numeric(gsub(",", "", finl.data$"2018"))
11  > finl.data$"2019" <- as.numeric(gsub(",", "", finl.data$"2019"))
12  > finl.data
13  # A tibble: 10 x 6
14    Metric                      `2015`    `2016`    `2017`    `2018`
      `2019`
15    <chr>                       <dbl>     <dbl>     <dbl>     <dbl>
      <dbl>
16   1 Net sales                  107006    135987    177866    232887    280522
17   2 Operating income            2233      4186      4106     12421     14541
18   3 Net income (loss)            596      2371      3033     10073     11588
19   4 Basic EPS                   1.28      5.01      6.32      20.7
      23.5
20   5 Diluted EPS                 1.25       4.9      6.15      20.1
      23.0
21   6 Wtd avg shares - Basic       467       474       480       487       494
22   7 wtd avg shares - Diulted     477       484       493       500       504
23   8 CF from operations         11909     17203     18365     30723     38514
24   9 Total assets               64747     83402    131310    162648    225248
25  10 LT obligations             17477     20301     45718     50708     75376
```

**Step 4: Convert to Data Frame**  The format of *finl.data* could be improved simply by converting the dataset into a data frame. The output above shows that the values appear to be aligned at the decimal place, whether the decimal is explicit or implicit. The output below shows that as a data frame object, the output includes two zeroes in the decimal place, which is what we would commonly see when presented with such data and everything looks aligned once more.

```
 1  > df.finl <- data.frame(finl.data)
 2  > str(df.finl)
 3  'data.frame': 10 obs. of  6 variables:
 4   $ Metric: chr  "Net sales" "Operating income" "Net income (loss)" ...
 5   $ X2015 : num  1.07e+05 2.23e+03 5.96e+02 1.28 1.25 ...
 6   $ X2016 : num  1.36e+05 4.19e+03 2.37e+03 5.01 4.90 ...
 7   $ X2017 : num  1.78e+05 4.11e+03 3.03e+03 6.32 6.15 ...
 8   $ X2018 : num  232887 12421 10073 20.7 20.1 ...
 9   $ X2019 : num  280522 14541 11588 23.5 23 ...
10  > df.finl
11                     Metric     X2015      X2016      X2017      X2018      X2019
12  1               Net sales 107006.00  135987.00  177866.00  232887.00  280522.00
13  2        Operating income   2233.00    4186.00    4106.00   12421.00   14541.00
14  3       Net income (loss)    596.00    2371.00    3033.00   10073.00   11588.00
15  4               Basic EPS      1.28       5.01       6.32      20.68      23.46
16  5             Diluted EPS      1.25       4.90       6.15      20.14      23.01
17  6  Wtd avg shares - Basic    467.00     474.00     480.00     487.00     494.00
18  7 wtd avg shares - Diulted   477.00     484.00     493.00     500.00     504.00
19  8      CF from operations  11909.00   17203.00   18365.00   30723.00   38514.00
20  9            Total assets  64747.00   83402.00  131310.00  162648.00  225248.00
```

21   10         LT obligations  17477.00  20301.00  45718.00  50708.00  75376.00

Now that we have this data converted into R, we can perform whatever analysis we would like to do. For example, if we wanted to show the historical sales of Amazon in a bar plot, we can extract the sales figures as a matrix. In this example, we divide the numbers by 1000 to make them easier to read. Since the numbers were already reported in millions of dollars, this adjustment would make the values that we report in billions of dollars. That is, the 2019 sales are $281 billion. Lines 13 and 14 add the values to the top of the bars. The 20 that is added to the y was identified through trial-and-error. The larger the number, the higher the position of the text.

```
1   > (sales <- as.matrix(round((df.finl[1, 2:6])/1000, 0)))
2     X2015 X2016 X2017 X2018 X2019
3   1   107   136   178   233   281
4   >
5   > x <- barplot(sales,
6   +     names.arg = seq(2015, 2019, 1),
7   +     ylim = c(0, 350),
8   +     ylab = "Net sales (bln)",
9   +     col = "blue",
10  +     border = 0,
11  +     main = "AMZN Net Sales, 2015 - 2019")
12  >
13  > y <- as.matrix(sales)
14  > text(x, y + 20, labels = as.character(y))
```

The output of the above code is shown in Fig. 8.1. As the chart shows, Amazon's sales increased from $107 billion in 2015 to $281 billion in 2019.

## 8.2  Projections

Shareholders are concerned with the free cash flows the firm is expected to generate in the future. The term "free" in free cash flows means cash flows that the firm can distribute without affecting its operations. For valuation purposes, we would like to use projections that are long enough such that by the final year of the projections, the company is considered to be in steady state. In steady state, the company's sales are growing at a sustainable rate, its margins are relatively stable, and it is able to reinvest funds from its profits to self-finance (i.e., no external financing) the growth of the business.

How long it takes for the firm to reach steady state should be determined on a case-by-case basis. For example, a mature firm could already be in steady state, while an early stage firm may take 10–15 years before it reaches steady state (assuming it ever does). Based on experience, we typically see projections that are generally from 5 to 15 years out. Some firms do not make projections, while some firms only make projections that are 1–3 years out.

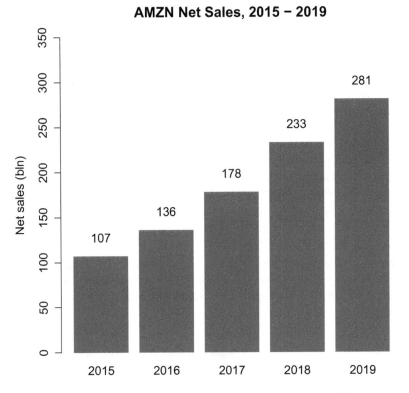

**Fig. 8.1**  Amazon's historical sales from 2015 to 2019. Source: Amazon 2019 10-K

The accuracy of projections decreases the further out we project. Intuitively, this is because we have less visibility the farther out we go and forecast errors get compounded. Therefore, more detailed projections are often more useful in the first few years of the projection period. Then, for the outer years of the projections, a more general model that is based on the evolution of key value drivers is more appropriate. In other words, it is unclear what gains, if any, a more detailed model provides in the outer years when the forecast error is large.

One common approach to developing projections is to forecast future revenues and then to base most of the other line items as some function of revenues. For example, if gross margin is a relatively constant percentage of revenue over time, then projected gross margin can be estimated using the appropriate gross margin percentage multiplied by the projected revenues. For purposes of our illustration below, we will use revenue projections as an example.

### 8.2.1   *Projecting Based on Historical Trends*

A simple way of developing projections is to use the historical trend of the company. This is particularly useful if a clear trend appears to exist, and such a trend is expected to continue. To use a real world company as an example, we will look at the historical sales for Barnes & Noble. From Barnes & Noble's 10-K, we are able to obtain the company's sales for fiscal years 2012 to 2019. Note that Barnes & Noble's fiscal year ends the last Saturday of April (e.g., April 28, 2018, April 27, 2019, etc.). We only use data beginning in 2012 because a review of the data shows that data prior to 2012 is not comparable with later period data. There are databases that could make this step easier, but we obtained the data by going on the SEC website. We then manually enter the sales figures using c().

```
1  > (sales <- c(5385.537, 5075.761, 4633.345, 4297.108, 4163.844,
2  +     3894.558, 3662.280, 3552.745))
3  [1] 5385.537 5075.761 4633.345 4297.108 4163.844 3894.558 3662.280 3552.745
```

Alternative to the manual entry above, we could have scraped the historical data from the web pages that contain them. The format of these tables is a little different, so we can use an alternative approach to scrape the data from these tables than the one used in the prior section.

**Step 1: Obtain Historical Sales Data** We would need to use data from two web pages to get the sales data from 2012 to 2019. The first URL is https://www.sec.gov/Archives/edgar/data/890491/000119312519176078/d703316dex131.htm. The second URL is https://www.sec.gov/Archives/edgar/data/890491/000119312516630221/d179245dex131.htm. Be sure to replace *url1* and *url2* below with the full URLs. The first URL gives us data for fiscal years 2015 to 2019 but in reverse chronological order (i.e., the $3.55 million is the 2019 sales figure). The second URL gives us data for fiscal years 2012 to 2016. Lines 8 and 18 remove the commas from the figures, so that R can convert these entries into numeric values.

```
1  > url1 <- "https://www.sec.gov/Archives/edgar/data/890491/ ..."
2  > barnes1 <- html_session(url1) %>%
3  +   html_node("table") %>%
4  +   html_table()
5  > (barnes1 <- barnes1[9, c(4, 8, 12, 16, 20)])
6            X4          X8          X12         X16         X20
7  9 3,552,745 3,662,280 3,894,558 4,163,844 4,297,108
8  > (sales1 <- as.numeric(gsub(",", "", barnes1)))
9  [1] 3552745 3662280 3894558 4163844 4297108
10 >
11 > url2 <- "https://www.sec.gov/Archives/edgar/data/890491/ ..."
12 > barnes2 <- html_session(url2) %>%
13 +   html_node("table") %>%
14 +   html_table()
15 > (barnes2 <- barnes2[9, c(4, 8, 12, 16, 20)])
16            X4          X8          X12         X16         X20
17 9 4,163,844 4,297,108 4,633,345 5,075,761 5,385,537
18 > (sales2 <- as.numeric(gsub(",", "", barnes2)))
```

19    [1] 4163844 4297108 4633345 5075761 5385537

**Step 2: Set Up Dataset for Plotting** We can take a quick look visually what the Barnes & Noble data looks like. Before doing so, we need to manipulate the data into shape. In our subsequent analyses, we would require the data in million, so we first create a *sales* variable in Line 1 that is in millions. Since there is overlapping data, we drop the first two columns of the *sales2* dataset, which are for the overlapping fiscal years of 2015 and 2016. After combining with the year data in *year*, we can see from the output of Line 5 that the sales figures are in reverse chronological order. We thus sort the data using   order(). Finally, we convert the sales figures into billions for plotting.

```
 1  > (sales <- c(sales1, sales2[-c(1:2)]) / 1000)
 2  [1] 3552.745 3662.280 3894.558 4163.844 4297.108 4633.345 5075.761 5385.537
 3  > (year <- seq(2019, 2012, -1))
 4  [1] 2019 2018 2017 2016 2015 2014 2013 2012
 5  > (sales <- data.frame(year, sales))
 6     year    sales
 7  1 2019 3552.745
 8  2 2018 3662.280
 9  3 2017 3894.558
10  4 2016 4163.844
11  5 2015 4297.108
12  6 2014 4633.345
13  7 2013 5075.761
14  8 2012 5385.537
15  > (sales <- sales[order(year), ])
16     year    sales
17  8 2012 5385.537
18  7 2013 5075.761
19  6 2014 4633.345
20  5 2015 4297.108
21  4 2016 4163.844
22  3 2017 3894.558
23  2 2018 3662.280
24  1 2019 3552.745
25  >
26  > (sales.plot <- round(sales$sales / 1000, 1))
27  [1] 5.4 5.1 4.6 4.3 4.2 3.9 3.7 3.6
```

In the above, we deleted overlapping data. However, the overlapping data could be helpful for us to check that we are combining values that are apples-to-apples. Here, the overlapping periods have identical values, which gives us comfort that we are using the same data and that we can confidently combine the two streams of revenues.

**Step 4: Plot Sales Data** By doing a quick bar plot as depicted in Fig. 8.2, we can see that Barnes & Noble's sales exhibits a decreasing trend. This is to be expected as bookstores, in general, and Barnes & Noble, in particular, have been disrupted in recent years by the growth of online book retailers/bookstores, namely Amazon. Therefore, if we were to project the next five years of sales based on this historical

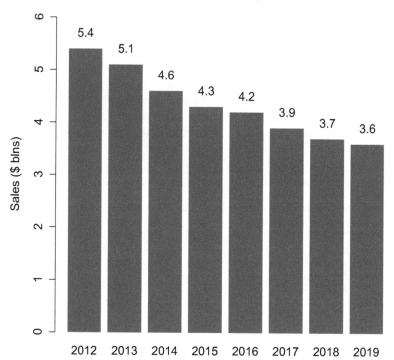

**Fig. 8.2** Barnes & Noble's historical sales. The chart shows historical Barnes & Noble's sales for FY 2012 to FY 2019. Source: Barnes & Noble's SEC filings

trend, we would expect Barnes & Noble's sales to continue this declining trend. One approach to quantify this declining trend going forward is through regression analysis. We go through an example of how to implement this approach below.

```
1  > x <- barplot(sales.plot,
2  +    ylim = c(0, 6),
3  +    ylab = "Sales ($ blns)",
4  +    col = "blue",
5  +    border = 0,
6  +    names.arg = seq(2012, 2019, 1),
7  +    main = "Barnes & Noble Sales, 2012—2019")
8  >
9  > y <- as.matrix(sales.plot)
10 > text(x, y + 0.30, labels = as.character(y))
```

**Step 5: Create a Dataset with the Historical Sales and a Trend Variable** We obtain the Barnes & Noble's historical sales data from the company's SEC filings. We obtain data for FY 2012 to 2019. We are projecting 5 years out (i.e., FY 2020 to 2024); thus, we need to add five extra observations at the end of the hist.sales variable, which we do using `rep()`. The first argument in `rep()` is the value that we wish to repeat, and the second argument is the number of times we wish to repeat that value. In this case, we told R to repeat zero 5x. Next, we created the `trend` variable, which takes on the values of 1, 2, 3, and so on until 13. To do this, we use `seq()`. We then combine the two vectors using `cbind()` and convert the object into a data frame using `data.frame()`.

```
1   > (hist.sales <- c(5385.537, 5075.761, 4633.345, 4297.108,
2   +    4163.844, 3894.558, 3662.280, 3552.745, rep(0, 5)))
3    [1] 5385.537 5075.761 4633.345 4297.108 4163.844 3894.558 3662.280 3552.745
4    [9]    0.000    0.000    0.000    0.000    0.000
5   > (trend <- seq(1, 13, 1))
6    [1]  1  2  3  4  5  6  7  8  9 10 11 12 13
7   > reg.data <- data.frame(cbind(hist.sales, trend))
8   > reg.data
9      hist.sales trend
10  1    5385.537     1
11  2    5075.761     2
12  3    4633.345     3
13  4    4297.108     4
14  5    4163.844     5
15  6    3894.558     6
16  7    3662.280     7
17  8    3552.745     8
18  9       0.000     9
19  10      0.000    10
20  11      0.000    11
21  12      0.000    12
22  13      0.000    13
```

**Step 6: Regress the Historical Data on the Trend Variable** In this step, we only want to use the data for FY 2012 to 2019. Thus, we only need the first eight observations of *reg.data*. We run the regression using `lm()`, and to subset *reg.data*, we add a square bracket with 1:8 to the right of the comma. As we expected, the regression results show that the coefficient of the trend variable is negative. Also, both the intercept and trend are statistically significant given the very small *p*-values.

```
1   > hist.reg <- lm(hist.sales ~ trend, data = reg.data[1:8, ])
2   > summary(hist.reg)
3
4   Call:
5   lm(formula = hist.sales ~ trend, data = reg.data[1:8, ])
6
7   Residuals:
8       Min      1Q Median      3Q     Max
9   -168.46 -55.26 -22.82   91.75  146.54
10
11  Coefficients:
12               Estimate Std. Error t value Pr(>|t|)
13  (Intercept)  5524.93      92.55   59.70 1.48e-09 ***
14  trend        -264.84      18.33  -14.45 6.88e-06 ***
```

```
15  ──
16  Signif. codes:  0 '***' 0.001 '**' 0.01 '*' 0.05 '.' 0.1 ' ' 1
17
18  Residual standard error: 118.8 on 6 degrees of freedom
19  Multiple R-squared:  0.9721, Adjusted R-squared:  0.9674
20  F-statistic: 208.8 on 1 and 6 DF,  p-value: 6.88e-06
```

**Step 7: Project Sales for FY 2020 to 2024**  We then use `predict()` to project the sales for the next 5 years. `predict()` requires two arguments: first, the regression output *hist.reg* from the prior step, and second, the data for the projection, which is observations 9 through 13 from *reg.data*. To use `predict()`, the second dataset needs to have the same variables as the data used in the regression. Thus, by taking a subset of the data we used in the regression, we make sure the dataset we use has the same variables.

```
1  > (proj.sales <- predict(hist.reg, reg.data[9:13, ]))
2        9        10       11       12       13
3  3141.367 2876.526 2611.686 2346.846 2082.006
```

**Step 8: Plot the Historical and Projected Sales**  In our example, it is easy to see that the trend during the projection period is also negative. The first year (FY 2020) sales are projected to be \$3.1 billion, while the last year (FY 2024) sales are projected to be \$2.1 billion. However, it may be easier and potentially more convincing to see this visually. We first have to do some data manipulation before we are able to create a bar plot of historical and projected sales. We first augment *proj.sales* to include eight 0s at the beginning.

```
1  > proj.sales <- c(rep(0, 8), proj.sales)
2  > names(proj.sales) <- seq(2012, 2024, 1)
3  > proj.sales
4      2012     2013     2014     2015     2016     2017     2018     2019
5     0.000    0.000    0.000    0.000    0.000    0.000    0.000    0.000
6      2020     2021     2022     2023     2024
7  3141.367 2876.526 2611.686 2346.846 2082.006
```

We then create a new dataset *forecast* that stacks the row vector *hist.sales* and *proj.sales*. We divide the sales by 1000 to convert them from millions to billions. We also rename the columns to make the output more comprehensible.

```
1  > forecast <- rbind(hist.sales, proj.sales) / 1000
2  > colnames(forecast) <- seq(2012, 2024, 1)
3  > forecast
4                  2012     2013     2014     2015     2016     2017    2018
5  hist.sales 5.385537 5.075761 4.633345 4.297108 4.163844 3.894558 3.66228
6  proj.sales 0.000000 0.000000 0.000000 0.000000 0.000000 0.000000 0.00000
7                  2019     2020     2021     2022     2023     2024
8  hist.sales 3.552745 0.000000 0.000000 0.000000 0.000000 0.000000
9  proj.sales 0.000000 3.141367 2.876526 2.611686 2.346846 2.082006
```

Next, we use `barplot()` to chart the historical and projected data. The y-axis range in Line 4 gives us extra room at the top of the chart to accommodate adding data labels to the top of the bars. To add the data labels to the top of bars, we first

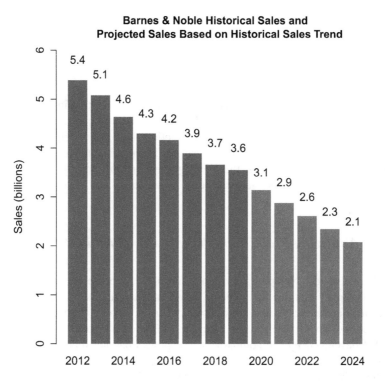

**Fig. 8.3** Barnes & Noble's historical and projected sales. The chart shows historical Barnes & Noble's sales for FY 2012 to FY 2019. The chart also shows projected sales for FY 2020 to 2024 based on the historical sales trend. Source: Barnes & Noble's SEC filings and author's calculations

create a *combo* dataset that combines the historical and projected sales into one row using colSums() in Line 10. Then, also in Line 10, we use to only show one decimal place. This is then used in Line 14. In Line 15, we add to the position in the y-axis an extra 0.4 (i.e., $400 million) to move the data labels higher up on the y-axis. Otherwise, the data labels would be too close to the top of the bars. The extra 0.4 was obtained using trial-and-error. Depending on our preferences as to the location of the data labels, we could use different values. The chart output is in Fig. 8.3.

```
1   > x <- barplot(forecast,
2   +     xlab = "",
3   +     ylab = "Sales (billions)",
4   +     ylim = c(0, 6),
5   +     col = c("blue", "red"),
6   +     border = 0,
7   +     main = "Barnes & Noble Historical Sales and
8   +     Projected Sales Based on Historical Sales Trend")
9   >
10  > (combo <- round(colSums(forecast), 1))
```

```
11   2012 2013 2014 2015 2016 2017 2018 2019 2020 2021 2022 2023 2024
12    5.4  5.1  4.6  4.3  4.2  3.9  3.7  3.6  3.1  2.9  2.6  2.3  2.1
13   >
14   > y <- as.matrix(combo)
15   > text(x, y + 0.4, labels = as.character(y))
```

Developing projections for a company requires not only training by the analyst but also knowledge about the business. Therefore, without support, simply following trends, like we did above, the resulting projections may not be reliable. The above is simply meant to illustrate how to use the historical trend to project sales of a company. Having said that, the above projections are not meant to be actual projections of Barnes & Noble's sales for FY 2020 to 2024.

## 8.2.2  Analyzing Projections Prepared by Third Parties

An alternative to developing our own projections is to use projections prepared by the company or third parties (e.g., advisors, analysts, etc.) to the extent such projections are available and we have access to those projections. For illustrative purposes, we can also use actual projections of Barnes & Noble's sales for FY 2020 to 2024 developed by the company. These projections are reported in Barnes & Noble's SEC Schedule 14D-9 dated July 26, 2019. Whether the projections were developed by the company or another third party, it makes sense for us to analyze such projections to gain a better understanding of the projections.

Although we do not require as much knowledge of the business to understand projections made by third parties compared to if we were going to develop the projections ourselves, we still need to have a decent understanding of the business to make reliable decisions and judgments about various trends and assumptions that were used. For example, we need to understand the sources of value. We would need to have some understanding of the competitive advantage of the firm in order to get a sense of whether there is demand that is generated by the company's products or services. We would need to have some understanding of the industry dynamics. What we need to gain an understanding of the firm's business would be different depending on the company we are analyzing.

However, we can perform some tests to help us ask the right questions. One of the things we can do is to analyze whether the projections are consistent with the historical trends. Before going through more complex analyses, a simple chart of the projections can give us a good sense of what to expect.

**Step 1: Obtain Projections Data**  We can manually enter the projections using c() or we can scrape the data from the web page that contains it. We show how to perform the latter here. The full URL of the Barnes & Noble projections data is https://www.sec.gov/Archives/edgar/data/890491/000119312519203182/ d770574dsc14d9a.htm. Be sure to use that in Line 1 below. Lines 3 to 7 use the same code we used at the beginning of this chapter to scrape the data from Amazon's 10-K.

```
1  > url.proj <- "https://www.sec.gov/Archives/edgar/data/890491/..."
2  >
3  > barnes.proj <- url.proj %>%
4  +    read_html() %>%
5  +    html_table() %>%
6  +    map_df(bind_cols) %>%
7  +    as_tibble()
8  > (barnes.proj <- barnes.proj[9, c(4, 8, 12, 16, 20)])
9  # A tibble: 1 x 5
10    X4         X8         X12        X16        X20
11    <chr>      <chr>      <chr>      <chr>      <chr>
12  1 3,475,850 3,555,666 3,657,419 3,772,115 3,909,922
13  > (barnes.proj <- as.numeric(gsub(",", "", barnes.proj)) / 1000)
14  [1] 3475.850 3555.666 3657.419 3772.115 3909.922
```

**Step 2: Set Up Chart Data** We will create a stacked bar chart below, so we have to create two series of data: historical sales and projected sales. For the historical sales data, we need it to have actual values from 2012 to 2019 and zeroes from 2020 to 2024. For the projected sales data, we need it to have zeroes from 2012 to 2019 and projected values from 2020 to 2024. We implement these in Lines 1 and 3 below. In Line 3, we divide the values in *barnes* by 1000 to convert them from millions to billions for charting purposes.

```
1  > hist <- c(sales$sales, rep(0, 5))
2  > proj <- c(rep(0, 8), barnes.proj)
3  > barnes <- rbind(hist, proj) / 1000
4  > colnames(barnes) <- seq(2012, 2024, 1)
5  > barnes[, 1:6]
6           2012      2013      2014      2015      2016      2017
7  hist 5.385537 5.075761 4.633345 4.297108 4.163844 3.894558
8  proj 0.000000 0.000000 0.000000 0.000000 0.000000 0.000000
9  > barnes[, 7:13]
10          2018     2019     2020     2021     2022     2023     2024
11 hist 3.66228 3.552745 0.00000 0.000000 0.000000 0.000000 0.000000
12 proj 0.00000 0.000000 3.47585 3.555666 3.657419 3.772115 3.909922
```

**Step 3: Plot the Data** Given the setup of the data above, a stacked bar chart, which is the default of `barplot()`, we can simply enter *barnes* in Line 1. This will stack the two data series we created, but each series only has a non-zero value for the appropriate period. This simplifies the code also because R will create two different colors for each series. In Line 5, we color the historical as blue bars and projected as red bars. We then use the same technique we used above to add data labels to the bar chart. The output of this code is Fig. 8.4.

```
1  > x <- barplot(barnes,
2  +    xlab = "",
3  +    ylab = "Sales (billions)",
4  +    ylim = c(0, 6),
```

```
5  +     col = c("blue", "red"),
6  +     border = 0,
7  +     names.arg = seq(2012, 2024, 1),
8  +     main = "Barnes & Noble Historical Sales and
9  +     Projected Sales Based on Company Projections")
10 >
11 > (combo <- round(colSums(barnes), 1))
12 2012 2013 2014 2015 2016 2017 2018 2019 2020 2021 2022 2023 2024
13  5.4  5.1  4.6  4.3  4.2  3.9  3.7  3.6  3.5  3.6  3.7  3.8  3.9
14 >
15 > y <- as.matrix(combo)
16 > text(x, y + 0.4, labels = as.character(y))
```

As Fig. 8.4 shows, the historical revenues have a downward trend, but the company appears to believe that there will be a reversal of that downward trend in FY 2020 to 2024. Although FY 2020 is slightly lower than FY 2019, the gray bars are sloping upward. In fact, the company is projecting that sales would grow 12.5% from $3.5 billion in FY 2020 to $3.9 billion in FY 2024. A break in the historical

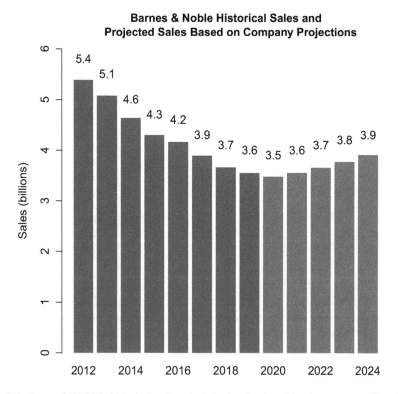

**Fig. 8.4** Barnes & Noble's historical and projected sales developed by the company. The chart shows historical Barnes & Noble's sales for FY 2012 to FY 2019. The chart also shows projected sales for FY 2020 to 2024 developed by the company. Source: Barnes & Noble's SEC filings

trends is not necessarily impossible, but it helps us formulate research tasks, such as whether there is a rationale for such a reversal and the likelihood of such a reversal.

### 8.2.3  Analyzing Growth Rates Embedded in Projections

We can also compare the average growth rates in the projection period with those of historical data. There are two common types of averaging methods: geometric average and arithmetic average. We first discuss the geometric average. The geometric average is also known as a **compound annual growth rate** (CAGR), which converts multi-year growth rates into an annual equivalent. We first create a dataset called *data* that contains the historical and projected sales data for Barnes & Noble.

```
1   > data <- data.frame(cbind(seq(2012, 2024, 1), company))
2   > names(data) <- c("Year", "Sales")
3   > data
4       Year    Sales
5    1  2012 5385.537
6    2  2013 5075.761
7    3  2014 4633.345
8    4  2015 4297.108
9    5  2016 4163.844
10   6  2017 3894.558
11   7  2018 3662.280
12   8  2019 3552.745
13   9  2020 3475.850
14  10  2021 3555.666
15  11  2022 3657.419
16  12  2023 3772.115
17  13  2024 3909.922
```

The projection period is from FY 2020 to FY 2024. Thus, to calculate the growth rate based on the level of the projections, we are looking at the change from the level in FY 2019 of $3.55 billion to the FY 2024 level of $3.91 billion. Mathematically, the calculation of the 5-year CAGR (i.e., geometric average) is as follows:

$$CAGR = \left(\frac{\$3.91}{\$3.55}\right)^{1/5} - 1. \tag{8.1}$$

As we can see, the CAGR over this 5-year period is 1.9%. We can perform a similar calculation over the 7-year historical sales period. Following the same calculation, we get a CAGR for historical sales of −5.8%. This means that, on average, sales fell 5.8% each year over the historical period, but the sales are expected to increase 1.9% each year during the projection period.

```
1   > (proj.cagr <- (data$Sales[13] / data$Sales[8])^(1/5) - 1)
2   [1] 0.01934409
3   > (hist.cagr <- (data$Sales[8] / data$Sales[1])^(1/7) -1)
4   [1] -0.05769669
```

Alternatively, we can also use the arithmetic average. We have to first calculate the annual growth rates by calculating the percentage change from one year to the next. We can calculate this percentage change using Delt(). We start the *growth* dataset in FY 2013 because we used the sales level in FY 2012 to calculate the growth from FY 2012 to FY 2013. Since we do not have FY 2011 data, the Growth would have equaled NA.

```
1  > growth <- Delt(data$Sales)
2  > growth <- growth[-1, ]
3  > growth <- data.frame(cbind(seq(2013, 2024, 1), growth))
4  > names(growth) <- c("Year", "Growth")
5  > growth
6      Year        Growth
7   1  2013 -0.05751998
8   2  2014 -0.08716250
9   3  2015 -0.07256895
10  4  2016 -0.03101249
11  5  2017 -0.06467245
12  6  2018 -0.05964168
13  7  2019 -0.02990896
14  8  2020 -0.02164383
15  9  2021  0.02296302
16 10 2022  0.02861714
17 11 2023  0.03135982
18 12 2024  0.03653309
```

We can then calculate the arithmetic average of the growth rates using mean(). From the above, we know that the annual growth in the historical period is in Rows 1 through 7 (Lines 7–13), and the annual growth in the projection period is in Rows 8 through 12 (Lines 14–18). We find that the arithmetic average growth rate during the historical period is -5.75% and the projection period is 1.96%.

```
1  > (hist.arith <- mean(growth[1:7, 2]))
2  [1] -0.05749815
3  > (proj.arith <- mean(growth[8:12, 2]))
4  [1] 0.01956585
```

Now, we can combine the geometric and arithmetic average growth rates we calculated, so we can easily compare them. As the output below shows, the growth rates calculated using either method yield similar results. A takeaway from this result, for example, could be that the expected turnaround in growth is approximately 8% (i.e., moving from the historical period annual growth of approximately $-5.7\%$ to a projection period annual growth of approximately 1.9%). Although it is possible for such a turnaround to occur, we have to evaluate whether, based on the facts and prospects of the company at the time, is an 8% turnaround in annual growth reasonable.

```
1  > growth.summary <- rbind(cbind(proj.cagr, hist.cagr),
2  +      cbind(proj.arith, hist.arith))
3  > colnames(growth.summary) <- c("Projected", "Historical")
4  > rownames(growth.summary) <- c("CAGR", "Arithmetic")
5  > growth.summary
```

```
6            Projected  Historical
7  CAGR      0.01934409 −0.05769669
8  Arithmetic 0.01956585 −0.05749815
```

In addition, note that having the CAGR and arithmetic average being very similar may not always be the case. Suppose you have sales of $1 million this year, $500,000 next year, and $1 million again the following year. Because you start at $1 million and end at $1 million, the CAGR or geometric average is 0%. However, the arithmetic average will take the simple average of a −50% growth and 100% growth, which yields an arithmetic average of 25%.

```
1  > temp.sales <- c(1, 0.5, 1)
2  > (cagr <- (1 / 1)^(1/2) − 1)
3  [1] 0
4  > temp.growth <- Delt(temp.sales)
5  > (temp.growth <- temp.growth[−1, ])
6  [1] −0.5  1.0
7  > mean(temp.growth)
8  [1] 0.25
```

### 8.2.4  Analyzing Projections Using Regression Analysis

One of the things that we can do to analyze the projections is to compare the projections developed by the company to the projections we developed based on the historical trend of the company's sales. First, we perform the analysis by visually looking at the projections.

**Step 1: Create a Dataset with the Two Projections**  We combine the historical sales and our projected sales in the *ours* dataset. We use `as.numeric()` in the projected sales to convert the data into a vector of numbers without a header. Otherwise, we will be combining `hist.sales`, which has no headers, with `proj.sales` that has headers, and we will see a bunch of blanks until we get to 2020 to 2024.

```
1  > (ours <- c(hist.sales[1:8], as.numeric(proj.sales[9:13])))
2   [1] 5385.537 5075.761 4633.345 4297.108 4163.844 3894.558 3662.280 3552.745
3   [9] 3141.367 2876.526 2611.686 2346.846 2082.006
4  > (company <- c(hist[1:8], proj[9:13]))
5   [1] 5385.537 5075.761 4633.345 4297.108 4163.844 3894.558 3662.280 3552.745
6   [9] 3475.850 3555.666 3657.419 3772.115 3909.922
7  > plot.data <- data.frame(cbind(ours, company))
8  > plot.data$year <- seq(2012, 2024, 1)
9  > rownames(plot.data) <- seq(1, 13, 1)
10 > plot.data
11        ours  company year
12 1  5385.537 5385.537 2012
13 2  5075.761 5075.761 2013
14 3  4633.345 4633.345 2014
15 4  4297.108 4297.108 2015
```

```
16   5   4163.844 4163.844 2016
17   6   3894.558 3894.558 2017
18   7   3662.280 3662.280 2018
19   8   3552.745 3552.745 2019
20   9   3141.367 3475.850 2020
21  10   2876.526 3555.666 2021
22  11   2611.686 3657.419 2022
23  12   2346.846 3772.115 2023
24  13   2082.006 3909.922 2024
```

**Step 2: Plot the Two Projections** What chart you use to plot the data and how you plot the data are a matter of preference. In this example, we use a line chart. In Fig. 8.5, the red line up to 2019 is actual historical sales data. The data from 2020 onward are projections and that is where we see the two projections diverge. The projections based on historical trends (i.e., the blue line) are downward sloping denoting that, if the historical trend is expected to continue, Barnes & Noble's sales from FY 2020 to 2024 will continue to decline. By contrast, the red line, which is the sales projections developed by the company, tells a different story. It shows Barnes & Noble's sales recovering during the projection period. Thus, the gap between the two projections widens as we go further out during the projection period. Determining why such a gap exists and why such a gap grows is critical to understanding whether the projections developed by the company are reliable.

```
 1   > plot(x = plot.data$year,
 2   +       xlab = "Years",
 3   +       y = plot.data$ours,
 4   +       ylab = "Sales (millions)",
 5   +       type = "l",
 6   +       lwd = 2,
 7   +       col = "blue",
 8   +       main = "Barnes & Noble Sales Projections
 9   +       Based on Historical Trend vs. Company Projections")
10   > lines(x = plot.data$year,
11   +       y = plot.data$company,
12   +       lwd = 2,
13   +       col = "red")
14   > legend("topright",
15   +       c("Historical Trend", "Company Projections"),
16   +       col = c("blue", "red"),
17   +       lwd = c(2, 2))
```

We can glean a decent amount of information by visually looking at the data. However, there are some additional ideas that may develop by using more sophisticated evaluation methods. A common technique that we can use is regression analysis. The point of regression analysis is that we separate out the historical period and the projection period and, after accounting for the trend in the data, determine whether the trend has changed or whether the sales level has shifted between the historical period and the projection period. Rather than giving an absolute answer as to which projection is right and which is wrong, the result of this analysis would

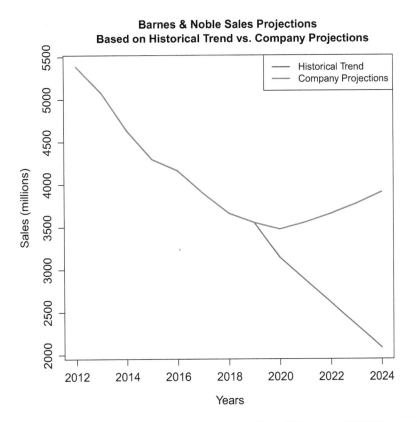

**Fig. 8.5** Comparing Barnes & Noble's sales projections. The solid line from FY 2012 to 2019 is historical sales data. The solid line from FY 2020 to 2024 is sales projections based on the trend in the company's historical sales. The dashed line from FY 2020 to 2024 is sales projections developed by the company. Source: Barnes & Noble's SEC filings and author's calculations

help us determine what questions to ask in order to help us get comfortable with the projections.

**Step 1: Set Up the Revenue and Trend Vectors** The *rev* vector is comprised of 13 elements with the first element equal to 1, the second element equal to 2, and so on.

```
1  > (rev <- c(hist[1:8], proj[9:13]))
2   [1] 5385.537 5075.761 4633.345 4297.108 4163.844 3894.558 3662.280 3552.745
3   [9] 3475.850 3555.666 3657.419 3772.115 3909.922
4  > (trend <- seq(1, 13, 1))
5   [1]  1  2  3  4  5  6  7  8  9 10 11 12 13
```

**Step 2: Create the Trend Change Vector** The *trend.chg* vector takes on the value of zero for the historical data and is an increasing sequence of numbers from 1 through 5 during the 5-year projection period.

```
1   > (trend.chg <- c(rep(0, 8), seq(1, 5, 1)))
2     [1] 0 0 0 0 0 0 0 0 1 2 3 4 5
```

**Step 3: Run a Regression to Determine Change in Trend**  We use lm() to run
an OLS regression of *rev* on *trend* and *trend.chg*. As we expected, the regression
results show that the trend is negative and significant. The output also shows that
the trend increased by a statistically significant amount. In other words, the trend
changes as we moved from historical sales to projected sales.

```
1   > reg1 <- lm(rev ~ trend + trend.chg)
2   > summary(reg1)
3
4   Call:
5   lm(formula = rev ~ trend + trend.chg)
6
7   Residuals:
8      Min      1Q  Median       3Q      Max
9   -165.31  -30.58  -11.70   33.95   162.31
10
11  Coefficients:
12               Estimate Std. Error t value Pr(>|t|)
13  (Intercept)   5534.39      70.57   78.42 2.78e-15 ***
14  trend         -267.99      13.01  -20.60 1.61e-09 ***
15  trend.chg      365.10      28.59   12.77 1.62e-07 ***
16  ---
17  Signif. codes:  0 '***' 0.001 '**' 0.01 '*' 0.05 '.' 0.1 ' ' 1
18
19  Residual standard error: 93.81 on 10 degrees of freedom
20  Multiple R-squared:  0.9804, Adjusted R-squared:  0.9765
21  F-statistic: 249.8 on 2 and 10 DF,  p-value: 2.907e-09
```

**Step 4: Create the Shift in Trend Vector**  The *shift* vector is comprised of 13
elements. The first eight elements representing the historical period are zero. The
last five elements representing the projection period have a value of 1. To create
these repeating values in the vector, we use  rep() with the first argument being
the value we want to repeat and the second argument being the number of times we
want the value to repeat.

```
1   > (shift <- c(rep(0, 8), rep(1, 5)))
2     [1] 0 0 0 0 0 0 0 0 1 1 1 1 1
```

**Step 5: Run a Regression to Determine Shift in Trend**  We use lm() to run an
OLS regression of *rev* on *trend* and  *shift*. As we expected, the regression results
show that the trend is negative and significant. The output also shows that the trend
did not shift (i.e., the level did not go up or down) by a statistically significant
amount. In other words, the projected level of sales did not differ from the level of
sales projected after accounting for the trend in the data.

```
1   > reg2 <- lm(rev ~ trend + shift)
2   > summary(reg2)
3
```

```
4  Call:
5  lm(formula = rev ~ trend + shift)
6
7  Residuals:
8       Min     1Q Median     3Q    Max
9   -584.45 -149.01 -72.78 259.98 621.83
10
11 Coefficients:
12              Estimate Std. Error t value Pr(>|t|)
13 (Intercept)  5201.88    249.63  20.838 1.44e-09 ***
14 trend        -193.05     48.27  -4.000  0.00252 **
15 shift         595.88    371.21   1.605  0.13952
16 ---
17 Signif. codes:  0 '***' 0.001 '**' 0.01 '*' 0.05 '.' 0.1 ' ' 1
18
19 Residual standard error: 348 on 10 degrees of freedom
20 Multiple R-squared:  0.7299, Adjusted R-squared:  0.6759
21 F-statistic: 13.51 on 2 and 10 DF,  p-value: 0.001437
```

The above analysis helps us determine what questions we need to ask to get comfortable with the projections by the company. For example, we would want to know what information supports the reversal in the declining sales trend.

Although we presented methodologies that could be used to test projections, the appropriate tests depend on the projections, availability of data, and economic factors that persist at the time the projections were made. Since the value of the firm depends significantly on the projections we use, it is important to get comfortable with the projections.

## 8.3 Equity Risk Premium

When one uses the capital asset pricing model (CAPM) to calculate the cost of equity, one of the required components is the **equity risk premium** (ERP). Although the ERP can be calculated in several ways, the most common way ERPs are presented in academic texts is by using historical data. In this section, we show how to use R to obtain ERP data from Professor Damodaran's website and calculate the ERP using the arithmetic mean and geometric mean. To do this calculation, we again use `tidyverse` and `rvest`, which we used at the start of this chapter.

**Step 1: Obtain ERP Data** We obtain the ERP data from Professor Damodaran's website, which contains a large table with several columns of data from 1928 to 2019. The full URL is http://pages.stern.nyu.edu/~adamodar/New_Home_Page/datafile/histretSP.html, which we should use in Line 1. Lines 4 to 8 read in the data on the web page.

```
 1  > url <- "http://pages.stern.nyu.edu/~adamodar/New_Home_Page/ . . .
 2  >
 3  > temp <-
 4  +    url %>%
 5  +    read_html() %>%
 6  +    html_table() %>%
 7  +    map_df(bind_cols) %>%
 8  +    as_tibble()
 9  > head.tail(temp)
10  # A tibble: 3 x 18
11    X1     X2    X3     X4     X5     X6     X7    X8    X9    X10    X11
12    <chr>  <chr> <chr>  <chr>  <chr>  <chr>  <chr> <chr> <chr> <chr>  <chr>
13  1 ""        "Annu~ "Annu~ "Annu~ "Annu~ Value~ Valu~ Valu~ Valu~ Annu~ Annu~
14  2 Year   S&P 5~ 3-mon~ US T.~ "Baa\~ S&P 5~ 3-mo~ US T~ "Baa~ "Sto~ "Sto~
15  3 1928   43.81% 3.08%  0.84%  3.22%  "$   ~ "$  ~ "$  ~ "$  ~ 40.7~ 42.9~
16  # ... with 7 more variables: X12 <chr>, X13 <chr>, X14 <chr>,
17  #    X15 <chr>, X16 <chr>, X17 <chr>, X18 <chr>
18  # A tibble: 3 x 18
19    X1     X2    X3     X4     X5     X6     X7    X8    X9    X10    X11
20    <chr>  <chr> <chr>  <chr>  <chr>  <chr>  <chr> <chr> <chr> <chr>  <chr>
21  1 2018   -4.23% 1.94% -0.0~  -2.7~  "$   ~ "$  ~ "$  ~ "$  ~ -6.1~  -4.2~
22  2 2019   31.22% 1.55%  9.64% 15.3~  "$   ~ "$  ~ "$  ~ "$  ~ 29.6~  21.5~
23  3 ""        ""     ""     ""     ""     ""    ""    ""    ""     ""
24  # ... with 7 more variables: X12 <chr>, X13 <chr>, X14 <chr>,
25  #    X15 <chr>, X16 <chr>, X17 <chr>, X18 <chr>
```

**Step 2: Manually Identify the Relevant Rows and Columns** Instead of using code to identify the relevant columns, we can simply look at the data and identify the rows and columns that we want to keep or delete. We can see from the output above, we can do with the first two rows. By viewing the dataset or going to the web page, we can see that we are interested in columns 1 (year), 10 (ERP based on Treasury Bills), and 11 (ERP based on Treasury Bonds).

```
 1  > erp <- data.frame(temp[-c(1,2), c(1, 10, 11)])
 2  > head.tail(erp)
 3       X1     X10     X11
 4  1 1928   40.73%  42.98%
 5  2 1929  -11.46% -12.50%
 6  3 1930  -29.67% -29.66%
 7       X1     X10     X11
 8  91 2018  -6.17%  -4.21%
 9  92 2019  29.67%  21.59%
10  93
```

**Step 3: Clean-up Dataset** As the output above shows, we may want to do some cleaning up of the data before we can start working on this. In Line 1, we rename the column headers. Note that we rename the ERP based on Treasury Bills, as "ST" for short-term. Accordingly, we rename the ERP based on Treasury Bonds, as "LT" for long-term. Treasury Bills are instruments that have maturities of 1 year or less, and Treasury Bonds are instruments that have maturities of 10 years or more. Line 2 converts the values in the Year column to numeric, while Lines 3 and 4 convert the

ERP values into numeric while simultaneously deleting the percentage sign using gsub(). Finally, Line 5 deletes the extraneous row 93, which as we see from the output above does not have any data.

```
1   > names(erp) <- c("Year", "ERP.ST", "ERP.LT")
2   > erp$Year <- as.numeric(erp$Year)
3   > erp$ERP.ST <- as.numeric(gsub("%", "", erp$ERP.ST))/100
4   > erp$ERP.LT <- as.numeric(gsub("%", "", erp$ERP.LT))/100
5   > erp <- erp[-93, ]
6   > head.tail(erp)
7      Year  ERP.ST  ERP.LT
8   1 1928   0.4073  0.4298
9   2 1929 -0.1146 -0.1250
10  3 1930 -0.2967 -0.2966
11     Year  ERP.ST  ERP.LT
12  90 2017  0.2067  0.1880
13  91 2018 -0.0617 -0.0421
14  92 2019  0.2967  0.2159
```

**Step 4: Calculate ERP Based on Arithmetic Mean**  With the above data, we can calculate the ERP based on the arithmetic mean from 1928 to 2019. To do this, we apply colMeans() to columns 2 and 3. This yields an ERP of 8.2% based on T-Bills and 6.4% based on T-Bonds.

```
1   > (arith.erp <- colMeans(erp[, 2:3]))
2       ERP.ST      ERP.LT
3   0.08175109 0.06425652
```

**Step 5: Calculate ERP Based on Geometric Mean**  The geometric mean calculation is a little more involved. We take this in several steps. First, in Lines 1 to 2, we calculate the gross annual excess market returns. We then use these gross annual excess market returns in Line 13, where we cumulate the product of these gross annual excess market returns.

```
1   > gross.erp <- erp
2   > gross.erp[, 2:3] <- 1 + gross.erp[, 2:3]
3   > head.tail(gross.erp)
4      Year ERP.ST ERP.LT
5   1 1928 1.4073 1.4298
6   2 1929 0.8854 0.8750
7   3 1930 0.7033 0.7034
8      Year ERP.ST ERP.LT
9   90 2017 1.2067 1.1880
10  91 2018 0.9383 0.9579
11  92 2019 1.2967 1.2159
12  >
13  > cum.erp <- apply(gross.erp[, 2:3], 2, cumprod)
14  > head.tail(cum.erp)
15      ERP.ST     ERP.LT
16  1 1.4073000 1.4298000
17  2 1.2460234 1.2510750
18  3 0.8763283 0.8800062
19      ERP.ST     ERP.LT
```

```
20   90 208.0378 35.53110
21   91 195.2019 34.03524
22   92 253.1183 41.38345
```

Next, we calculate the number of years of returns we are calculating. In this case, we take the difference between 1928 and 2019, but we add 1. The reason is that the 1928 figure is a return, which means that the starting value of the return calculation is 1927. Then, in Lines 4 and 6, we calculate the geometric ERP based on Treasury Bills and Bonds, respectively. The output shows that the ERP based on T-Bills is 6.2% and the ERP based on T-Bonds is 4.1%.

```
1  > (num.years <- gross.erp[nrow(gross.erp), 1] - gross.erp[1, 1] + 1)
2  [1] 92
3  >
4  > (geo.erp.st <- cum.erp[92, 1]^(1 / num.years) - 1)
5  [1] 0.06199649
6  > (geo.erp.lt <- cum.erp[92, 2]^(1 / num.years) - 1)
7  [1] 0.04129601
8  >
9  > (geo.erp <- cbind(geo.erp.st, geo.erp.lt))
10      geo.erp.st geo.erp.lt
11  [1,] 0.06199649 0.04129601
```

Finally, we create a summary table, so we can easily compare the ERPs we calculated. As we can see, the geometric ERP is lower than the arithmetic ERP. Note also that the long-term ERP is smaller than the short-term ERP. Thus, we have to remember to be consistent with what ERP we are applying and what risk-free rate we are using. In particular, if we are using a T-Bill rate in the CAPM, then we have to use the ERP based on T-Bills. If we are using a T-Bond rate in the CAPM, then we have to use the ERP based on T-Bonds.

```
1  > (erp.summary <- rbind(arith.erp, geo.erp))
2                ERP.ST      ERP.LT
3  arith.erp 0.08175109 0.06425652
4            0.06199649 0.04129601
5  > rownames(erp.summary) <- c("Arithmetic", "Geometric")
6  > erp.summary
7                ERP.ST      ERP.LT
8  Arithmetic 0.08175109 0.06425652
9  Geometric  0.06199649 0.04129601
```

## 8.4   Unlevering Betas

Another critical component for the discount rate calculated using the CAPM is the use of the appropriate equity beta. In some instances, the beta of the subject firm is estimated using the betas of peer companies. These peer companies will likely have different leverage than the subject firm, so we have to adjust the beta for the differential in leverage. A common formula that is used under the assumption that the debt beta is equal to zero is the following formula developed by Bob Hamada:

$$\beta_U = \frac{\beta_L}{1 + (1 - t)D/E}, \tag{8.2}$$

where $\beta_U$ is the unlevered beta, $\beta_L$ is the levered beta, $t$ is the tax rate, and $D/E$ is the debt-to-equity ratio of the firm.

Let us look at an example to see how to use the above formula. Suppose the firm has a regression-based beta of 1.2, the average debt-to-equity ratio during the regression period is 40%, and the tax rate during the regression period is 20%. These inputs result in an unlevered beta of 0.91.

```
1  > reg.beta <- 1.2
2  > avg.de <- 0.40
3  > tax <- 0.20
4  > target.de <- 0.50
5  >
6  > (beta.unl <- reg.beta / (1 + (1 - tax) * avg.de))
7  [1] 0.9090909
```

Now, assuming a target debt-to-equity ratio of 50%, we find that the levered beta is equal to 1.27. This makes sense because, all else equal, more leverage implies more risk.

```
1  > (beta.lev <- beta.unl * (1 + (1 - tax) * target.de))
2  [1] 1.272727
```

For betas estimated using historical data and regression analysis, I argue that we should use the average regression period leverage ratio to unlever betas. I discuss this issue in more detail in [4]. The reason is that the beta estimate from a regression is based on the average returns over the period, which is then based on the average leverage during the period. Therefore, using the regression beta as is implicitly assumes that the average regression period leverage ratio is the same as your target leverage ratio. An alternative way of thinking about this is, by using the regression beta as is, we are not assuming that the current leverage ratio is the target leverage ratio.

This problem extends to when we use betas from peer companies. A typical analysis we observe when using betas of peer companies goes as follows. Each peer company's beta is estimated using a regression over some historical period. Each peer's regression beta is then unlevered using that peer's current leverage ratio. The mean or median unlevered beta of the peer companies is then relevered using a target leverage ratio for the subject firm to arrive at the subject firm's levered beta. However, the problem is that the use of the current leverage ratio does not properly unlever each peer company's regression beta. Instead, we should use each peer's average regression period leverage ratio to unlever each peer company's regression beta.

The above formula assumes that debt betas are zero. If the debt beta is not zero, a formula that incorporates debt betas would need to be used. Because the beta of a portfolio is equal to the weighted average of the betas of the components of the portfolio, then the unlevered (asset) beta is equal to the weighted average of the debt beta and the equity beta. That is,

$$\beta_{unlevered} = \beta_{debt}\frac{D}{D+E} + \beta_{equity}\frac{E}{D+E}. \qquad (8.3)$$

This is the formula used in the popular corporate finance textbook by Brealey, Myers, and Allen. I discuss in the further reading section several other formulas that incorporate debt betas.

## 8.5   Sensitivity Analysis

When performing valuations, we make judgments about inputs and assumptions. Some of those inputs are more critical drivers of value than others. Although we may have a basis when we came up with those assumptions, we could have recognized that there is a reasonable range of outcomes that those inputs or assumptions could also fall. Thus, we may want to see how the value changes if we are off in our judgment. The technique used to perform such tests is called a **sensitivity analysis**.

Here, we use as an example a simplified version of a valuation exercise in which we value a $100 annuity $c$ that grows into perpetuity at a rate $g$ and being discounted at a rate $k$. Mathematically, we define the value $V$

$$V = \frac{c(1+g)}{k-g}. \qquad (8.4)$$

Now, assume that we initially assume a perpetuity growth rate of 4% and cost of equity of 10%. Under this scenario, we have a value of $1733 [= $100 * (1 + 4%) / (10% - 4%)].

```
1   > c <- 100
2   > g = 0.04
3   > k = 0.10
4   > (V = c * (1 + g) / (k - g))
5   [1] 1733.333
```

Now, suppose we want to test the sensitivity of our valuation if $g$ is either 3% or 5% and the discount rate is either 8% or 12%. Below, we show how to go about performing this sensitivity.

**Step 1: Set Up Range for k and g**  We are going to develop a sensitivity table that is a 3 × 3 matrix. This means that there will be nine values in total, which are combinations of the various discount rate and growth rate values we assume.

```
1   > (k.range <- seq(0.08, 0.12, 0.02))
2   [1] 0.08 0.10 0.12
3   > (g.range <- seq(0.03, 0.05, 0.01))
4   [1] 0.03 0.04 0.05
```

**Step 2: Construct the Value Function**  In this example, we are calculating the present value of a growing annuity. Thus, the value function is simple. However, we can easily extend this analysis regardless of whether we have a simple or complex value function as long as the drivers of the value function that we allow to change are the discount rate and growth rate.

```
1  > val.fx <- function(c, k, g) {
2  +      value <- c * (1 + g) / (k - g)
3  + }
```

**Step 3: Create a Temporary Matrix** We need to create a temporary $3 \times 3$ matrix for the nine values we are going to calculate. We do this by using `matrix()`. The first argument of 999 is somewhat arbitrary, but it helps make sure that the program runs properly because the number is quite odd that it is unlikely to be a correctly calculated sensitivity.

```
1  > (val.table <- matrix(999, ncol = 3, nrow = 3))
2       [,1] [,2] [,3]
3  [1,]  999  999  999
4  [2,]  999  999  999
5  [3,]  999  999  999
```

**Step 4: Calculate the Values for the Sensitivities** We use two for-loops to calculate the value in the sensitivity. The first loop goes through each of the three discount rate values, and the second loop goes through each of the three perpetuity growth rate values. At each of the nine runs, we replace the values in the $3 \times 3$ matrix *val.table* with the value calculated by the value function we specified above. We then clean up the table in Lines 22 and 23 so that the column and row headers make sense.

```
1  > for (i in 1:3) {
2  +      k = k.range[i]
3  +      for (j in 1:3){
4  +          g = g.range[j]
5  +          val.table[i, j] = val.fx(c, k, g)
6  +      }}
7  > val.table
8            [,1]      [,2] [,3]
9  [1,] 2060.000 2600.000 3500
10 [2,] 1471.429 1733.333 2100
11 [3,] 1144.444 1300.000 1500
12 >
13 > colnames(val.table) <- c("3%", "4%", "5%")
14 > rownames(val.table) <- c("8%", "10%", "12%")
15 > val.table
16           3%       4%   5%
17 8%  2060.000 2600.000 3500
18 10% 1471.429 1733.333 2100
19 12% 1144.444 1300.000 1500
```

As we can see from the output, the central value of $k = 10\%$ and $g = 4\%$ is \$1733, which is what we calculated above as our base scenario. Then, we can see how the value changes as we change either the discount rate or the perpetuity growth rate or both from the base values of $k = 10\%$ and $g = 4\%$. The output above shows that the value could range from \$1144 to \$3500.

Note that the above shows the mechanics of how to calculate a sensitivity table in R. The above assumes that the discount rate and the perpetuity growth rate are independent of one another. In reality, those two inputs are not completely independent. The perpetuity growth rate is a function of the percentage of your earnings during the terminal period you plowback (i.e., reinvestment rate) and the return on new invested capital (i.e., return you expect to earn on the amount reinvested). However, in competitive markets, the return on new invested capital should converge to the discount rate over the long-term because the competition will eat away at any excess profits. Therefore, there is a link between the discount rate and the perpetuity growth rate and those have implications for the terminal value.

## 8.6   Relative Valuation Using Regression Analysis

Relative valuation is a valuation method that estimates the value of the subject firm based on the value of comparable firms. The analysis typically starts by identifying comparable firms. These are firms that are affected by the same economic forces as the subject firm. As such, we often start with firms in the same industry, but we typically have to make adjustments to ensure that we have reasonably comparable firms. For example, a lot of firms competing against the subject firm may not be comparable based on factors such as size, profitability, and risk.

Once the comparable firms have been identified, we would then select a multiple that best reflects these characteristics. For example, a common multiple used is the Price-to-Earnings or P-E ratio. The P-E Ratio is equal to the market price divided by the earnings per share. We can calculate a P-E Ratio for each comparable firm and decide what percentile of the range of comparable company's multiples the subject firm belongs to. In theory, if we selected the comparable firms appropriately, the subject firm's multiple would be approximately equal to the median or average comparable company's multiple, as the cross-sectional averaging would eliminate firm-specific factors that could affect the individual peer multiples.

We then use that selected P-E Ratio and apply the subject firm's earnings, to arrive at a value for the subject firm. This approach is called **relative valuation** because the value derived from this approach is "relative" to the valuation of the comparable companies. If those firms are overvalued, then the resulting multiples and subject firm's valuation are also overvalued. Conversely, if those firms are undervalued, then the resulting multiples and subject firm's valuation are also undervalued.

The selection of the percentile within the comparable companies' multiples range typically requires a mix of qualitative and quantitative analyses. The quantitative component analyzes various metrics, such as those that could be proxy for size, profitability, leverage, growth, etc. Then, based on some assessment, we could make a determination as to whether the subject firm is more like the median or average comparable company or it is more like the 25th percentile or 75th percentile comparable company or some other percentile.

An alternative to the above is to use regression analysis when selecting the multiple based on various known determinants of that multiple. In our example below, we regress the P-E Ratio on the earnings per share (EPS) growth rate, dividend payout ratio, and beta.

**Step 1: Obtain Data for Comparable Companies** We assume that Amazon has the following Internet peer group: Apple (AAPL), Alibaba (BABA), Baidu (BIDU), Booking.com (BKNG), ebay (EBAY), Expedia (EXPE), Facebook (FB), Groupon (GRPN), Alphabet (GOOG), Netflix (NFLX), Tripadvisor (TRIP), and Twitter (TWTR). We go to Yahoo Finance and obtain data on these firms' P-E Ratio, long-term EPS growth rate, dividend payout ratio, and beta. You would have to manually enter the data below because we cannot get these historical values if we look at Yahoo Finance today.

```
 1  > aapl <- c(19.43, 0.0986, 0.2523, 1.23)
 2  > baba <- c(24.59, .0366, 0, 2.25)
 3  > bidu <- c(18.02, -0.0123, 0, 1.77)
 4  > bkng <- c(18.09, 0.12, 0, 1.1)
 5  > ebay <- c(12.50, 0.1284, 0.19, 1.26)
 6  > expe <- c(15.63, 0.0880, 0.3881, 0.99)
 7  > fb <- c(22.71, 0.1146, 0, 1.06)
 8  > grpn <- c(10.22, 0.0948, 0, 1.35)
 9  > goog <- c(25.07, 0.1291, 0, 1.02)
10  > nflx <- c(60.70, 0.4168, 0, 1.30)
11  > trip <- c(16.14, 0.1102, 0, 1.30)
12  > twtr <- c(36.26, 0.1390, 0, 0.56)
13  >
14  > reg.data <- rbind(aapl, baba, bidu, bkng, ebay, expe, fb,
15  +      grpn, goog, nflx, trip, twtr)
16  > reg.data <- data.frame(reg.data)
17  > names(reg.data) <- c("P.E", "Growth", "Div.Pay", "Beta")
18  > reg.data
19           P.E  Growth Div.Pay Beta
20  aapl 19.43  0.0986  0.2523 1.23
21  baba 24.59  0.0366  0.0000 2.25
22  bidu 18.02 -0.0123  0.0000 1.77
23  bkng 18.09  0.1200  0.0000 1.10
24  ebay 12.50  0.1284  0.1900 1.26
25  expe 15.63  0.0880  0.3881 0.99
26  fb   22.71  0.1146  0.0000 1.06
27  grpn 10.22  0.0948  0.0000 1.35
28  goog 25.07  0.1291  0.0000 1.02
29  nflx 60.70  0.4168  0.0000 1.30
30  trip 16.14  0.1102  0.0000 1.30
31  twtr 36.26  0.1390  0.0000 0.56
```

**Step 2: Run Regression on Peer Data** We run a regression of the P-E Ratio on the EPS growth rate, dividend payout ratio, and beta. We run the regression using lm()

and make sure that we save the results in *reg.results*, as we will need that dataset in
the next step to construct the prediction.

```
1   > reg.result <- lm(P.E ~ Growth + Div.Pay + Beta,
2   +     data = reg.data)
3   > summary(reg.result)
4
5   Call:
6   lm(formula = P.E ~ Growth + Div.Pay + Beta, data = reg.data)
7
8   Residuals:
9       Min     1Q  Median     3Q     Max
10  -11.634  -6.126   1.392   4.170  12.043
11
12  Coefficients:
13              Estimate Std. Error t value Pr(>|t|)
14  (Intercept)    7.091    10.560   0.671  0.52084
15  Growth       110.382    26.428   4.177  0.00309 **
16  Div.Pay      -18.883    19.794  -0.954  0.36803
17  Beta           3.185     6.519   0.489  0.63830
18  ---
19  Signif. codes:  0 '***' 0.001 '**' 0.01 '*' 0.05 '.' 0.1 ' ' 1
20
21  Residual standard error: 8.374 on 8 degrees of freedom
22  Multiple R-squared:  0.7249, Adjusted R-squared:  0.6218
23  F-statistic: 7.028 on 3 and 8 DF,  p-value: 0.01244
```

**Step 3: Predict Amazon P-E Ratio Using Regression Results**  We obtain Amazon's P-E Ratio (69.19x), long-term EPS growth rate (28%), dividend payout ratio (0%), and beta (1.52) from Yahoo Finance. We then use `predict()` to predict what Amazon's P-E Ratio would be based on the regression results stored in *reg.result*. Based on the regression results of the peers and Amazon's long-term EPS growth rate, dividend payout ratio, and beta, Amazon's P-E Ratio is predicted to be 42.84x, which is 38% lower than Amazon's actual 69.19x P-E Ratio . This result suggests that Amazon stock could be overvalued based on an analysis of our selected peers.

```
1   > amzn <- c(69.19, 0.28, 0, 1.52)
2   > predicted <- data.frame(t(amzn))
3   > names(predicted) <- c("P.E", "Growth", "Div.Pay", "Beta")
4   > predicted
5       P.E Growth Div.Pay Beta
6   1 69.19   0.28       0 1.52
7   >
8   > rhat <- predict(reg.result, predicted)
9   > rhat
10         1
11  42.8383
```

## 8.7   Identifying Significant Shifts in Stock Returns

We sometimes want to determine whether a particular stock's characteristics have changed between certain periods. For example, we may want to know whether the firm's returns are higher or lower by a significant amount between two periods. We may also want to know if the relationship between the firm's stock return and certain factors has changed after an acquisition or divestiture. Below, we discuss several statistical tools that could help us make that determination.

### 8.7.1   t-Test: Testing Difference in Average Returns

A **t-test** could be used to test the difference in average returns between two periods. Here, we use t.test() on AMZN stock returns.

**Step 1: Calculate Returns** We first import AMZN stock price data using load.data(). Then, we calculate log returns by first taking the log of the AMZN adjusted close price using log() and then take the difference of those log prices using diff(). We then remove the first observation (December 31, 2014), which we only use to calculate the return on the first trading day of 2015.

```
1  > data.amzn <- load.data("AMZN Yahoo.csv", "AMZN")
2  > rets <- diff(log(data.amzn$AMZN.Adjusted))
3  > rets <- rets[-1, ]
4  > names(rets) <- "AMZN"
5  > head.tail(rets)
6                    AMZN
7  2015-01-02 -0.005914077
8  2015-01-05 -0.020730670
9  2015-01-06 -0.023098010
10                   AMZN
11 2019-12-27  0.0005510283
12 2019-12-30 -0.0123283480
13 2019-12-31  0.0005142195
```

**Step 2: Separate the Data into Two Periods** Since we are testing the difference in average returns between two periods, we have to identify those two periods. For our exercise, we compare the average return from 2015 to 2017 to the average return from 2018 to 2019. We can create two numeric vectors by using xts-style subsetting using dates and as.numeric(). Note that in the xts-style subsetting, we can ignore the date to the left of the slash if we want R to keep all the dates on or prior to the date we define after the slash. Similarly, if we only have a date to the right of the slash but no date to the right of the slash, then R keeps all the dates on or before the date defined to the left of the slash.

```
1  > rets.A <- as.numeric(rets["/2017-12-31"])
2  > head.tail(rets.A)
3  [1] -0.005914077 -0.020730670 -0.023098010
```

```
4  [1]  0.004662962  0.003242724 −0.014119964
5  >
6  > rets.B <− as.numeric(rets["2018−01−01/"])
7  > head.tail(rets.B)
8  [1] 0.016570407 0.012694369 0.004466026
9  [1]  0.0005510283 −0.0123283480  0.0005142195
```

**Step 3: Run t-Test** We run t.test() on *rets.A* and *rets.B*. We test whether the difference between the two returns mu is equal to zero. The alternative hypothesis is thus two sided (i.e., that the average return could be positive or negative). We also assume that the variances in the sample are not equal and that this is not a paired t-test because we assume the two samples are independent. Finally, we use a 95% confidence level for our test.

```
1  > (diff.mean <− t.test(rets.A, rets.B,
2  +    mu = 0,
3  +    alternative = "two.sided",
4  +    var.equal = FALSE,
5  +    paired = FALSE,
6  +    conf.level = 0.95))
7
8    Welch Two Sample t−test
9
10  data:  rets.A and rets.B
11  t = 0.79525, df = 1021.6, p−value = 0.4267
12  alternative hypothesis: true difference in means is not equal to 0
13  95 percent confidence interval:
14   −0.001243893  0.002939126
15  sample estimates:
16     mean of x    mean of y
17  0.0017570930 0.0009094766
```

As the output above shows, the p-value is 43%, which means that we cannot reject the null that there is no difference in the average returns for AMZN in both periods.

### 8.7.2  Identifying Breakpoints

Time series of returns data may exhibit breaks because of structural changes in the company. Thus, it could be helpful for us to be able to determine whether there are breakpoints in the data that we are using. From a practical perspective, the presence of a breakpoint could mean that we cannot use a pooled regression (i.e., one regression over the entire period), but it would be better to capture the relationship in the data if we split the data into separate periods given by the breakpoints. Sometimes, we may be able to identify the breakpoints by eyeballing the data. However, this is not usually the case, and we may have to resort to methods that allow us to identify the breakpoints using some quantitative criteria. In this section, we discuss two such methods: using **F-statistics** and using the **Bayesian Information Criterion** (BIC).

**Step 1: Combine Returns Data**  We will identify the breakpoint in the relationship between AMZN and SPY, so we have to first calculate the log returns of both securities. Since we already calculated the log returns for AMZN and stored them in *rets*, we only need to calculate the log returns for SPY. To do so, we start by importing the SPY data using load.data(). Then, we calculate the log returns by first taking the log of the SPY adjusted close price using log() and then taking the difference of those log returns using diff(). We then remove the first observation (December 31, 2014), which we only use to calculate the return on the first trading day of 2015.

```
1  > data.spy <- load.data("SPY Yahoo.csv", "SPY")
2  > MKT <- diff(log(data.spy$SPY.Adjusted))
3  > MKT <- MKT[-1, ]
4  > names(MKT) <- "MKT"
5  > head.tail(MKT)
6                     MKT
7  2015-01-02 -0.0005354247
8  2015-01-05 -0.0182246098
9  2015-01-06 -0.0094636586
10                    MKT
11 2019-12-27 -0.0002478074
12 2019-12-30 -0.0055284735
13 2019-12-31  0.0024263490
14 >
15 > returns <- cbind(rets, MKT)
16 > head.tail(returns)
17                 AMZN          MKT
18 2015-01-02 -0.005914077 -0.0005354247
19 2015-01-05 -0.020730670 -0.0182246098
20 2015-01-06 -0.023098010 -0.0094636586
21                 AMZN          MKT
22 2019-12-27  0.0005510283 -0.0002478074
23 2019-12-30 -0.0123283480 -0.0055284735
24 2019-12-31  0.0005142195  0.0024263490
```

**Step 2: Run Regression of AMZN on SPY**  We establish the relationship over the entire period by using lm() to run a linear regression model of AMZN returns on SPY returns.

```
1  > returns.df <- data.frame(returns)
2  > reg.amzn <- lm(AMZN ~ MKT, data = returns.df)
```

**Step 3: Identify Breakpoint Using F-Statistic**  To identify the breakpoints using the F-Statistic, we use Fstats() and breakpoints() in the strucchange package. The output shows that the breakpoint is at observation number 811, which is March 22, 2018. Graphically, we can see in Fig. 8.6 where the breakpoint is. The plot uses relative time, which makes it difficult to figure out what the dates are, but we know by manually identifying observation number 811 which date the breakpoint is.

# Choosing Breakpoint Using F Statistic

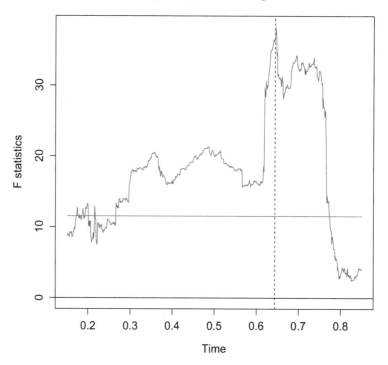

**Fig. 8.6** Breakpoint in relationship between AMZN and SPY from 2015 to 2019 based on F-Statistics. The vertical line denotes a breakpoint at observation number 811, which is March 22, 2018. Data source: Price data reproduced with permission of CSI ©2020. www.csidata.com

```
1   > library(strucchange)
2   > fs.amzn <- Fstats(returns.df$AMZN ~ returns.df$MKT)
3   > breakpoints(fs.amzn)
4
5      Optimal 2-segment partition:
6
7   Call:
8   breakpoints.Fstats(obj = fs.amzn)
9
10  Breakpoints at observation number:
11  811
12
13  Corresponding to breakdates:
14  0.6438792
15  > returns.df[811, ]
16                  AMZN           MKT
17  2018-03-22 -0.0236292 -0.02531506
18  > plot(fs.amzn,
19  +     col = "blue",
```

```
20  +      main = "Choosing Breakpoint Using F Statistic")
21  > lines(breakpoints(fs.amzn))
```

**Step 4: Identify Breakpoint Using BIC** As an alternative, we can use the Bayesian Information Criterion (BIC) to identify the breakpoint. As the output shows, using BIC also identifies observation number 811 as the breakpoint.

```
1   > bp.amzn <- breakpoints(returns.df$AMZN ~ returns.df$MKT)
2   > breakpoints(bp.amzn)
3
4     Optimal 2-segment partition:
5
6   Call:
7   breakpoints.breakpointsfull(obj = bp.amzn)
8
9   Breakpoints at observation number:
10  811
11
12  Corresponding to breakdates:
13  0.6446741
```

## 8.7.3   Chow Test

Once we have identified a breakpoint, we can test whether the coefficients of the linear regression are different prior to the breakpoint and after the breakpoint. The formal test to do this is called the **Chow Test**. We know from the above that a breakpoint in the returns data is at observation number 811. We can then use sctest(), which is also in the strucchange package, to run a Chow Test. If the F-Statistic in the Chow Test is greater than the critical value, we reject the null hypothesis that there is no breakpoint in the data. As we can see below, the p-value associated with the F-Statistic of 10 is very small and, thus, we reject the null hypothesis that observation number 811 is not a breakpoint.

```
1   > sctest(AMZN ~ MKT, data = returns.df,
2   +      type = "Chow", points = 811)
3
4     Chow test
5
6   data:  AMZN ~ MKT
7   F = 10.031, p-value = 4.765e-05
```

This implies that to better model the relationship of AMZN and SPY over this period, we may want to consider splitting the data into a pre-March 22, 2018 period and a post-March 22, 2018 period.

### *8.7.4    Test Equality of Two Betas*

The Chow Test above tests whether the coefficients (i.e., intercept and beta) change between the two periods. In some instances, we may want to only test whether the betas (i.e., slope of a regression line) during the two periods are the same. To do this, we use a standard **z-test**.

**Step 1: Calculate Value and Standard Error of Beta in Period A**  Using lm(), we run a regression of AMZN returns on MKT returns from the period 2015 to 2017. We then extract the estimate of beta and the standard error of beta from the summary results. These values are elements number 2 and 4 of the  coefficients list.

```
1  > reg.A <- lm(AMZN ~ MKT, data = returns["/2017-12-31"])
2  > summary(reg.A)
3
4  Call:
5  lm(formula = AMZN ~ MKT, data = returns["/2017-12-31"])
6
7  Residuals:
8        Min        1Q     Median        3Q        Max
9  -0.107743 -0.006844 -0.000577  0.005784  0.141576
10
11  Coefficients:
12              Estimate Std. Error t value Pr(>|t|)
13  (Intercept) 0.0012733  0.0005605    2.272   0.0234 *
14  MKT         1.1342683  0.0721784   15.715   <2e-16 ***
15  ---
16  Signif. codes:  0 '***' 0.001 '**' 0.01 '*' 0.05 '.' 0.1 ' ' 1
17
18  Residual standard error: 0.01538 on 753 degrees of freedom
19  Multiple R-squared:  0.247,  Adjusted R-squared:  0.246
20  F-statistic:   247 on 1 and 753 DF,  p-value: < 2.2e-16
21
22  > (beta.A <- summary(reg.A)$coefficients[2])
23  [1] 1.134268
24  > (se.A <- summary(reg.A)$coefficients[4])
25  [1] 0.07217837
```

**Step 2: Calculate Value and Standard Error of Beta in Period B**  Similarly, we extract the estimate of beta and the standard error of beta in the second period, which is from 2018 to 2019. To save space, we do not output the regression summary output like we did above, but we know that elements number 2 and 4 of the coefficients list are the values we want.

```
1  > reg.B <- lm(AMZN ~ MKT, data = returns["2018-01-01/"])
2  > (beta.B <- summary(reg.B)$coefficients[2])
3  [1] 1.506851
4  > (se.B <- summary(reg.B)$coefficients[4])
5  [1] 0.05982215
```

**Step 3: Calculate Difference in Betas**   The beta in the first period is 1.13, and the beta in the second period is 1.51, which yields a raw difference of 0.37 with the beta in the second period being higher than the beta in the first period. However, like many values that are estimated, whether the size of the difference is large enough to be statistically significant depends on the standard errors.

```
1  > (diff.beta <- beta.A - beta.B)
2  [1] -0.3725827
```

**Step 4: Perform z-Test on the Difference in Betas**   The z-test is calculated as

$$\frac{\beta_1 - \beta_2}{\sqrt{\sigma_1^2 + \sigma_2^2}}, \tag{8.5}$$

where $\beta$ is the beta of the period and $\sigma$ is the standard error of the period. As the output of the **z-test** shows, the difference in betas during the two periods has a z-score of $-3.97$ and a $p$-value that is very small. In other words, this difference in betas is significantly different from zero.

```
1  > (z <- diff.beta / sqrt(se.A^2 + se.B^2))
2  [1] -3.974363
3  > options(scipen = 999)
4  > (p.value <- 2 * pnorm(-abs(z)))
5  [1] 0.00007056775
6  > options(scipen = 0)
```

## 8.8   Further Reading

Two excellent valuation books are [6] and [10]. Aside from providing a general discussion on many valuation topics, [10] has an excellent discussion of how to develop projections, and [6] has a discussion of multiples and its determinants.

The terminal value in a DCF typically comprises a big portion of a firm's valuation. However, we often observe that many terminal values change substantially when the discount rate and perpetuity growth rate are changed. In [3], I explain how this should not be the case, as the terminal value, discount rate, and perpetuity growth rates should all be linked.

With respect to unlevering betas, the original paper on the Hamada formula is [8]. Note that the above formula by Brealey and Myers (see, for example, [5]) assumes a debt beta that is not zero. There are also alternative formulas that unlever equity betas that account for a non-zero debt beta, such as ones by [11], [9], and [7].

I have written about several practical issues commonly observed in practice with regard to betas. In [2], I show why one should consider cash-adjusting betas especially when a firm has a lot of excess cash. The reason is that firms with a lot of excess cash have observed betas that understate the beta of the operating assets. Thus, we do not properly measure the risk of the operations of the company, which is the cash flows that the cost of capital is discounting. In [4], we show that the appropriate leverage ratio to unlever betas obtained by running a regression on historical data is the average leverage ratio of the regression period. We find that there we cannot know beforehand the direction and magnitude of the error unless we perform the correct calculation. Finally, in many instances, practitioners simply take a debt beta estimated in academic articles or they find published somewhere. However, this debt beta is not estimated with the same return frequency and estimation period as the equity beta used in the valuation. In [1], I show how to estimate debt betas that are consistent with the methodology we use to estimate the equity beta.

# References

1. Ang, C. (2017). Estimating debt betas and beta unlevering formulas. *NACVA QuickRead*. http://quickreadbuzz.com/2017/02/08/estimating-debt-betas-beta-unlevering-formulas/
2. Ang, C. (2018). Why you may want to consider cash-adjusting CAPM betas. *Bloomberg CFA Blog*. https://www.bloombergprep.com/blog-posts/2018/6/8/why-you-may-want-to-consider-cash-adjusting-capm-betas-1
3. Ang, C. (2019). Terminal values and runaway valuations. *NACVA QuickRead*. https://quickreadbuzz.com/2019/11/20/business-valuation-clifford-ang-terminal-values-in-dcfs/
4. Ang, C., Lin, A. (2020). The valuation impact f using the wrong leverage ratio to unlever betas. *NACVA QuickRead*. http://quickreadbuzz.com/2020/04/29/business-valuation-ang-lin-the-valuation-impact-of-using-the-wrong-leverage-ratio-to-unlever-betas/
5. Brealey, R., Myers, S., Allen, F. (2011). *Principles of corporate finance* (10th ed.). New York: McGraw-Hill.
6. Damodaran, A. (2013) *Investment valuation* (3rd ed.). Hoboken: Wiley.
7. Fernandez, P. (2004) The value of tax shields is NOT equal to the present value of tax shields. *Journal of Financial Economics, 73*, 145–165.
8. Hamada, R. (1972). The effect of the firm's capital structure on the systematic risk of common stock. *Journal of Finance, 37*, 435–452.
9. Harris, R., Pringle, J. (1985). Risk-adjusted discount rates extensions from the average-risk case. *Journal of Financial Research, 8*, 237–244.
10. Koller, T., Goedhart, M., & Wessels, D. (2020). *Valuation: Measuring and managing the value of companies* (7th ed.). Hoboken: Wiley.
11. Miles, J., Ezzell, J. (1980). The weighted average cost of capital, perfect capital markets and project life: a clarification. *Journal of Financial and Quantitative Analysis, 15*, 719–730.

# Chapter 9
# Fixed Income

Up to this point our analysis has been focused primarily on equity securities. In this chapter, we will discuss **fixed income** securities. The simplest example of a fixed income security is a bond with fixed rate coupon payments and a finite term. For example, a 5% coupon bond due in 5 years. If the principal for this bond is $1000 and it pays semi-annual coupon payments, we would expect to receive $25 every 6 months for the next 5 years. At the end of the 5 years, we get our principal of $1000 back. The fixed income space includes many variations of these types of securities and, among other things, includes securities with floating rate coupon payments and bonds with embedded options.

Based on size, the global stock market is actually smaller than the global fixed income market. According to the Securities Industry and Financial Market Association (SIFMA), in 2018, the global equities market capitalization was $74.7 trillion, while the global bond market outstanding was $102.8 trillion. As such, it is worth spending time going through an analysis of fixed income securities and fixed income models.

Although the applications in this chapter are implemented on fixed income securities, many of the techniques we will go through are equally applicable to equity securities or other types of data.

We begin our analysis of fixed income securities by showing how to obtain and analyze economic data in Sect. 9.1. Many investments, including fixed income instruments, have a relationship with macroeconomic variables, which makes analyzing economic data important to understanding how our investments may perform during our holding period. We then demonstrate an analysis of **US Treasury** yields in Sect. 9.2. Since some fixed income instruments rely on a premium over the rates of Treasury securities, being able to analyze the Treasury yield curve is essential to making sound fixed income investment decisions. We also show how to look at the real yield of Treasury securities and observe the decline in real yields in recent years. Then, we demonstrate how to identify mean reverting patterns in Treasury securities.

© The Author(s), under exclusive license to Springer Nature Switzerland AG 2021    265
C.S. Ang, *Analyzing Financial Data and Implementing Financial Models Using R*,
Springer Texts in Business and Economics,
https://doi.org/10.1007/978-3-030-64155-9_9

Next, in Sect. 9.3, we use **principal components analysis** or PCA to verify that most of the variation in interest rates is a result of the level of interest rates. Then, in Sect. 9.4, we analyze the time series of spreads between corporates of different **investment grade** ratings. We show how such an analysis can reveal the widening or tightening of credit spreads. This can be viewed as a measure of investor's appetite for credit risk.

We then show how to implement **bond valuation** using discounted cash flow analysis in Sect. 9.5. We show how plain vanilla bonds are valued as well as how to find the yield of a bond if its price and cash flows are known. In Sect. 9.6, we demonstrate how to calculate **duration** and **convexity** of bonds, which are tools used to manage **interest rate risk** and show when these techniques can provide reasonable estimates of the change in the bond's price. This method is helpful when managing a portfolio of bonds, as we do not need to perform a full valuation of each bond in the portfolio.

We end this chapter with a discussion of **short rate** models in Sect. 9.7. In particular, we look at the models by Vasicek [9] and Cox et al. [2].

## 9.1  Economic Analysis

When making investments, understanding how the economy is doing is essential to the success of our investment strategy. For example, when interest rates rise, demand for fixed income securities may increase. As a practical matter, there are too many economic indicators to discuss in this text and different indicators are more important for some sectors compared to others. Thus, for illustrative purposes, we only analyze three economic indicators: Real GDP, unemployment, and inflation. The sources and techniques used in this section can be applied to other economic data as well.

### 9.1.1  Real GDP

We first discuss **real gross domestic product** (GDP), which is an indicator of the strength of the economy. Most securities tend to move with the overall economy. When the economy does well, prices of securities tend to increase as well. Conversely, when the economy does poorly, prices of securities tend to decrease too. Hence, it may be helpful to know how well or poorly the economy is doing to determine the best investments that fit our strategy.

To analyze the growth in real GDP, we will construct a **bar chart** of historical and projected real US GDP growth using data retrieved from the IMF website. In particular, we obtain the data from the October 2019 World Economic Outlook. This database contains much more information that what we use here and is a very useful source of economic data. As of early 2019, the real GDP growth data contains actual historical results from 1980 through 2018 and projected results from 2019 through 2024. From a presentation perspective, a key element for this analysis is to distinguish between historical results and projected results. For that purpose, we will construct a chart that uses different colored bars for the historical data and for the projected data.

The April 2020 WEO data does not have a forecast beyond 2020. Thus, I have saved the IMF forecast data we will use below into a CSV file labeled IMF WEO US RGDP.csv, which can be downloaded from my website http://www.cliffordang. com.

**Step 1: Import Real GDP Growth Data from the IMF Website**  We import the IMF real GDP forecast data using read.csv() (Line 1). The IMF real GDP data is from 1980 to 2024. The data for 1980–2018 are historical data and the data for 2019–2024 are projected data. We only keep the data in the third row, which is the data on real GDP. We also drop the first column, which is the label. We then use as.numeric() to convert the real GDP data from being by R as a character to numeric (Line 8).

```
1  > (raw.rgdp <- read.csv("IMF WEO US RGDP.csv", header = FALSE)[3, -1])
2       V2  V3   V4  V5  V6  V7  V8  V9 V10 V11 V12  V13 V14 V15 V16 V17 V18 V19
3    3 -0.3 2.5 -1.8 4.6 7.2 4.2 3.5 3.5 4.2 3.7 1.9 -0.1 3.5 2.8   4 2.7 3.8 4.4
4       V20 V21 V22 V23 V24 V25 V26 V27 V28 V29  V30  V31 V32 V33 V34 V35 V36 V37
5    3 4.5 4.8 4.1   1 1.7 2.9 3.8 3.5 2.9 1.9 -0.1 -2.5 2.6 1.6 2.2 1.8 2.5 2.9
6       V38 V39 V40 V41 V42 V43 V44 V45 V46
7    3 1.6 2.4 2.9 2.4 2.1 1.7 1.6 1.6 1.6
8  > Value <- as.numeric(as.character(raw.rgdp))
9  > str(Value)
10   num [1:45] -0.3 2.5 -1.8 4.6 7.2 4.2 3.5 3.5 4.2 3.7 ...
11 > head.tail(Value)
12   [1] -0.3  2.5 -1.8
13   [1] 1.6 1.6 1.6
```

Note that we did not read in the first row of the CSV file, which is the year label. Although we need the year label, we can easily add that later using seq() because it is simply each year between 1980 and 2024. We then combine the two objects into one data frame using data.frame() (Line 7).

```
1  > (Year <- seq(1980, 2024, 1))
2   [1] 1980 1981 1982 1983 1984 1985 1986 1987 1988 1989 1990 1991 1992 1993
3   [15] 1994 1995 1996 1997 1998 1999 2000 2001 2002 2003 2004 2005 2006 2007
4   [29] 2008 2009 2010 2011 2012 2013 2014 2015 2016 2017 2018 2019 2020 2021
5   [43] 2022 2023 2024
6  >
7  > us.rgdp <- data.frame(Year, Value)
8  > head.tail(us.rgdp)
9    Year Value
10 1 1980  -0.3
11 2 1981   2.5
12 3 1982  -1.8
13    Year Value
```

```
14   43 2022   1.6
15   44 2023   1.6
16   45 2024   1.6
```

**Step 2: Separate Historical From Projected Data**  The *us.rgdp* dataset contains a single column **Value** that contains both the historical and projected real GDP data. We use `ifelse()` to separate the data. For the *historical* dataset, we tell R to copy the value in **Value** if the **Year** is less than or equal to 2018 (i.e., 1980–2018). Otherwise, the value in *historical* will equal zero. For the *projected* dataset, we tell R to copy the value in **Value** if the **Year** is greater than 2018 (i.e., 2019 and later). Otherwise, the value in those years will equal zero. We then use `data.frame()` to combine *Year*, *historical*, and *projected* into one dataset called *chart.data* (Line 11).

```
1  > historical <- ifelse(us.rgdp$Year <= 2018, us.rgdp$Value, 0)
2  > head.tail(historical)
3  [1] -0.3  2.5 -1.8
4  [1] 0 0 0
5  >
6  > projected <- ifelse(us.rgdp$Year > 2018, us.rgdp$Value, 0)
7  > head.tail(projected)
8  [1] 0 0 0
9  [1] 1.6 1.6 1.6
10 >
11 > chart.data <- data.frame(Year, historical, projected)
12 > head.tail(chart.data)
13    Year historical projected
14 1 1980      -0.3         0
15 2 1981       2.5         0
16 3 1982      -1.8         0
17    Year historical projected
18 43 2022        0        1.6
19 44 2023        0        1.6
20 45 2024        0        1.6
```

**Step 3: Transpose Chart Data**  The data in *chart.data* has 3 columns and 45 rows. For our **bar chart**, we would like to have 45 columns and 3 rows. To do this, we use `t()` to transpose the data.

```
1  > tchart.data <- t(chart.data)
2  > head(tchart.data)
3                [,1]   [,2]   [,3]   [,4]   [,5]   [,6]   [,7]   [,8]   [,9]
4  Year       1980.0 1981.0 1982.0 1983.0 1984.0 1985.0 1986.0 1987.0 1988.0
5  historical   -0.3    2.5   -1.8    4.6    7.2    4.2    3.5    3.5    4.2
6  projected     0.0    0.0    0.0    0.0    0.0    0.0    0.0    0.0    0.0
7               [,10]  [,11]  [,12]  [,13]  [,14] [,15]  [,16]  [,17]  [,18]
8  Year       1989.0 1990.0 1991.0 1992.0 1993.0  1994 1995.0 1996.0 1997.0
9  historical    3.7    1.9   -0.1    3.5    2.8     4    2.7    3.8    4.4
10 projected     0.0    0.0    0.0    0.0    0.0     0    0.0    0.0    0.0
11              [,19]  [,20]  [,21] [,22]  [,23]  [,24]  [,25]  [,26]  [,27]
```

```
12  Year       1998.0 1999.0 2000.0  2001 2002.0 2003.0 2004.0 2005.0 2006.0
13  historical   4.5    4.8    4.1      1    1.7    2.9    3.8    3.5    2.9
14  projected    0.0    0.0    0.0      0    0.0    0.0    0.0    0.0    0.0
15             [,28]  [,29]  [,30]  [,31]  [,32]  [,33]  [,34]  [,35]  [,36]
16  Year       2007.0 2008.0 2009.0 2010.0 2011.0 2012.0 2013.0 2014.0 2015.0
17  historical   1.9   −0.1   −2.5    2.6    1.6    2.2    1.8    2.5    2.9
18  projected    0.0    0.0    0.0    0.0    0.0    0.0    0.0    0.0    0.0
19             [,37]  [,38]  [,39]  [,40]  [,41]  [,42]  [,43]  [,44]  [,45]
20  Year       2016.0 2017.0 2018.0 2019.0 2020.0 2021.0 2022.0 2023.0 2024.0
21  historical   1.6    2.4    2.9    0.0    0.0    0.0    0.0    0.0    0.0
22  projected    0.0    0.0    0.0    2.4    2.1    1.7    1.6    1.6    1.6
```

**Step 4: Create Chart of Historical and Projected GDP** We first want to make sure we have enough space in the y-axis. Earlier, we would manually enter the range. Here, in Line 3, we try a different technique. To get the minimum value of the range and round it down to the next smallest integer, we can use a combination of floor() and min() (Line 3). The minimum value of the real GDP data is -2.5%. Applying the combination of floor() and min() yields a value of -3%. Similarly, we can find the maximum value of the range and round it up to the next largest integer, we can use a combination of ceiling() and max() (Line 4). The maximum value of the real GDP data is 7.2%. Applying the combination of ceiling() and max() yields a value of 8.0%. We see this range for the y-axis in Line 5. We then use barplot() to plot a bar chart. By default, this creates a stacked bar chart. Since there is no value during the projection period when there is a historical period and vice versa, stacking does not affect the results but makes what we are doing slightly easier. The result of this plot is shown in Fig. 9.1.

```
1  > range(tchart.data[2:3, ])
2  [1] −2.5  7.2
3  > (y.range <- c(floor(min(tchart.data[2:3, ])),
4  +     ceiling(max(tchart.data[2:3, ]))))
5  [1] −3  8
6  barplot(tchart.data[2:3, ],
7      col = c("blue", "red"),
8      border = 0,
9      ylab = "Real GDP Growth (%)",
10     ylim = y.range,
11     names.arg = seq(1980, 2024, 1),
12     las = 2,
13     main = "United States Real GDP Growth
14     Historical (1980–2018) and Projected (2019–2024)")
15 legend("topright",
16     c("Historical", "Projected"),
17     col = c("blue", "red"),
18     pch = c(15, 15))
```

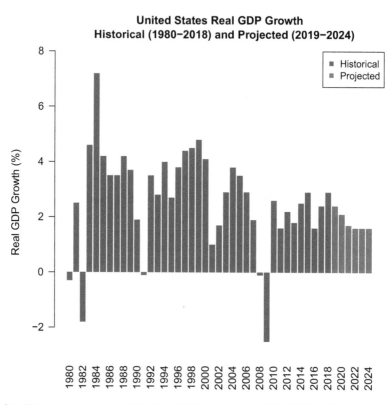

**Fig. 9.1**  Historical and projected US Real GDP growth rate, 1980–2024. IMF data reproduced with permission. Data Source: International Monetary Fund World Economic Outlook October 2019

### Retrieving and Importing FRED Data

In what follows, we will use a good amount of economic data obtained from the Federal Reserve Electronic Database (FRED) in this chapter. Thus, it would be worthwhile to efficiently obtain the data that we need as well as set up a function to import data into R.

**Retrieving Data**  The website for FRED data is at https://fred.stlouisfed.org/. On the top-right portion of the page, there is a search box. In that box, we can enter the ticker symbol of the data series we would like to download. For example, we can enter **DGS3MO** for the 3-month constant maturity Treasury data. As we are typing a valid symbol, we can see the available series for that symbol. For DGS3MO the only series available is daily data. By clicking on that series, we go to the page for DGS3MO that shows a

(continued)

chart of the time series of 3-month Treasury rates. For our purposes, we can change the date range of the chart to the maximum date range. Then, we can click the download button on the upper right side of the screen. Clicking the download button shows a menu of download options, from which we choose to download the data as a CSV file. A CSV file labeled DGS3MO.csv will be downloaded in the browser's download directory. We can move that CSV file into our R working directory from there. Note that it is easier to read in R and we will write the function below based on importing a CSV file.

**Symbols to Download** Once we are on the DGS3MO page, we can go to the search box on the top-right portion of the page to enter other symbols. We then download a CSV file of the maximum data for these other symbols as well. The symbols of the 18 data series we will use in this book are as follows: DGS3MO, DGS6MO, DGS1, DGS3, DGS5, DGS7, DGS10, DGS20, DGS30, WGS10, DFII10, UNRATE, CPIAUCNS, GDPC1, VIXCLS, SP500, AAA, and BBB. The FRED data for all the symbols except for UNRATE, CPIAUCNS, and GDPC1 was downloaded on January 4, 2020 because those already contained data through December 2019 on that date. UNRATE and CPIAUCNS were downloaded on January 19, 2020, as data through December 2019 was available as of that date. Finally, GDPC1 data was also downloaded on January 19, 2020 but only data through July 2019 was available at that time.

**Importing FRED Data Function** Since we will be importing a good amount of FRED data, it will be good to walk through how to create a function that allows us to simplify the import process. For this example, we will use the DGS3MO data because the Treasury yield data as we downloaded contains missing data represented as dots. Thus, it creates an additional layer of complication that we have to deal with. The other FRED data without such a complication can still use the code we will develop based on DGS3MO. Following the steps above, the DGS3MO data should be saved as DGS3MO.csv in our R working directory.

**Step 1: Set up Variables Needed Later in the Function** The function we will create takes on two arguments: the filename of the data ($x$) and some symbol of the data ($symbol$) we will use going forward.

```
1  > x <- "DGS3MO.csv"
2  > symbol <- "DGS3MO"
```

**Step 2: Create a Temporary Dataset to Import the CSV File** We use read.csv() to import the filename represented by $x$. We use header = TRUE to tell R that the first row is the header. This will read in values of DATE and DGS3MO as headers to the columns of data.

(continued)

```
 1  > temp <- read.csv(x, header = TRUE)
 2  > head.tail(temp)
 3            DATE DGS3MO
 4  1 1982-01-04  11.87
 5  2 1982-01-05  12.20
 6  3 1982-01-06  12.16
 7            DATE DGS3MO
 8  9912 2019-12-31   1.55
 9  9913 2020-01-01      .
10  9914 2020-01-02   1.54
```

Since we have the dot for the January 1, 2020 observation, which is a holiday, we need to add an extra step to clean up that data to leave only numbers. We do this by subsetting the data to anything that is not a dot. We use != to let R know that we want data that is not equal to a dot. We use the column reference here because we will use the same convention when we create a function to import FRED data later.

```
 1  > temp <- subset(temp, temp[, 2] != ".")
 2  > head.tail(temp)
 3            DATE DGS3MO
 4  1 1982-01-04  11.87
 5  2 1982-01-05  12.20
 6  3 1982-01-06  12.16
 7            DATE DGS3MO
 8  9911 2019-12-30   1.57
 9  9912 2019-12-31   1.55
10  9914 2020-01-02   1.54
```

**Step 3: Create Date Variable** When we look at the structure of the *temp* dataset, we can see that the date variable is considered a Factor by R. The easiest way to convert this variable to a date variable is using the lubridate package. In that package, we use ymd() because the format of DATE is year-month-day. As we can see at the bottom of the code, the date variable in this object is now recognized by R as a date. This will be important when we need to subset the data. In the last line of the code, we delete the DATE variable in *temp* because we no longer need it as we have our new date variable.

```
 1  > str(temp)
 2  'data.frame': 9502 obs. of  2 variables:
 3   $ DATE  : Factor w/ 9914 levels "1982-01-04","1982-01-05",..: 1 2 3 ...
 4   $ DGS3MO: Factor w/ 1136 levels ".","0.00","0.01",..: 279 290 288 ...
 5  > library(lubridate)
 6  > date <- ymd(temp$DATE)
 7  > head.tail(date)
 8  [1] "1982-01-04" "1982-01-05" "1982-01-06"
 9  [1] "2019-12-30" "2019-12-31" "2020-01-02"
10  > str(date)
```

(continued)

```
11   Date[1:9502], format: "1982-01-04" "1982-01-05" "1982-01-06" ...
```

**Step 4: Create a Numeric Vector of Values** As the output of the structure for *temp* in Step 3 shows, the yield data is also considered a factor. We thus have to convert this data to numeric values. We do this by nesting an `as.character()` inside `as.numeric()` and apply those to the yield data (Column 2) of the *temp* directory.

```
1   > value <- as.numeric(as.character(temp[, 2]))
2   > head.tail(value)
3   [1] 11.9 12.2 12.2
4   [1] 1.57 1.55 1.54
```

**Step 5: Create xts Object of the Combined Data** We load the `xts` package in case it or the `quantmod` package is not loaded in the R environment. In Line 2, we create an xts object using `xts()` in Line 2. We then rename the variable in Line 3 using the symbol we defined in Step 1. As the bottom part of the code shows, this data is now an xts object. We will then be able to do similar things to it as if we used `getSymbols()` and used that to import the FRED data.

```
1   > library(xts)
2   > temp2 <- xts(value, order.by = date)
3   > names(temp2) <- symbol
4   > head.tail(temp2)
5               DGS3MO
6   1982-01-04   11.9
7   1982-01-05   12.2
8   1982-01-06   12.2
9               DGS3MO
10  2019-12-30   1.57
11  2019-12-31   1.55
12  2020-01-02   1.54
13  >
14  > str(temp2)
15  An 'xts' object on 1982-01-04/2020-01-02 containing:
16    Data: num [1:9502, 1] 11.9 12.2 12.2 12.2 12 ...
17    - attr(*, "dimnames")=List of 2
18     ..$ : NULL
19     ..$ : chr "DGS3MO"
20    Indexed by objects of class: [Date] TZ: UTC
21    xts Attributes:
22  NULL
```

**Step 5: Create a Function Using Above Code** We now put all of the above together to create a function called `load.fred()`.

```
1   > load.fred <- function(rawdata, symbol){
```

(continued)

```
 2   +    temp <- read.csv(rawdata, header = TRUE)
 3   +    temp <- subset(temp, temp[, 2] != ".")
 4   +    library(lubridate)
 5   +    date <- ymd(temp$DATE)
 6   +    value <- as.numeric(as.character(temp[, 2]))
 7   +    library(xts)
 8   +    temp2 <- xts(value, order.by = date)
 9   +    names(temp2) <- symbol
10   +    return(temp2)
11   + }
```

We can then test this function using the same 3-month US Treasury yield data. We can compare the output below to the output in Step 5 and see that the first three and last three observations are identical. We can now use this function to import all the FRED data we will use in this book.

```
 1   > t3mo <- load.fred("DGS3MO.csv", "DGS3MO")
 2   > head.tail(t3mo)
 3                DGS3MO
 4   1982-01-04   11.9
 5   1982-01-05   12.2
 6   1982-01-06   12.2
 7                DGS3MO
 8   2019-12-30   1.57
 9   2019-12-31   1.55
10   2020-01-02   1.54
```

**Check Against `getSymbols()` Output** The last step is to check the output we generated from the above with the output from using `getSymbols()`. We have identical results as the above data was downloaded a few minutes before we imported the data using `getSymbols()`. As we can see from the `getSymbols()` code below, we do not use any arguments to control the date range (unlike when we download Yahoo Finance data) because the date range argument is not applicable to FRED data. The data below includes the missing observation for January 1, 2020, which we cleaned up above in Step 2.

```
 1   > check <- getSymbols("DGS3MO", src = "FRED", auto.assign = FALSE)
 2   > head.tail(check)
 3                DGS3MO
 4   1982-01-04   11.87
 5   1982-01-05   12.20
 6   1982-01-06   12.16
 7                DGS3MO
 8   2019-12-31   1.55
 9   2020-01-01   NA
10   2020-01-02   1.54
11   > str(check)
12   An 'xts' object on 1982-01-04/2020-01-02 containing:
```

(continued)

```
13   Data: num [1:9914, 1] 11.9 12.2 12.2 12.2 12 ...
14   - attr(*, "dimnames")=List of 2
15   ..$ : NULL
16   ..$ : chr "DGS3MO"
17   Indexed by objects of class: [Date] TZ: UTC
18   xts Attributes:
19  List of 2
20   $ src    : chr "FRED"
21   $ updated: POSIXct[1:1], format: "2020-01-04 13:14:53"
```

## 9.1.2 Unemployment Rate

In this section, we will use US unemployment rate data from FRED. Specifically, we will analyze the **unemployment rate** over the last 50 years and identify its peaks and long-term average. To emphasize the peaks and long-term average, we create an **annotated chart**.

**Step 1: Obtain Unemployment Rate Data from 1970 to 2019** We are interested in the Civilian Unemployment Rate with symbol UNRATE from the US Department of Labor. This is a monthly and seasonally adjusted series. Following the steps described above, we saved the FRED unemployment rate data as **UNRATE.csv**. Using the `load.fred()` function we created above, we import UNRATE. That yields data from 1948 to 2019. Since we only want the last 50 years of data, we use xts-style date subsetting to limit the data to between January 1970 and December 2019 (Line 12). In Line 23, using `dim()`, we can see the data has 600 observations, which is what we would expect for 50 years of monthly data (i.e., 60 observations = 50 years * 12 months per year).

```
1  > unrate <- load.fred("UNRATE.csv", "unrate")
2  > head.tail(unrate)
3              unrate
4  1948-01-01   3.4
5  1948-02-01   3.8
6  1948-03-01   4.0
7              unrate
8  2019-10-01   3.6
9  2019-11-01   3.5
10 2019-12-01   3.5
11 >
12 > unrate <- unrate["1970-01-01/2019-12-31"]
13 > head.tail(unrate)
14              unrate
15 1970-01-01   3.9
16 1970-02-01   4.2
```

```
17   1970–03–01     4.4
18                unrate
19   2019–10–01     3.6
20   2019–11–01     3.5
21   2019–12–01     3.5
22   >
23   > dim(unrate)
24   [1] 600    1
```

**Step 2: Calculate Long-Term Average** Using mean(), we calculate the monthly average unemployment rate over the last 50 years. We find that the long-term average is 6.2%.

```
1   > (lt.avg <- mean(unrate$unrate))
2   [1] 6.2
```

**Step 3: Plot Unemployment Rate Data** We use plot() to create a line chart of the unemployment rate data. We need some room to add annotations, so although the range of unemployment rates is 3.5% to 10.8%, we choose the y-axis to have a range of 2% to 12% (Line 7). There is no magic to 2% or 12% and you can change that range to suit your preference.

```
 1   > range(unrate$unrate)
 2   [1]   3.5 10.8
 3   > plot(x = index(unrate),
 4   +       y = unrate,
 5   +       xlab = "Date (Quarters)",
 6   +       ylab = "Unemployment Rate (%)",
 7   +       ylim = c(2, 12),
 8   +       type = "l",
 9   +       col = "blue",
10   +       main = "US Unemployment Rate From 1970 to 2019")
```

We then overlay the annotations onto the chart. There are two things that we want to mark. First, we want to add a green horizontal line that spans the chart to denote the long-term average over the period of 6.2%. We do this using abline() with the h option for a horizontal line. To make it more visible, we want to make this line thicker, so we use lwd = 2. We then use text() to add labels that will show up on the chart. We split our label into two lines. The first line will say Long-Term and the second line will say Avg. = 6.2%. We also want to create the arrow pointing downwards to the green horizontal line denoting the long-term average. We create the arrow using arrow(). All the inputs to the various arguments for the text() and arrow() I found using trial-and-error and some of it (e.g., positioning) could be a matter of preference as well. The date of January 1, 2001 was selected by eyeballing where on the chart is there room for us to drop the label.

```
1   > abline(h = lt.avg, lwd = 2, col = "darkgreen")
2   > text(as.Date("2001–01–01"), 8.0, "Long–Term", col = "red")
3   > text(as.Date("2001–01–01"), 7.5, "Avg. = 6.2%", col = "red")
4   > arrows(x0 = as.Date("2001–01–01"),
5   +        y0 = 7.0,
```

```
 6   +      x1 = as.Date("2001-01-01"),
 7   +      y1 = 6.4,
 8   +      col = "red",
 9   +      code = 2,
10   +      length = 0.10)
```

We then add a black dot for the unemployment rate as of October 2009, which is
the local peak during the 2008/2009 financial crisis. The dot is created using pch
= 16 and we change the size of the dot using cex. Here, we add a point on the
chart using points(). We then also do use a similar combination of text()
and arrows() to add text and an arrow pointing to the dot. Again, the size of the
marker, text location, and arrow locations are based on trial-and-error for the most
part and is also dependent on one's preferences.

```
1   > points(as.Date("2009-10-01"), 10, pch = 16, cex = 1.5)
2   > text(as.Date("2010-01-01"), 11.6, "October 2009", col = "red")
3   > arrows(x0 = as.Date("2009-10-01"),
4   +      y0 = 11,
5   +      x1 = as.Date("2009-10-01"),
6   +      y1 = 10.3,
7   +      col = "red",
8   +      code = 2,
9   +      length = 0.10)
```

Figure 9.2 shows the annotated unemployment rate chart.

### 9.1.3   Inflation Rate

Another example of key economic data available on FRED is the US **Consumer
Price Index** (CPI) data, which is commonly used to calculate the **inflation rate**. The
inflation rate is one of the most important factors investors consider in determining
their expected rates of return. A higher inflation rate would lead to a higher required
expected rate of return, as a dollar tomorrow is worth less than a dollar is worth
today. The Bureau of Labor Statistics of the US Department of Labor reports the
CPI data every month. The year-over-year changes in the CPI are typically used
as a common measure of the inflation rate. The year-over-year change means that
the inflation rate for a particular month is equal to the percentage change in the
CPI for that same month last year. For our example, we look at the time series
of inflation rates based on year-over-year changes in CPI over the last 50 years. To
show the inflation rate during recessionary and non-recessionary periods, we overlay
the recession periods as identified by the National Bureau of Economic Research
(NBER).

**Step 1: Import CPI Data from FRED**   The measure of CPI we use is the CPI for
all urban consumers, which has the symbol CPIAUCNS on FRED. This is monthly
data that is not seasonally adjusted. Note that the data begins in 1913. We save
the FRED data as CPIAUCNS FRED.csv, which as of January 2020 contains data

through December 2019. Since the data on FRED gets continuously updated, pulling the data at a later date will result in retrieving data that includes data for more recent periods. We use the `load.fred()` function we created above to import the data.

```
1  > us.cpi <- load.fred("CPIAUCNS.csv", "us.cpi")
2  > head.tail(us.cpi)
3               us.cpi
4   1913-01-01    9.8
5   1913-02-01    9.8
6   1913-03-01    9.8
7               us.cpi
8   2019-10-01    257
9   2019-11-01    257
10  2019-12-01    257
```

**Step 2: Calculate a 12-Month Lag Variable**  Since the inflation rate is calculated as year-over-year changes in CPI, we create a variable that takes on the value of the CPI during the same month the year before. As we have monthly data, this requires us to take a 12-month lag of the data. As such, we use `Lag()` with `k = 12` to

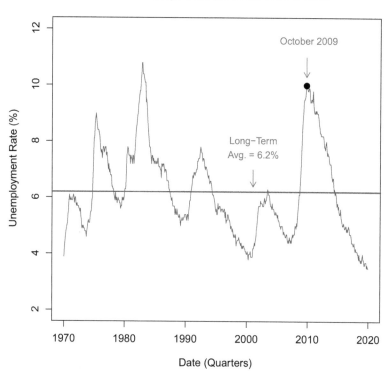

**Fig. 9.2**  US inflation rate, 1970–2019. Data Source: Federal Reserve Electronic Database

denote the 12-period lag. As we can see, the first 12 observations do not have any lag values. The 13th observation for January 2014 has 9.8%, which is the value for January 2013 that we can see above.

```
1  > lag.cpi <- Lag(us.cpi$us.cpi, k = 12)
2  > lag.cpi[10:14, ]
3            Lag.12
4  1913-10-01    NA
5  1913-11-01    NA
6  1913-12-01    NA
7  1914-01-01   9.8
8  1914-02-01   9.8
```

**Step 3: Combine Data and Subset to 1990–2019** We use cbind() to combine both datasets. In Line 11, we then subset the data to January 1990 to December 2019 (i.e., last 50 years) using the xts-style date subsetting. Since we were going to subset the data to a much later time period, there was no need to clean up the NAs generated by lagging in the earlier periods because they will be excluded anyway.

```
1  > us.cpi <- cbind(us.cpi, lag.cpi)
2  > head.tail(us.cpi)
3              us.cpi Lag.12
4  1913-01-01    9.8     NA
5  1913-02-01    9.8     NA
6  1913-03-01    9.8     NA
7              us.cpi Lag.12
8  2019-10-01    257    253
9  2019-11-01    257    252
10 2019-12-01    257    251
11 > us.cpi <- us.cpi["1990-01-01/2019-12-31"]
12 > head.tail(us.cpi)
13             us.cpi Lag.12
14 1990-01-01    127    121
15 1990-02-01    128    122
16 1990-03-01    129    122
17             us.cpi Lag.12
18 2019-10-01    257    253
19 2019-11-01    257    252
20 2019-12-01    257    251
```

**Step 4: Calculate Inflation Rate** We can then calculate the inflation rate as the percentage change between us.cpi and lag.cpi for each period. We then multiply the resulting change, which is in decimals, by 100 to convert it to a percentage point change. This means that the January 1990 inflation rate of 5.2 is 5.2%. The use of percentage points makes reading graph labels easier.

```
1  > us.cpi$inflation <- (us.cpi$us.cpi / us.cpi$Lag.12 - 1) * 100
2  > head.tail(us.cpi)
3             us.cpi Lag.12 inflation
4  1990-01-01    127    121      5.20
5  1990-02-01    128    122      5.26
6  1990-03-01    129    122      5.23
```

```
7                        us.cpi Lag.12 inflation
8    2019–10–01     257    253       1.76
9    2019–11–01     257    252       2.05
10   2019–12–01     257    251       2.29
```

**Step 5: Plot Inflation Rate Time Series**  We then create a line chart of the inflation rate data using plot(). We first create the base plot, which is the line chart of the inflation rate time series.

```
1   > plot(x = index(us.cpi),
2   +      y = us.cpi$inflation,
3   +      xlab = "Date",
4   +      ylab = "Inflation Rate (%)",
5   +      type = "l",
6   +      col = "blue",
7   +      main = "US Inflation Rates From 1990 to 2019
8   +      Based on Year Over Year Changes in CPI")
```

Next, we shade three recessionary periods that our time period covers. We use recession period dates identified by the NBER Dating Committee. The shading is added in two steps. First, we par("usr") to report the coordinates of the chart. Then, we use rect() to draw a rectangle, where the coordinates are given by (1) our first date, (2) the coordinates for the top of the chart given by the second element of *shade* (shade[2]), (3) our last date, and (4) the coordinates for the bottom of the chart given by the third element of *shade* (shade[3]). We use box() to redraw the borders of the chart that were partially overwritten when we shaded the recession periods.

```
1    > (shade <- par("usr"))
2    [1]  6867.960000 18668.040000    −2.432640      6.625288
3    > rect(as.Date("2007–12–01"), shade[2],
4    +      as.Date("2009–06–01"), shade[3],
5    +      col = "pink", lty = 0)
6    > rect(as.Date("2001–03–01"), shade[2],
7    +      as.Date("2001–11–01"), shade[3],
8    +      col = "pink", lty = 0)
9    > rect(as.Date("1990–07–01"), shade[2],
10   +      as.Date("1991–03–01"), shade[3],
11   +      col = "pink", lty = 0)
12   > box(which = "plot", lty = 1)
```

Note that the recession period shading overlaps with the inflation rate plot. However, we needed to plot the inflation rate first to get the boundaries of the plot so that we can create the rectangular boxes needed by the recession period shading. As such, we fix the issue by redrawing the inflation rate line using lines(). Finally, we add a horizontal line to denote a 0% inflation rate as a frame of reference when interpreting the chart.

```
1    > lines(x = index(us.cpi), y = us.cpi$inflation, col = "blue")
2    > abline(h = 0)
```

Figure 9.3 shows the inflation chart from 1990 to 2019. As we can see, over the last 50 years, inflation was highest during two out of the three recession periods with the recession from 1990 to 1991 resulting in inflation of over 6%. We also see a brief period of negative inflation around the end of the 2009 financial crisis.

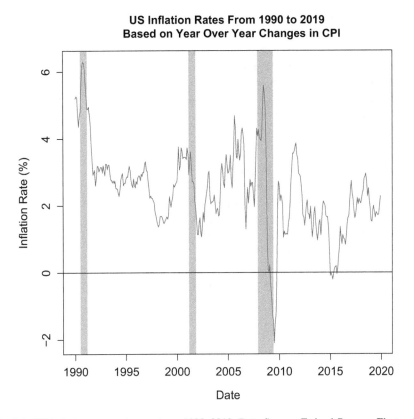

**Fig. 9.3** US inflation rate and recessions, 1990–2019. Data Source: Federal Reserve Electronic Database and NBER Business Cycle Dates

## 9.2 US Treasuries

### 9.2.1 Shape of the US Treasury Yield Curve

The rates of US Treasury securities form the benchmark for the rates of other fixed income instruments. As such, knowledge of Treasury yields is required to determine whether the yields of other fixed income securities are large enough for the incremental risk we are taking on. In addition, trading strategies can be constructed around the differential between short rates and long-term rates. If T-bill rates for 3-months substantially lower than 6-month rates, an investor who expects to hold a T-Bill for 3 months may instead purchase a 6-month T-Bill and sell it in 3 months to take advantage of the higher yields on the longer dated instrument. This strategy is known as **riding the yield curve**.

The **Treasury yield curve** is a curve comprised of rates on key maturities of US Treasury securities. The yield curve can take on many shapes. It can be **upward sloping**, which means that short-term rates are lower than long-term rates. An upward sloping yield curve is also called *normal*, which implies this is the most typical shape of the yield curve we observe. The yield curve can be **inverted**, which means that the short-term rates are higher than long-term rates. The yield curve can also be **flat**, which means that the short-term rates are almost the same as the long-term rates. In this section, we demonstrate how to find examples of these different shapes of the yield curve and plot these examples.

**Step 1: Obtaining US Treasury Yields for Key Maturities** To construct the Treasury yield curve, we have to first obtain the yields of Treasury securities of different maturities. For our example, we use the following key maturities: 3 months (DGS3MO), 6 months (DGS6MO), 1 year (DGS1), 2 years (DGS2), 3 years (DGS3), 5 years (DGS5), 7 years (DGS7), 10 years (DGS10), 20 years (DGS20), and 30 years (DGS30). The text in the parentheses are the FRED symbols for these Treasury securities. We import the data for each of these Treasury securities using the load.fred() function we created above.

```
 1   > t3mo <- load.fred("DGS3MO.csv", "DGS3MO")
 2   > head.tail(t3mo)
 3                DGS3MO
 4   1982-01-04   11.87
 5   1982-01-05   12.20
 6   1982-01-06   12.16
 7                DGS3MO
 8   2019-12-30    1.57
 9   2019-12-31    1.55
10   2020-01-02    1.54
11   >
12   > t6mo <- load.fred("DGS6MO.csv", "DGS6MO")
13   > head.tail(t6mo)
14                DGS6MO
15   1982-01-04   13.16
16   1982-01-05   13.41
17   1982-01-06   13.46
18                DGS6MO
19   2019-12-30    1.60
20   2019-12-31    1.60
21   2020-01-02    1.57
22   >
23   > t1yr <- load.fred("DGS1.csv", "DGS1")
24   > head.tail(t1yr)
25                DGS1
26   1962-01-02   3.22
27   1962-01-03   3.24
28   1962-01-04   3.24
29                DGS1
30   2019-12-30   1.57
31   2019-12-31   1.59
32   2020-01-02   1.56
```

```
33  >
34  > t2yr <-load.fred("DGS2.csv", "DGS2")
35  > head.tail(t2yr)
36              DGS2
37  1976-06-01 7.26
38  1976-06-02 7.23
39  1976-06-03 7.22
40              DGS2
41  2019-12-30 1.58
42  2019-12-31 1.58
43  2020-01-02 1.58
44  >
45  > t3yr <- load.fred("DGS3.csv", "DGS3")
46  > head.tail(t3yr)
47              DGS3
48  1962-01-02 3.70
49  1962-01-03 3.70
50  1962-01-04 3.69
51              DGS3
52  2019-12-30 1.59
53  2019-12-31 1.62
54  2020-01-02 1.59
55  >
56  > t5yr <- load.fred("DGS5.csv", "DGS5")
57  > head.tail(t5yr)
58              DGS5
59  1962-01-02 3.88
60  1962-01-03 3.87
61  1962-01-04 3.86
62              DGS5
63  2019-12-30 1.68
64  2019-12-31 1.69
65  2020-01-02 1.67
66  >
67  > t7yr <- load.fred("DGS7.csv", "DGS7")
68  > head.tail(t7yr)
69              DGS7
70  1969-07-01 6.88
71  1969-07-02 6.89
72  1969-07-03 6.85
73              DGS7
74  2019-12-30 1.81
75  2019-12-31 1.83
76  2020-01-02 1.79
77  >
78  > t10yr <- load.fred("DGS10.csv", "DGS10")
79  > head.tail(t10yr)
80              DGS10
81  1962-01-02  4.06
82  1962-01-03  4.03
83  1962-01-04  3.99
84              DGS10
85  2019-12-30  1.90
86  2019-12-31  1.92
```

```
87   2020-01-02  1.88
88   >
89   > t20yr <- load.fred("DGS20.csv", "DGS20")
90   > head.tail(t20yr)
91                 DGS20
92   1993-10-01  6.12
93   1993-10-04  6.10
94   1993-10-05  6.12
95                 DGS20
96   2019-12-30  2.21
97   2019-12-31  2.25
98   2020-01-02  2.19
99   >
100  > t30yr <- load.fred("DGS30.csv", "DGS30")
101  > head.tail(t30yr)
102                 DGS30
103  1977-02-15  7.70
104  1977-02-16  7.67
105  1977-02-17  7.67
106                 DGS30
107  2019-12-30  2.34
108  2019-12-31  2.39
109  2020-01-02  2.33
```

**Step 2: Combine Yield Data Into One Dataset** Using merge(), we combine the 10 separate yield datasets into one dataset. Because different yield series have different date ranges, we use all = TRUE to force the datasets to merge based on the superset of available dates. In other words, all dates are kept in the combined series and, if a particular series does not have a value on a date, the value for that series will be an NA. As we can see from the output, for a number of the series, there is no data in 1962. However, by the time we reach the end of 2019, all 10 series have yields.

```
1   > treasury <- merge(t3mo, t6mo, t1yr, t2yr, t3yr, t5yr,
2   +     t7yr, t10yr, t20yr, t30yr, all = TRUE)
3   > head.tail(treasury)
4              DGS3MO DGS6MO DGS1 DGS2 DGS3 DGS5 DGS7 DGS10 DGS20 DGS30
5   1962-01-02    NA     NA 3.22   NA 3.70 3.88   NA  4.06    NA    NA
6   1962-01-03    NA     NA 3.24   NA 3.70 3.87   NA  4.03    NA    NA
7   1962-01-04    NA     NA 3.24   NA 3.69 3.86   NA  3.99    NA    NA
8              DGS3MO DGS6MO DGS1 DGS2 DGS3 DGS5 DGS7 DGS10 DGS20 DGS30
9   2019-12-30  1.57   1.60 1.57 1.58 1.59 1.68 1.81  1.90  2.21  2.34
10  2019-12-31  1.55   1.60 1.59 1.58 1.62 1.69 1.83  1.92  2.25  2.39
11  2020-01-02  1.54   1.57 1.56 1.58 1.59 1.67 1.79  1.88  2.19  2.33
```

**Step 3: Select Last 30 Years of Data (1990–2019) with Data for 3-Month, 10-Year, and 30-Year Treasuries** We subset the *treasury* dataset to only include observations from 1990 to 2019 (Line 1). Then, in Line 11, we use na.omit() to remove any observation that has an NA in either DGS3MO, DGS10, or DGS30. Note that this deletes a whole chunk of data from February 18, 2002 through

February 9, 2006 because the 30-year Treasury was not available during that period.
After this procedure, we end up with over 6500 observations.

```
1  > extreme <- treasury["1990-01-01/2019-12-31"]
2  > head.tail(extreme)
3           DGS3MO DGS6MO DGS1 DGS2 DGS3 DGS5 DGS7 DGS10 DGS20 DGS30
4  1990-01-02   7.83   7.89 7.81 7.87 7.90 7.87 7.98  7.94    NA  8.00
5  1990-01-03   7.89   7.94 7.85 7.94 7.96 7.92 8.04  7.99    NA  8.04
6  1990-01-04   7.84   7.90 7.82 7.92 7.93 7.91 8.02  7.98    NA  8.04
7           DGS3MO DGS6MO DGS1 DGS2 DGS3 DGS5 DGS7 DGS10 DGS20 DGS30
8  2019-12-27   1.57   1.59 1.51 1.59 1.60 1.68 1.80  1.88  2.18  2.32
9  2019-12-30   1.57   1.60 1.57 1.58 1.59 1.68 1.81  1.90  2.21  2.34
10 2019-12-31   1.55   1.60 1.59 1.58 1.62 1.69 1.83  1.92  2.25  2.39
11 > extreme <- na.omit(extreme[, c(1, 8, 10)])
12 > head.tail(extreme)
13           DGS3MO DGS10 DGS30
14 1990-01-02   7.83  7.94  8.00
15 1990-01-03   7.89  7.99  8.04
16 1990-01-04   7.84  7.98  8.04
17           DGS3MO DGS10 DGS30
18 2019-12-27   1.57  1.88  2.32
19 2019-12-30   1.57  1.90  2.34
20 2019-12-31   1.55  1.92  2.39
21 > dim(extreme)
22 [1] 6511     3
```

**Step 4: Identify Examples of Different Shapes of the Yield Curve** The yield
curve can be upward sloping (normal), flat, or downward sloping (inverted). We
identify examples of each shape of the yield curve below. We define how steep
the slope is by taking the difference between DGS30 and DGS3MO. We call this
variable sign.diff (Line 1). We then identify the days with the steepest upward and
downward slopes as well as the day with the flattest slope. We use subset() to
extract the days we identify. The day with the most negative sign.diff is November
24, 2000 and that is our example of the inverted (downward sloping) yield curve.
The day with the largest sign.diff is January 11, 2010 and that is our example of the
normal (upward sloping) yield curve.

```
1  > extreme$sign.diff <- extreme$DGS30 - extreme$DGS3MO
2  > head.tail(extreme)
3           DGS3MO DGS10 DGS30 sign.diff
4  1990-01-02   7.83  7.94  8.00      0.17
5  1990-01-03   7.89  7.99  8.04      0.15
6  1990-01-04   7.84  7.98  8.04      0.20
7           DGS3MO DGS10 DGS30 sign.diff
8  2019-12-27   1.57  1.88  2.32      0.75
9  2019-12-30   1.57  1.90  2.34      0.77
10 2019-12-31   1.55  1.92  2.39      0.84
11 > (inverted <- subset(extreme, sign.diff == min(sign.diff)))
12           DGS3MO DGS10 DGS30 sign.diff
13 2000-11-24   6.36  5.63  5.67     -0.69
14 > (normal <- subset(extreme, sign.diff == max(sign.diff)))
```

```
15                   DGS3MO DGS10 DGS30 sign.diff
16   2010-01-11   0.04  3.85  4.74          4.7
17   > (flat <- subset(extreme, sign.diff == min(abs(sign.diff))))
18                   DGS3MO DGS10 DGS30 sign.diff
19   2000-04-13   5.81  5.94  5.81             0
20   2006-03-02   4.62  4.64  4.62             0
21   2006-07-20   5.08  5.03  5.08             0
22   2006-07-21   5.10  5.05  5.10             0
23   2006-07-25   5.13  5.07  5.13             0
24   2006-07-28   5.07  5.00  5.07             0
25   2006-08-14   5.12  5.00  5.12             0
26   2019-09-04   1.97  1.47  1.97             0
```

We then attempted to identify the day with the smallest sign.diff in absolute value terms. This should give us a sense of the flattest yield curve. However, as the above output shows, there are 8 days when the DGS3MO and DGS30 yields were the same. Thus, we have to come up with a second criteria to determine what is the flattest yield curve. We select as our second criteria the difference between the DGS10 and DGS30 yields. By finding the smallest sign.diff2 in absolute value terms, we find our flattest yield curve on March 2, 2006.

```
1    > flat$sign.diff2 <- flat$DGS30 - flat$DGS10
2    > flat
3                    DGS3MO DGS10 DGS30 sign.diff sign.diff2
4    2000-04-13   5.81  5.94  5.81          0      -0.13
5    2006-03-02   4.62  4.64  4.62          0      -0.02
6    2006-07-20   5.08  5.03  5.08          0       0.05
7    2006-07-21   5.10  5.05  5.10          0       0.05
8    2006-07-25   5.13  5.07  5.13          0       0.06
9    2006-07-28   5.07  5.00  5.07          0       0.07
10   2006-08-14   5.12  5.00  5.12          0       0.12
11   2019-09-04   1.97  1.47  1.97          0       0.50
12   > (flat <- subset(flat, abs(sign.diff2) == min(abs(sign.diff2))))
13                    DGS3MO DGS10 DGS30 sign.diff sign.diff2
14   2006-03-02   4.62  4.64  4.62          0      -0.02
```

### Step 5: Extract Full Yield Curve on Days Identified and On December 31, 2019

We can now create variables that hold the dates with our examples of an inverted, flat, and upward sloping (i.e., normal) yield curve. Then, we can extract these dates from the *treasury* dataset. We also extract the yield curve on December 31, 2019, which as of the time of this writing gives us the current shape of the yield curve.

```
1    > invert.date <- index(inverted)
2    > normal.date <- index(normal)
3    > flat.date <- index(flat)
4    > current.date <- as.Date("2019-12-31")
5    >
6    > tyld.curve<-subset(treasury,
7    +      index(treasury) == invert.date |
8    +      index(treasury) == flat.date |
9    +      index(treasury) == normal.date |
10   +      index(treasury) == current.date)
11   > tyld.curve
```

```
12                DGS3MO DGS6MO DGS1 DGS2 DGS3 DGS5 DGS7 DGS10 DGS20 DGS30
13  2000-11-24    6.36    6.34 6.12 5.86 5.74 5.63 5.70  5.63  5.86  5.67
14  2006-03-02    4.62    4.75 4.74 4.72 4.72 4.68 4.66  4.64  4.80  4.62
15  2010-01-11    0.04    0.13 0.35 0.95 1.55 2.58 3.32  3.85  4.64  4.74
16  2019-12-31    1.55    1.60 1.59 1.58 1.62 1.69 1.83  1.92  2.25  2.39
```

**Step 6: Prepare Data For Plotting** The data we have has the yields in columns and dates in rows. To chart this properly, we first transpose the data in the *yield.curve* dataset using t (). Then, we change the names of the rows to numbers that represent the maturity date of the different instruments. For example, the DGS3MO is a 3-month Treasury, so the row name for that maturity is 0.25 years.

```
1   > class(tyld.curve)
2   [1] "xts" "zoo"
3   > tyld.curve <- t(tyld.curve)
4   > rownames(tyld.curve) <- paste(c(0.25, 0.5, 1, 2, 3, 5, 7, 10, 20, 30))
5   > colnames(tyld.curve) <- paste(c("inverted", "flat", "normal", "current"))
6   > tyld.curve
7           inverted flat normal current
8   0.25       6.36 4.62   0.04    1.55
9   0.5        6.34 4.75   0.13    1.60
10  1          6.12 4.74   0.35    1.59
11  2          5.86 4.72   0.95    1.58
12  3          5.74 4.72   1.55    1.62
13  5          5.63 4.68   2.58    1.69
14  7          5.70 4.66   3.32    1.83
15  10         5.63 4.64   3.85    1.92
16  20         5.86 4.80   4.64    2.25
17  30         5.67 4.62   4.74    2.39
```

**Step 7: Plot the Yield Curve on the Four Dates** Before we plot the data, we generate two vectors. First, the *TTM* vector that we will use as the x-axis labels denote the different times to maturity (in years) of the Treasury securities in our dataset. Second, the *y.range* vector that we will use as the range of the y-axis so we do not truncate the plot. We can see that the lowest yield across all four dates is 0.04% and the highest yield is 6.36%. In Line 7, the type = "o" denotes a line chart with markers as we want to identify the points on the curve that we have actual Treasury yield data. The points in between the markers are interpolated (i.e., the two markers are connected by a straight line). Figure 9.4 shows the output of our code.

```
1   > (TTM <- c(0.25, 0.5, 1, 2, 3, 5, 7, 10, 20, 30))
2    [1]  0.25  0.50  1.00  2.00  3.00  5.00  7.00 10.00 20.00 30.00
3   > (y.range<-range(tyld.curve))
4   [1] 0.04 6.36
5   > plot(x = TTM,
6   +      y = tyld.curve[, 1],
7   +      type = "o",
8   +      ylim = y.range,
9   +      xlab = "Years to Maturity",
10  +      ylab = "Yield (Percent)",
```

```
11  +      col = "black",
12  +      main = "Shapes of the Treasury Yield Curve")
13  > lines(x = TTM, y = tyld.curve[, 2],
14  +      type = "o", col = "blue")
15  > lines(x = TTM, y = tyld.curve[, 3],
16  +      type = "o", col = "red")
17  > lines(x = TTM, y = tyld.curve[, 4],
18  +      type = "o", col = "darkgreen")
19  > legend("bottomright", legend = c("Inverted (11/24/2000)",
20  +        "Flat (03/02/2006)","Normal (01/11/2010)",
21  +        "December 31, 2019"),
22  +      cex = 0.75,
23  +      lty = c(1, 1, 1, 1),
24  +      col = c("black", "blue", "red", "darkgreen"))
```

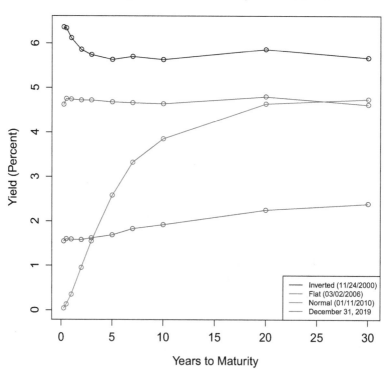

**Fig. 9.4** Different shapes of the US treasury yield curve. Data Source: Federal Reserve Electronic Database

## 9.2.2   Slope of the US Treasury Yield Curve

The **slope** of the Treasury yield curve has been known as a leading indicator of future economic activity and rising inflation expectations. The slope is calculated as the difference between the yields of long-term and short-term Treasury securities. In prior recessions, short-term interest rates rose above long-term rates, which is the opposite of the more conventional pattern of long-term interest rates being higher than short-term interest rates. For more details, see, for example, [4]. In this section, we show how to calculate a time series of the slope of the yield curve using the 3-Month Treasury Bill (DGS3MO) and 30-Year Treasury Bond (DGS30) from 2007 to 2019.

**Step 1: Obtain Data for Three Month and 30-Year Treasury from 2007 to 2019**
We already imported into R the data for DGS3MO (*t30mo*) and DGS30 (*t30yr*). Thus, we now use merge() to combine the two datasets. The all = TRUE argument tells R to merge the data regardless of one dataset not having a date available in the other dataset. If one dataset does not have a value for a particular date, R will enter a value of NA. We can see this in the output below for DGS3MO on the first 3 days of 1977 (Lines 4–6). Since we only need data from 2007 to 2019, we subset the data to that time period. The final *slope* dataset has 3253 observations. In the last three lines of the code, we tried to delete any day in which either the DGS3MO or DGS30 had an NA using na.omit(). As we can see from the output, there were no observations during 2007–2019 in which either the DGS3MO or DGS30 had an NA.

```
1    > slope <- merge(t3mo, t30yr, all = TRUE)
2    > head.tail(slope)
3              DGS3MO DGS30
4    1977-02-15    NA  7.70
5    1977-02-16    NA  7.67
6    1977-02-17    NA  7.67
7              DGS3MO DGS30
8    2019-12-30  1.57  2.34
9    2019-12-31  1.55  2.39
10   2020-01-02  1.54  2.33
11   >
12   > slope <- slope["2007-01-01/2019-12-31"]
13   > head.tail(slope)
14             DGS3MO DGS30
15   2007-01-02  5.07  4.79
16   2007-01-03  5.05  4.77
17   2007-01-04  5.04  4.72
18             DGS3MO DGS30
19   2019-12-27  1.57  2.32
20   2019-12-30  1.57  2.34
21   2019-12-31  1.55  2.39
22   > dim(slope)
23   [1] 3253    2
24   > slope <- na.omit(slope)
```

```
25  > dim(slope)
26  [1] 3253    2
```

**Step 2: Plot Yield Data with Shading for 2008–2009 Financial Crisis Period**
Sometimes we may want to create a chart and highlight the portion of the data that
occurs during a certain time period. For example, below, we show how to create a
shaded box during the financial crisis period to highlight the crisis period yields. To
mark the date range of the most recent recession, we will add a shaded rectangular
box to the chart to denote December 1, 2007 to June 1, 2009, which are the dates
identified by the NBER Dating Committee as the official start and end dates of
the last recession. The first step in doing this is to use par("usr") to report the
coordinates of the chart (Line 12). Then, we use rect() to draw a rectangle (Lines
14–19), where the coordinates are given by (1) our first date, (2) the coordinates for
the top of the chart (shade[2]), (3) our last date, and (4) the coordinates for the
bottom of the chart (shade[3]). Finally, using box() we redraw the original
chart border because the upper border gets covered by the pink shaded rectangle
(Line 17). Then, in Lines 20 and 21, we redraw the lines for the DGS3MO and
DGS30 yields in order for those lines to show up on top of the shaded area. We add
a legend using the code in Lines 22–25. Figure 9.5 shows the result of our plot. As
the chart shows, DGS3MO yields at the end of 2019 were much higher than the low
yields coming out of the crisis in mid-2009, while DGS30 yields are lower at the
end of 2019 than they were coming out of the financial crisis in mid-2009.

```
1   > (y.range <- range(slope))
2   [1] 0.00 5.35
3   >
4   > plot(x = index(slope),
5   +      xlab = "Date",
6   +      y = slope$DGS30,
7   +      ylab = "Yield (Percent)",
8   +      ylim = y.range,
9   +      type = "l",
10  +      col = "blue",
11  +      main = "Yields on 3-Month and 30-Year Treasuries, 2007 - 2019")
12  > (shade <- par("usr"))
13  [1] 13325.160 18450.840    -0.214     5.564
14  > rect(as.Date("2007-12-01"),
15  +      shade[2],
16  +      as.Date("2009-06-01"),
17  +      shade[3],
18  +      col="pink",
19  +      lty=0)
20  > box(which = "plot", lty = 1)
21  > lines(x = index(slope), y = slope$DGS30, col = "blue")
22  > lines(x = index(slope), y = slope$DGS3MO, col = "darkgreen")
23  > legend("topright",
24  +      c("3-Month Treasury", "30-Year Treasury"),
25  +      col = c("blue", "darkgreen"),
26  +      lty = 1)
```

**Fig. 9.5** Chart of yields on short-term and long-term Treasury securities with shading for the financial crisis, 2007–2019. Data Source: Federal Reserve Electronic Database

**Step 3: Calculate Slope of the Yield Curve** Another interesting chart for us to look at is the slope of the yield curve (i.e., difference between DGS30 and DGS3MO) over the same period. In this exercise, we multiply the yields by 100 in order to represent the spread in basis points (bps), which is a common convention when reporting yields. Note that 100bps is equal to 1%.

```
1   > slope$slope <- (slope$DGS30 - slope$DGS3MO) * 100
2   > head.tail(slope)
3              DGS3MO DGS30 slope
4   2007-01-02   5.07  4.79   -28
5   2007-01-03   5.05  4.77   -28
6   2007-01-04   5.04  4.72   -32
7              DGS3MO DGS30 slope
8   2019-12-27   1.57  2.32    75
9   2019-12-30   1.57  2.34    77
10  2019-12-31   1.55  2.39    84
```

**Slope of the US Treasury Yield Curve from 2007 to 2019
Based on the Spread Between the
3−Month and 30−Year Treasury Yields**

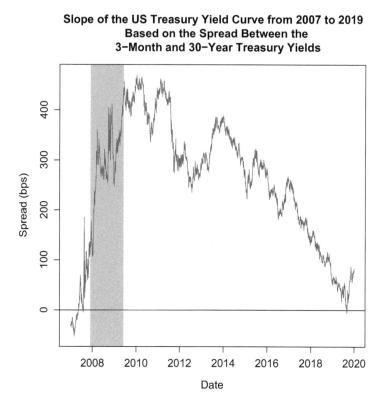

**Fig. 9.6** Slope of treasury yield curve based on spread of 3-month and 30-year treasuries, 2007–2019. Data Source: Federal Reserve Electronic Database and author's calculations

**Step 4: Plot Slope of the Yield Curve** Using the same technique as in Step 3, we plot the slope of the yield curve with the financial crisis period shaded. Figure 9.6 shows the output of the code below. The figure shows that the long-term rates were pretty high relative to short-term rates coming out of the crisis with the difference exceeding 400 basis points during certain times in 2010 and 2011. Over the next 8 years, the difference in yields between DGS30 and DGS3MO exhibited a declining trend. By the end of 2019, that difference in yields has diminished and in 2019 there was a period in which the difference in the yields was virtually zero.

```
 1  > plot(x = index(slope),
 2  +     xlab = "Date",
 3  +     y = slope$slope,
 4  +     ylab = "Spread (bps)",
 5  +     type = "l",
 6  +     col = "blue",
 7  +     main = "Slope of the US Treasury Yield Curve from 2007 to 2019
 8  +     Based on the Spread Between the
 9  +     3-Month and 30-Year Treasury Yields")
10  > (shade <- par("usr"))
```

```
11   [1] 13325.16 18450.84   −72.88    490.88
12   > rect(as.Date("2007−12−01"),
13   +      shade[2],
14   +      as.Date("2009−06−01"),
15   +      shade[3],
16   +      col = "pink",
17   +      lty = 0)
18   > box(which = "plot")
19   > abline(h = 0, lty = 1)
20   > lines(x = index(slope), y = slope$slope, col = "blue")
```

### 9.2.3   Real Yields on US Treasuries

The **real yield** or inflation-adjusted yield on Treasuries is an indicator of economic growth. The real yield falls when economic growth falls, and the real yield rises when economic growth rises. Real yields can be observed by looking at the yields on Treasury Inflation Protected Securities (TIPS). The inflation protection comes from the fact that the principal and coupon of the TIPS are tied to the consumer price index, the change in which is commonly used as the rate of inflation. Another important reason to look at real yields is to observe how the real interest rate is moving. Recall that the real yield is equal to the nominal yield minus the inflation rate. Therefore, when we see the nominal yield decline, it could be a result of the real yield declining or the inflation rate declining or both. Consider the case in which we observe the nominal yield decline from 5% to 3%, which is a 2% decline. If the inflation rate declined from 3% to 1%, the real yield in both instances would be the same at 2% (i.e., $5\% - 3\% = 2\%$ and $3\% - 1\% = 2\%$). This can be revealed by analyzing the real yields.

**Step 1: Obtain 10-Year TIPS from 2007 to 2019** We use the load.fred() function we created earlier to import the 10-year TIPS data from FRED labeled DFII10.csv. We then subset the TIPS data to the period January 1, 2007 to December 31, 2019.

```
1    > tips <− load.fred("DFII10.csv", "DFII10")
2    > head.tail(tips)
3                DFII10
4    2003−01−02   2.43
5    2003−01−03   2.43
6    2003−01−06   2.46
7                DFII10
8    2019−12−30   0.15
9    2019−12−31   0.15
10   2020−01−02   0.08
11   >
12   > tips <− tips["2007−01−01/2019−12−31"]
13   > head.tail(tips)
14                DFII10
```

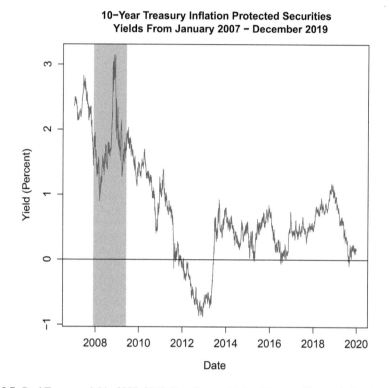

**Fig. 9.7**  Real Treasury yields, 2007–2019. Data Source: Federal Reserve Electronic Database and author's calculations

```
15    2007–01–02    2.38
16    2007–01–03    2.36
17    2007–01–04    2.36
18                  DFII10
19    2019–12–27    0.15
20    2019–12–30    0.15
21    2019–12–31    0.15
```

**Step 2: Plot the Real Yield Data**  Using the same methodology we previously used to plot a chart with shading for the financial crisis period, we plot the TIPS data. Figure 9.7 shows the output of our code. As the chart shows, real yields began a sharp decline after the crisis and turned negative for parts of 2012 and 2013. Even though real yields are generally positive again, the real yield levels through 2019 were still lower than their pre-crisis levels.

```
1   > plot(x = index(tips),
2   +     xlab = "Date",
3   +     y = tips$DFII10,
4   +     ylab = "Yield (Percent)",
5   +     type = "l",
```

```
 6   +      col = "blue",
 7   +      main = "10–Year Treasury Inflation Protected Securities
 8   +      Yields From January 2007 – December 2019")
 9   > (shade <– par("usr"))
10   [1] 13325.1600 18450.8400    –1.0308      3.3108
11   > rect(as.Date("2007–12–01"),
12   +      shade[2],
13   +      as.Date("2009–06–01"),
14   +      shade[3],
15   +      col = "pink",
16   +      lty = 0)
17   > box(which = "plot", lty = 1)
18   > lines(x = index(tips), y = tips$DFII10, col = "blue")
19   > abline(h = 0, lty = 1)
```

### 9.2.4   Expected Inflation Rates

The **breakeven rate** of TIPS is equal to the yield on conventional Treasury securities less the yield on TIPS. This difference can be interpreted as the market's **expected inflation** over the term of the TIPS/Treasury security. Below, we use 10-year TIPS and 10-year US Treasury Bond yields to infer the market's expected inflation over the next 10 years.

**Step 1: Obtain 10-Year TIPS and Treasury Data** We previously imported the TIPS data (*tips*) and 10-Year Treasury data (*t10yr*) into R. If they are no longer in the R environment, we can use the load.data() function we created earlier to import them again. Then, using merge() we combine the two yield series. We then use na.omit() to delete all the observations in which there is at least an NA in either the TIPS or Treasury yield. This results in a combined data that begins in January 2007. The last two lines of code show that the final dataset has 3253 observations. Given we are looking at 13 years of data and assuming 252 trading days, we can expect to see 3276 observations in the dataset. Over such a long period, having 23 fewer observations than what we would expect does not in and of itself give us pause.

```
 1   > exp.infl <– merge(tips, t10yr, all = TRUE)
 2   > head.tail(exp.infl)
 3                    DFII10 DGS10
 4   1962–01–02      NA  4.06
 5   1962–01–03      NA  4.03
 6   1962–01–04      NA  3.99
 7                    DFII10 DGS10
 8   2019–12–30    0.15  1.90
 9   2019–12–31    0.15  1.92
10   2020–01–02      NA  1.88
11   >
12   > exp.infl <– na.omit(exp.infl)
13   > head.tail(exp.infl)
```

```
14                  DFII10 DGS10
15   2007–01–02   2.38  4.68
16   2007–01–03   2.36  4.67
17   2007–01–04   2.36  4.62
18                  DFII10 DGS10
19   2019–12–27   0.15  1.88
20   2019–12–30   0.15  1.90
21   2019–12–31   0.15  1.92
22   > dim(exp.infl)
23   [1] 3253    2
```

**Step 2: Calculate Inflation Expectations** We calculate the market's expected inflation by subtracting the 10-Year TIPS yield from the 10-Year Treasury yield. Since we are performing this calculation using 10-year nominal and real yields, this difference can be interpreted as the market's average annual inflation expectation over the next 10 years.

```
1   > exp.infl$infl.10yr <- exp.infl$DGS10 − exp.infl$DFII10
2   > head.tail(exp.infl)
3                  DFII10 DGS10 infl.10yr
4   2007–01–02   2.38  4.68      2.30
5   2007–01–03   2.36  4.67      2.31
6   2007–01–04   2.36  4.62      2.26
7                  DFII10 DGS10 infl.10yr
8   2019–12–27   0.15  1.88      1.73
9   2019–12–30   0.15  1.90      1.75
10  2019–12–31   0.15  1.92      1.77
```

**Step 3: Plot Expected Inflation** Using a similar methodology from our previous few charts, we plot the expected 10-year inflation rate and shade the crisis period. Figure 9.8 shows the market's inflation expectation from 2007 to 2019. The chart shows that, during the last recession, the 10-year expected inflation fell sharply to almost zero. Coming out of the recession, the 10-year expected inflation was right around its post-crisis average of approximately 2%. In addition, we can also compare the expected inflation implied by these market yields to other sources of inflation projections. A common source of inflation expectations is the Livingston Survey by the Federal Reserve Bank of Philadelphia. The Livingston Survey is the oldest continuous survey of economists' expectations. In the December 2019 Livingston Survey, the long-term outlook for average annual inflation for the next 10 years is 2.22%, which is above the market's 10-year inflation expectation in December 2019 of below 2.0%.

```
1   > (avg.infl <- mean(exp.infl$infl.10yr["2010–01–01/"]))
2   [1] 1.98924
3   > plot(x = index(exp.infl),
4   +      y = exp.infl$infl.10yr,
5   +      xlab = "Date",
6   +      ylab = "Inflation Rate (%)",
7   +      type = "l",
8   +      col = "blue",
9   +      main = "Market's Expectation of 10–Year Inflation (2007 − 2019)")
```

## Market's Expectation of 10-Year Inflation (2007 − 2019)

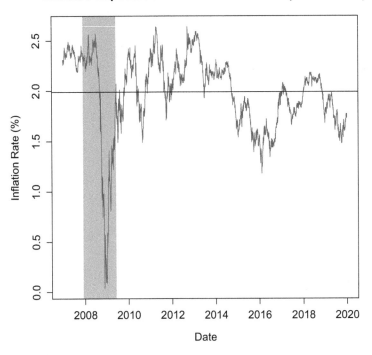

**Fig. 9.8** Expected 10-year inflation rates based on breakeven rate of TIPS, 2007–2019. Data Source: Federal Reserve Electronic Database and author's calculations

```
10   > (shade <- par("usr"))
11   [1] 13325.160 18450.840    -0.064     2.744
12   > rect(as.Date("2007-12-01"),
13   +      shade[2],
14   +      as.Date("2009-06-01"),
15   +      shade[3],
16   +      col = "pink",
17   +      lty = 0)
18   > box(which = "plot", lty = 1)
19   > lines(x = index(exp.infl), y = exp.infl$infl.10yr, col = "blue")
20   > abline(h = avg.infl)
```

**Step 4: Compare Expected Inflation with Actual Inflation** The expected inflation is a forward-looking measure of inflation. As such, it can give very different numbers from historical inflation, such as those calculated based on changes in the CPI. Below, we create a chart that compares the historical inflation with the expected inflation. To calculate the historical inflation, we use the changes in CPI data we calculated earlier (*us.cpi*) and subset the data to the period 2007–2019. We do this so that both datasets show inflation over the same time period. The inflation based on changes in CPI is in basis points, so we convert that to percentage points by dividing the values by 100. Then, we plot the historical inflation and expected inflation by stacking one chart on top of the other for ease of comparison. We do this using the mfrow = c(2, 1) in Line 12. We use the code in Line 19 to revert any future charts back to the typical one chart on one page layout. If we do not run the code in Line 19, all subsequent charts will assume the 2 × 1 stacked layout we use in this example.

```
1   > cpi.sub <- us.cpi$inflation["2007-01-01/2019-12-31"]
2   > head.tail(cpi.sub)
3                inflation
4   2007-01-01  2.075643
5   2007-02-01  2.415199
6   2007-03-01  2.778779
7                inflation
8   2019-10-01  1.764043
9   2019-11-01  2.051278
10  2019-12-01  2.285130
11  >
12  > par(mfrow = c(2, 1))
13  > plot(cpi.sub,
14  +     col = "blue",
15  +     main = "Inflation Based on Changes in CPI")
16  > plot(exp.infl$infl.10yr,
17  +     col = "red",
18  +     main = "10-Year Inflation Based on Market Expectations")
19  > par(mfrow = c(1, 1))
```

The result of the above code is shown in Fig. 9.9. We can see that the expected inflation ranges from slightly above zero around the start of 2009 to a high of 12.6%. By contrast, historical inflation fell to −2% in mid-2009 and was 5.6% in 2008. Therefore, historical inflation had a much wider range of values during the crisis compared to the values implied by looking at the TIPS. By the end of 2019, the difference is much closer. During that time, historical inflation is around 2% while the expected inflation is a little less than 2%. Note, however, we have to be a little careful with a temporal comparison between these two series because the expected inflation calculated from the 10-Year TIPS is the inflation expected over the 10-year period of the security.

### 9.2.5  Mean Reversion

When yields are predictable, there is opportunity for investors to make money. One indicator of return predictability is when yields are **mean reverting**. This means that if yields today are higher than some historical average, yields would have a tendency to decline towards the mean in the future. Conversely, if yields today are lower than some historical average, yields would have a tendency to rise towards the mean in the future. We implement an analysis to determine whether yields for

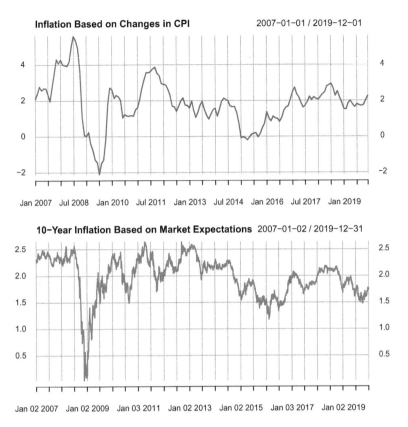

**Fig. 9.9** Expected and historical US inflation rates, 2007–2019. Data Source: Federal Reserve Electronic Database

the 10-Year Constant Maturity Treasury exhibit mean reversion. In this analysis, we will use a 200-week moving average of yields.

**Step 1: Obtain 10-Year Weekly Treasury Data** We use the `load.fred()` function to import the data in **WGS10YR.csv** into R.

```
 1   > cmt <- load.fred("WGS10YR.csv", "WGS10YR")
 2   > head.tail(cmt)
 3                WGS10YR
 4   1962-01-05    4.03
 5   1962-01-12    4.06
 6   1962-01-19    4.11
 7                WGS10YR
 8   2019-12-13    1.84
 9   2019-12-20    1.91
10   2019-12-27    1.90
```

**Step 2: Calculate 200-Week Moving Average** Using `rollmean()` and `k = 200`, we calculate the 200-week moving average of the yields. Since we are doing a

200-week moving average, the first 199 observations of roll will have NAs because it will not have 200 weeks of data to calculate a moving average.

```
1  > cmt$roll <- rollmeanr(cmt$WGS10YR, k = 200, fill = NA)
2  > cmt[195:205, ]
3              WGS10YR roll
4  1965-09-24    4.29   NA
5  1965-10-01    4.34   NA
6  1965-10-08    4.33   NA
7  1965-10-15    4.32   NA
8  1965-10-22    4.36   NA
9  1965-10-29    4.39 4.09
10 1965-11-05    4.43 4.09
11 1965-11-12    4.46 4.09
12 1965-11-19    4.45 4.09
13 1965-11-26    4.45 4.09
14 1965-12-03    4.50 4.10
```

**Step 3: Subset Data from 1970 to 2019**  We subset the data to only include data from January 1970 to December 2019. We use the xts-style subsetting of the dates between January 1970 and December 2019.

```
1  > class(cmt)
2  [1] "xts" "zoo"
3  > cmt <- cmt["1970-01-01/2019-12-31"]
4  > head.tail(cmt)
5              WGS10YR roll
6  1970-01-02    7.94 5.62
7  1970-01-09    7.93 5.64
8  1970-01-16    7.82 5.65
9              WGS10YR roll
10 2019-12-13    1.84 2.32
11 2019-12-20    1.91 2.32
12 2019-12-27    1.90 2.32
```

**Step 4: Plot Yield and Moving Average**  We use the plot() to create a line chart of the 10-year Treasury yield. Then, we add a dashed line for the 200-week moving average. Next, we add a legend. The output of our plot is shown in Fig. 9.10. As the chart shows, the 10-Year Treasury yields follow closely its 200-week moving average.

```
1  > (y.range <- range(cmt))
2  [1]  1.38 15.68
3  >
4  > pdf(file = "fig-bonds-mean_reversion.pdf")
5  > plot(x = index(cmt),
6  +      y = cmt$WGS10YR,
7  +      ylim = y.range,
8  +      type = "l",
9  +      col = "blue",
10 +      main = "10-Year Treasury Yield & 200-Week Moving Average Yield
11 +      1970 - 2019")
12 > lines(x = index(cmt), y = cmt$roll, col = "red")
```

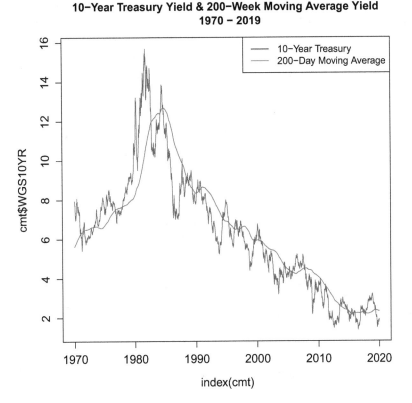

**Fig. 9.10** Mean reversion in 10-year Treasury yields, 1970–2019. Data Source: Federal Reserve Electronic Database and author's calculations

```
13  > legend("topright",
14  +     c("10-Year Treasury", "200-Day Moving Average"),
15  +     lty = c(1, 1),
16  +     col = c("black", "red"))
```

## 9.3   Principal Components Analysis

One of the common results in fixed income analysis is that virtually all of the variation in interest rates is a result of the level of interest rates. The slope of the term structure and curvature of the yield curve explain the remaining variation. This result can be verified through the use of **principal components analysis** (PCA).

**Step 1: Create Dataset of Key Rates** We use the *treasury* dataset we obtained before. It is important for this analysis to have a complete series for all Treasury securities. As such, we will choose DGS3MO, DGS6MO, DGS1, DGS3, DGS5,

DGS7, and DGS10 and we will perform the analysis over the 30-year period from the period 1990 to 2019. Thus, we tell R to keep columns 1 to 3 and columns 5 to 8. We can then check using sum() and is.na() to count the number of NAs in the dataset and, as confirmed by the output, we have zero NAs in *key.rates*.

```
1  > key.rates <- subset(treasury, index(treasury) >= "1990-01-01")
2  > key.rates <- key.rates[, c(1:3, 5:8)]
3  > sum(is.na(key.rates))
4  [1] 0
5  > head.tail(key.rates)
6            DGS3MO DGS6MO DGS1 DGS3 DGS5 DGS7 DGS10
7  1990-01-02  7.83   7.89 7.81 7.90 7.87 7.98  7.94
8  1990-01-03  7.89   7.94 7.85 7.96 7.92 8.04  7.99
9  1990-01-04  7.84   7.90 7.82 7.93 7.91 8.02  7.98
10           DGS3MO DGS6MO DGS1 DGS3 DGS5 DGS7 DGS10
11 2019-12-30  1.57   1.60 1.57 1.59 1.68 1.81  1.90
12 2019-12-31  1.55   1.60 1.59 1.62 1.69 1.83  1.92
13 2020-01-02  1.54   1.57 1.56 1.59 1.67 1.79  1.88
```

**Step 2: Run the PCA Analysis** Using prcomp() in the stats package, we can calculate the principal components. We use the argument scale. = TRUE, which tells R to scale the variables to have unit variance before the analysis takes place. Note that there is a dot (.) after scale.

```
1  > library(stats)
2  > pca <- prcomp(key.rates, scale. = TRUE)
3  > summary(pca)
4  Importance of components:
5                          PC1     PC2    PC3     PC4     PC5     PC6
   PC7
6  Standard deviation     2.5913 0.52052 0.1059 0.04766 0.02040 0.01684 0.01402
7  Proportion of Variance 0.9592 0.03871 0.0016 0.00032 0.00006 0.00004 0.00003
8  Cumulative Proportion  0.9592 0.99794 0.9996 0.99987 0.99993 0.99997 1.00000
```

As the output above shows, the proportion of variance (Line 7) explained by the first principal component is 95.9%. The second principal component adds 3.9%. Essentially, the first two components capture 99.8% of the variation of the series. The third principal component adds 0.2%. Thus, the first three principal components explain virtually all of the variation in the interest rate series.

**Step 3: Interpret Principal Components** We output the eigenvectors using pca$rotation. We interpret the signs as the opposite from what is reported in the output (i.e., negatives are positives and positives are negatives). Thus, we add a negative sign to the beginning of the code to make interpretation more consistent with the output. The first principal component has approximately equal weights at all maturities. Thus, it is often thought to capture the level of interest rates. The second principal component has negative factors in the short maturities, becomes more neutral (i.e., closer to zero) for the intermediate maturities, and becomes positive for the long maturities. Thus, it is believed the second component captures the slope of the term structure. Finally, the third principal component has positive weights in the short and long end of maturities but has negative weights in the intermediate maturity. Thus, the third component is said to capture the curvature of the yield curve.

```
1  > options(digits = 3)
2  > -pca$rotation
3          PC1      PC2      PC3      PC4      PC5      PC6      PC7
4  DGS3MO 0.375 -0.44454  0.4760   0.513   0.3538 -0.0919 -0.1961
5  DGS6MO 0.377 -0.41878  0.1572 -0.187  -0.4871  0.3159  0.5350
6  DGS1   0.380 -0.33425 -0.2351 -0.579  -0.0294 -0.2763 -0.5257
7  DGS3   0.385  0.00952 -0.5970  0.098   0.4929 -0.1340  0.4740
8  DGS5   0.382  0.25555 -0.3017  0.293  -0.1602  0.6485 -0.4065
9  DGS7   0.377  0.39847  0.0372  0.287  -0.5054 -0.5988  0.0291
10 DGS10  0.370  0.53954  0.4946 -0.434   0.3357  0.1357  0.0907
11 > options(digits = 7)
```

## 9.4 Investment Grade Bond Spreads

Treasury securities, like those we looked at in the previous section, are often referred to as default risk-free instruments or, simply, risk-free instruments. However, fixed income securities issued by companies, such as corporate bonds, have varying levels of default risk. Rating agencies, such as Standard & Poor's, Fitch, and Moody's, provide credit ratings to corporate fixed income instruments. Using the Moody's definition, the highest credit rating is Aaa and the lowest investment grade credit rating is Baa. This means that firms or bond issues with less than a Baa rating from Moody's is considered a high-yield bond or a junk bond.

### 9.4.1 Time Series of Spreads

A common analysis is to evaluate the size of the spread between investment grade bonds. As an example, we analyze the period the spread of Aaa and Baa bonds from 1990 to 2019. A below investment grade firm may have limited access to investors' funds, as some types of investors cannot invest in below investment grade debt as a matter of investment policy. The rationale for this is that, empirically, default risk jumps substantially between the lowest investment grade and the highest junk bond. The default risk between investment grade securities are different but generally IG bonds are reasonably safe. As such, Aaa and Baa rated issues (i.e., investment grade bonds) are fighting for access to a similar pool of capital and investors may decide to invest in one over the other depending on the then-existing risk-return trade off. Therefore, evaluating the spread between the yields of Aaa and Baa bonds is a good indicator of the size of **credit risk premium** and, based on the investors' own analysis, the investor can determine whether the spread is sufficiently large or sufficiently small to entice her into making an investment.

**Step 1: Obtain Moody's Aaa and Baa Corporate Index Yield Data for 1990–2019** We downloaded the Moody's Aaa Corporate Bonds Index data as AAA.csv and Baa Corporate Bond Index data as BAA.csv. Then, using load.data() we import the data in these two CSV files into R. We use merge() to combine the two datasets and then subset the data to the period January 1, 1990 to December 31, 2019.

```
1    > aaa <- load.fred("AAA.csv", "AAA")
2    > head.tail(aaa)
3                AAA
4    1919-01-01 5.35
5    1919-02-01 5.35
6    1919-03-01 5.39
7                AAA
8    2019-10-01 3.01
9    2019-11-01 3.06
10   2019-12-01 3.01
11   >
12   > baa <- load.fred("BAA.csv", "BAA")
13   > head.tail(baa)
14                BAA
15   1919-01-01 7.12
16   1919-02-01 7.20
17   1919-03-01 7.15
18                BAA
19   2019-10-01 3.92
20   2019-11-01 3.94
21   2019-12-01 3.88
22   >
23   > moodys <- merge(aaa, baa)
24   > head.tail(moodys)
25                AAA  BAA
26   1919-01-01 5.35 7.12
27   1919-02-01 5.35 7.20
28   1919-03-01 5.39 7.15
29                AAA  BAA
30   2019-10-01 3.01 3.92
31   2019-11-01 3.06 3.94
32   2019-12-01 3.01 3.88
33   >
34   > moodys <- moodys["1990-01-01/2019-12-31"]
35   > head.tail(moodys)
36                AAA   BAA
37   1990-01-01 8.99  9.94
38   1990-02-01 9.22 10.14
39   1990-03-01 9.37 10.21
40                AAA  BAA
41   2019-10-01 3.01 3.92
42   2019-11-01 3.06 3.94
43   2019-12-01 3.01 3.88
```

**Step 2: Set up Data for Plotting** We first create a *date* vector. Using `data.frame()`, we then combine into a data frame object this *date* vector and the *Moody's* dataset.

```
 1  > date <- as.Date(index(moodys))
 2  > moodys <- data.frame(date, moodys)
 3  > rownames(moodys) <- seq(1, nrow(moodys), 1)
 4  > head.tail(moodys)
 5          date AAA    BAA
 6  1 1990-01-01 8.99  9.94
 7  2 1990-02-01 9.22 10.14
 8  3 1990-03-01 9.37 10.21
 9          date AAA  BAA
10  358 2019-10-01 3.01 3.92
11  359 2019-11-01 3.06 3.94
12  360 2019-12-01 3.01 3.88
```

**Step 3: Plot Spread Data** To shade the area in between the two lines, we use `polygon()`. In addition, because we are already doing the shading for the area between the two yields, we should not do shading as well to denote the time period of the financial crisis. As an alternative, we use `abline()` to draw vertical lines and by using `v =`. The output of this code is shown in Fig. 9.11. To see the individual pieces get added, we can split running the following code. We can run Lines 4–12 first, Lines 13–17 second, Lines 18–19 third, then Lines 20–23 last.

```
 1  > (y.range = range(moodys[, 2:3]))
 2  [1]  2.98 10.74
 3  >
 4  > plot(x = moodys$date,
 5  +      xlab = "",
 6  +      y = moodys$AAA,
 7  +      ylab = "Yield (%)",
 8  +      ylim = y.range,
 9  +      type = "l",
10  +      col = "blue",
11  +      main = "Moody's Aaa and Baa Index Yields, 1990 to 2019")
12  > lines(x = moodys$date, y = moodys$BAA, col = "red")
13  > polygon(c(moodys$date, rev(moodys$date)),
14  +      c(moodys$BAA, rev(moodys$AAA)),
15  +      col = "pink",
16  +      density = 20,
17  +      border = NA)
18  > abline(v = c(as.Date("2007-12-01"),
19  +      as.Date("2009-06-01")), col = "darkgreen")
20  > legend("topright",
21  +      c("Aaa", "Baa"),
22  +      lwd = 1,
23  +      col = c("blue", "red"))
```

Figure 9.11 shows the widening of the yields during the crisis period. This is because the yield reflects the cost of borrowing of companies, which increased for firms with lower credit ratings during the crisis.

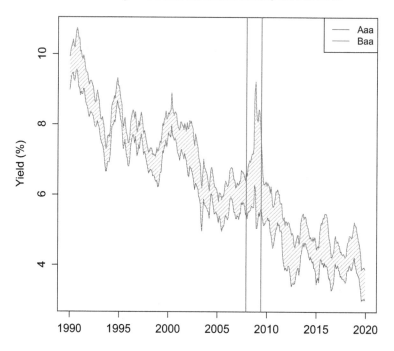

**Fig. 9.11** Spread between investment grade bonds, 1990–2019. Data Source: Federal Reserve Electronic Database

### 9.4.2 Spreads and Real GDP Growth

We again look at the investment grade bond market, but this time in relation to the economic cycle. During times of stress, we expect that higher-rated bonds would perform better than lower-rated bonds. To analyze this relationship, we look at the spreads between investment grade bonds and the real GDP growth rate. In our example, we look at this relationship over the last 30 years from 1990 to 2019. As of the writing of this book, the real GDP yield from FRED only has data through July 2019. Thus, we will perform the analysis below from January 1990 through June 2019.

**Step 1: Calculate the Aaa and Baa Spread from January 1989 to June 2019**
We are using the same *aaa* and *baa* dataset in this analysis as we did in the analysis we did earlier. We merge the two datasets, subset the data to the January 1989 to December 2019 period, and take the difference of the two spreads. When we calculate the spread, we subtract the smaller yield (i.e., Aaa) from the larger yield (i.e., Baa). We also multiply the difference by 100 to format the spread as basis points.

```
1  > spread <- merge(aaa, baa)
2  > spread <- spread["1989-01-01/2019-06-30"]
3  > spread$spread <- (spread$BAA - spread$AAA) * 100
4  > head.tail(spread)
5              AAA  BAA spread
6  1989-01-01 9.62 10.6    103
7  1989-02-01 9.64 10.6     97
8  1989-03-01 9.80 10.7     87
9              AAA  BAA spread
10 2019-04-01 3.69 4.70    101
11 2019-05-01 3.67 4.63     96
12 2019-06-01 3.42 4.46    104
```

**Step 2: Obtain Real GDP Data For January 1989 to June 2019** We can download real GDP data from FRED. The symbol is GPDC1 and save the file as GDPC1.csv. We then subset the data to the period January 1989 through June 2019.

```
1  > rgdp <- load.fred("GDPC1.csv", "GDPC1")
2  > rgdp <- rgdp["1989-01-01/2019-06-30"]
3  > head.tail(rgdp)
4             GDPC1
5  1989-01-01  9102
6  1989-04-01  9171
7  1989-07-01  9239
8             GDPC1
9  2018-10-01 18784
10 2019-01-01 18927
11 2019-04-01 19022
```

**Step 3: Calculate Year-Over-Year GDP Growth** The real GDP data we retrieved is quarterly, so to calculate a year-over-year GDP growth rate we have to create a 4-period lag variable using the Lag() command with k = 4. We then combine the lag variable into *rgdp*, which as we can see from the output shows that we have correctly done the correct number of lags. We then calculate the year-over-year change (in percentage points) as the percentage change between the value under GDPC1 and the value under Lag.GDP multiplied by 100. Using percentage points makes it easier for charting purposes.

```
1  > rgdp$Lag <- Lag(rgdp$GDPC1, k = 4)
2  > head.tail(rgdp)
3             GDPC1 Lag
4  1989-01-01  9102  NA
5  1989-04-01  9171  NA
6  1989-07-01  9239  NA
7             GDPC1   Lag
8  2018-10-01 18784 18322
9  2019-01-01 18927 18438
10 2019-04-01 19022 18598
11 >
12 > rgdp$Rgdp.Growth <- ((rgdp$GDPC1 - rgdp$Lag) / rgdp$Lag) * 100
13 > head.tail(rgdp)
```

```
14                GDPC1 Lag Rgdp.Growth
15   1989-01-01 9102  NA           NA
16   1989-04-01 9171  NA           NA
17   1989-07-01 9239  NA           NA
18                GDPC1   Lag Rgdp.Growth
19   2018-10-01 18784 18322        2.52
20   2019-01-01 18927 18438        2.65
21   2019-04-01 19022 18598        2.28
```

**Step 4: Subset Data to 1990–2019**   The reason we use January 1989 as the starting point earlier was that to make sure the relevant period we are looking at (i.e., January 1990 through December 2019) will not have any NAs because we have to take the lag of certain variables. But at this stage, we can subset the data by limiting it to data between January 1990 to December 2019 using the xts-style subsetting based on dates.

```
1   > rgdp <- rgdp["1990-01-01/2019-12-31"]
2   > head.tail(rgdp)
3                 GDPC1  Lag Rgdp.Growth
4   1990-01-01   9358 9102        2.82
5   1990-04-01   9392 9171        2.41
6   1990-07-01   9398 9239        1.73
7                 GDPC1   Lag Rgdp.Growth
8   2018-10-01 18784 18322        2.52
9   2019-01-01 18927 18438        2.65
10  2019-04-01 19022 18598        2.28
```

**Step 5: Plot 2-Axis Chart**   Because we need the RGDP y-axis to go the opposite direction (i.e., negative at the top and positive at the bottom), we need to construct a 2-axis chart. We first set up two ranges for the two y-axes. Using range(), we find that the values in *spread* go from 3.28 to 338 bp. As such, we set the *y1.range* equal to 0–350 bp. For the range of real GDP growth rates (*y2.range*), we use ceiling() and floor() to get the next biggest integer and the next smallest integer, respectively. Note also the order with which we apply the range. We start with the positive number (+6%) and end with the negative number (−4%). This allows us to invert the axis when charting.

```
1   > range(spread)
2   [1]    3.28 338.00
3   > (y1.range <- c(0, 350))
4   [1]    0 350
5   > (y2.range <- c(ceiling(max(range(rgdp$Rgdp.Growth))),
6   +      floor(min(range(rgdp$Rgdp.Growth)))))
7   [1]   6 -4
```

Let us break the charting itself into several steps. First, we create the plot of the spread as though it was a normal single chart plot, but we add the line  par(mar= c(5, 5, 5, 5)). This creates enough space for us to add a second axis later. The rest of the code, including the shading for the two recession periods (March 2001 to November 2001 and December 2007 to June 2009), follows the same technique as our prior plotting codes.

```
1  > par(mar = c(5, 5, 5, 5))
2  > plot(x = index(spread),
3  +      xlab = "Date",
4  +      y = spread$spread,
5  +      ylab = "Baa − Aaa Spread (bps)",
6  +      ylim = y1.range,
7  +      type = "l",
8  +      col = "darkgreen",
9  +      main = "Moody's Baa Minus Aaa Spread and Real GDP Growth
10 +      1990 − 2019")
11 > (shade <- par("usr"))
12 [1]  6495.68 18492.32  −14.00   364.00
13 > rect(as.Date("2007−12−01"), shade[2],
14 +      as.Date("2009−06−01"), shade[3], col = "pink", lty = 0)
15 > rect(as.Date("2001−03−01"), shade[2],
16 +      as.Date("2001−11−01"), shade[3], col = "pink", lty = 0)
17 > lines(x = index(spread), y = spread$spread, type = "l", col = "darkgreen")
```

Next, we now plot the data for the second axis. The par(new = TRUE) tells
R that we are adding to an existing plot. The xaxt = "n" and yaxt = "n"
suppress the axes labels generated by this second plot command. We manually add
the second axis label using the axis(4) and mtext commands. The side = 4
denotes the right side of the chart and line = 3 starts the text on the third line
from the margin.

```
1  > par(new = TRUE)
2  > plot(x = index(rgdp),
3  +      xlab = " ",
4  +      xaxt = "n",
5  +      y = rgdp$Rgdp.Growth,
6  +      ylab = " ",
7  +      yaxt = "n",
8  +      ylim = y2.range,
9  +      type = "l",
10 +      col = "blue")
11 > axis(4)
12 > mtext("Year−over−Year Real GDP Growth (%)", side = 4, line = 3)
```

Finally, we add a legend to the plot. The legend is quite long, so it overlaps with
the top of the recession period shading. The legend default is to have a transparent
background, so the recession shading actually penetrates the legend making it hard
to read the text in the legend. To prevent this, we add the bg = "white" argument
to fill-in the legend background with the color white.

```
1  > legend("topleft",
2  +      c("Baa−Aaa Spread (Left)", "RGDP Growth (Right, Inverted)"),
3  +      col = c("darkgreen", "blue"),
4  +      lty = 1,
5  +      cex = 0.75,
6  +      bg = "white")
```

Figure 9.12 shows that bond spreads spike shortly after (i.e., with a lag) the recession
began. The biggest spike in spreads occurred during the 2008/2009 recession, where

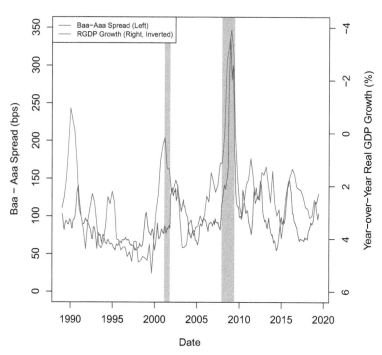

**Fig. 9.12** Spread between investment grade bonds and the economic cycle, 1990–2019. Data Source: Federal Reserve Electronic Database

GDP growth was the most negative. Recall that the right axis was intentionally constructed to be inverted, so that a "spike" in real GDP growth is actually a huge negative year-over-year GDP growth.

## 9.5   Bond Valuation

The value of a bond is the present value of the coupon and principal payments bondholders expect to receive from owning the bond discounted at the appropriate risk-adjusted discount rate. Therefore, to value a bond, we would need to know the bond's promised cash flows and its yield to maturity (YTM). On the other hand, if we know the price of the bond and the bond's cash flows, we can then find the bond's YTM. In this section, we demonstrate how to implement these two calculations.

### 9.5.1 Valuing Bonds with Known Yield to Maturity

We can calculate the value of a bond as the present value of its promised cash flows discounted using the yield to maturity. That is,

$$V_{Bond} = \sum_{t=1}^{Tf} \frac{C/f}{(1+y/f)^t} + \frac{FV}{(1+y/f)^{Tf}}, \qquad (9.1)$$

where $C$ is the coupon payment, $f$ is the frequency of the coupon payment, $y$ is the yield to maturity of the bond, $FV$ is the face value or par value of the bond, and $T$ is the time to maturity of the bond (in years). This formula works best in the context of a plain vanilla bond (i.e., a bond with a fixed rate, fixed maturity, and no embedded options).

Equation (9.1) works well when we value the bond immediately after a coupon payment is made. The reason is that the promised future cash flows are in equally spaced intervals and there is no accrued interest. This means that the investor who purchases a bond that settles on that day will get the full coupon on the next coupon payment date. The price that an investor pays for the bond, including the calculation of the accrued interest, is from the **settlement date** and not the trade date. The settlement date is usually the trade date plus three trading days (T+3) for US corporates. However, the number of days after the trade date can vary substantially. For our purposes, we will assume the trade date and settlement date are the same, but we should keep in mind that the trade date and settlement date can be different.

Let us go through an example of how to use Eq. (9.1) to value a bond. Consider the Wal-Mart Bond due on December 19, 2030 that pays semi-annual coupons with a rate of 5.75%. According to Fixed Income Investor (UK), as of December 26, 2019, the price of that bond was £1441.98 and its annual yield was 1.391%. Below, we will value this bond assuming a valuation date of December 19, 2019.

**Step 1: Set up Variables for Bond Characteristics** Since we are starting a new analysis in this section, we may want to consider clearing the R environment. For purposes of our analysis, we assume we have formed an expectation as to what the yield on this bond should be and, based on that belief, we calculate the bond price. We will assume the yield is 1.391%, which is the yield reported by Fixed Income Investor.

```
1  > rm(list = ls())
2  > coupon <- 0.0575
3  > maturity <- 11
4  > yield <- 0.01391
5  > par <- 1000
6  > coupon.freq <- 2
```

Note that for purposes of this section, we will refer to the price or value of a bond in relation to the £1000 par value. This is not the typical way bond prices are quoted. In general, bond prices are quoted as a percentage of the par value. So a price of

99 means that we would have to pay £990 to purchase a bond with a par value of £1000. For simplicity, we will use the term price to refer to the £990 in the above example.

**Step 2: Create Variables Accounting For the Periodicity of the Coupon Payments** First, we create a *period* variable that takes the value of the time to maturity multiplied by the coupon frequency. With semi-annual coupon payments, the 11-year time to maturity of the Wal-Mart bond means that there will be 22 coupon payment dates. Second, *coupon.period* refers to the coupon payment made per period as a percentage of par. Given the 5.75% coupon and semi-annual coupon payments, the per period coupon payment is 2.875% of par. Third, *yield.period* denotes the yield to maturity per period, so we divide the 1.391% annual yield by 2 for a yield per 6-month period of 0.6955%.

```
1  > (periods <- maturity * coupon.freq)
2  [1] 22
3  > (coupon.period <- coupon / coupon.freq)
4  [1] 0.02875
5  > (yield.period <- yield / coupon.freq)
6  [1] 0.006955
```

**Step 3: Construct a Vector of Per Period Coupon Rates** We create a vector labeled *bond.coupon* that contains the coupon payments for each period. We do this by using `rep()`, which repeats the per period coupon of 2.875% 22 times.

```
1  > bond.coupon <- rep(coupon.period, periods)
2  > bond.df <- as.data.frame(bond.coupon)
3  > head.tail(bond.df)
4     bond.coupon
5  1      0.02875
6  2      0.02875
7  3      0.02875
8     bond.coupon
9  20     0.02875
10 21     0.02875
11 22     0.02875
```

**Step 4: Calculate the Present Value of the Bond's Periodic Cash Flows** We now set up four additional variables: First, in the cf column, we enter the periodic dollar coupon payments of £28.75 for all but the last payment period. The coupon payment of £28.75 every 6 months is calculated by multiplying the periodic coupon rate of 2.875% by the par value of £1000. On the last payment period, we get both the coupon of £28.75 plus the principal of £1000. Second, in the ytm column, we enter the periodic yield of 0.6955%. Third, in the period column, we enter the payment number. We use this when calculating the number of periods to discount the cash flows. Fourth, in the pv.cf column, we enter the present value of each cash flow. For example, the first payment 6 months from the valuation date of £28.75 is equal to a present value of £28.55 as of December 19, 2019. We repeat this calculation for the other 21 payments.

```
1  > bond.df$cf <- coupon.period * par
2  > bond.df$cf[periods] <- bond.df$cf[periods] + par
3  > bond.df$ytm <- yield.period
4  > bond.df$period <- c(1:periods)
5  > bond.df$pv.cf <- bond.df$cf / ((1 + bond.df$ytm)^bond.df$period)
6  > head.tail(bond.df)
7    bond.coupon    cf    ytm period    pv.cf
8  1      0.02875 28.75 0.006955      1 28.55142
9  2      0.02875 28.75 0.006955      2 28.35422
10 3      0.02875 28.75 0.006955      3 28.15838
11    bond.coupon      cf    ytm period      pv.cf
12 20     0.02875   28.75 0.006955     20   25.02860
13 21     0.02875   28.75 0.006955     21   24.85573
14 22     0.02875 1028.75 0.006955     22 883.25981
```

**Step 5: Calculate Value of the Bond** The value of the bond is the sum of the present value of its expected cash flows. We apply sum() to the present value of each period's cash flows to get a value of £1443.18 for the Wal-Mart bond. This is consistent with the price of the bond reported by Fixed Income Investor on December 26, 2019 (i.e., 1 week later) of £1441.98.

```
1  > (value <- sum(bond.df$pv.cf))
2  [1] 1443.184
```

In this section, we discussed the use of yield to maturity to price a bond. In many instances, yield-to-worst is often typically quoted in the market. Yield-to-worst is the same calculation as yield to maturity except that it is performed across all dates in which the bond can be called and the maturity date. As such, the yield-to-worst for a non-callable bond will equal its yield to maturity. It may also be possible that the highest yield is the one that is at the maturity date, so in those situations the yield-to-worst will also equal the yield to maturity.

### 9.5.2   Bond Valuation Function

In the previous section, we went through how to calculate the value of a bond in R by laying out all the bond's cash flows. In this section, we will create a bond valuation function bondprc() that simplifies the way we calculate bond values so that we can value many bonds easily. The structure of the code and the variable names are tied closely to the analysis in the prior section to assist in understanding how this function works.

```
1  > bondprc <- function(coupon, maturity, yield, par, coupon.freq) {
2  +    periods <- maturity * coupon.freq
3  +    coupon.period <- coupon / coupon.freq
4  +    yield.period <- yield / coupon.freq
5  +
6  +    bond.coupon <- rep(coupon.period, periods)
7  +    bond.df <- as.data.frame(bond.coupon)
```

```
 8  +
 9  +      for (i in 1:periods) {
10  +          bond.df$cf[i] <- par * coupon.period
11  +          bond.df$ytm[i] <- yield.period
12  +          bond.df$period[i] <- i
13  +          }
14  +      bond.df$cf[periods] <- bond.df$cf[periods] + par
15  +      bond.df$pv.cf <- bond.df$cf / ((1 + bond.df$ytm)^bond.df$period)
16  +      value <- sum(bond.df$pv.cf)
17  +      return(value)
18  + }
```

We now apply the above function to calculate the value of the Wal-Mart bond. We know the Wal-Mart bond has a coupon rate of 5.75%, time to maturity of 11 years, yield of 1.391%, par of £1000, and semi-annual coupon payments (i.e., coupon frequency of 2). The bondprc() function yields a value of £1443.184, which is identical to the bond value we calculated above.

```
1  > bondprc(0.0575, 11, 0.01391, 1000, 2)
2  [1] 1443.184
```

### 9.5.3   Finding the Yield to Maturity

In the prior sections, we calculated the value of the bond when we know the yield to maturity. In this section, we will estimate the yield to maturity of the bond when we know the market price of the bond. The yield of the bond is important because it is a proxy for the cost of debt. Different bonds can have different yields to maturity, as the yield to maturity is a function of the bond's features such as time to maturity and coupon rate. Unlike calculating the value of the bond using a formula, we cannot do the same when finding the yield to maturity. Therefore, the only way to find the YTM is through trial-and-error. Fortunately, there are algorithms that we can use to automate this trial-and-error process. To find the YTM, we use the uniroot() function. We also create a bondval() function that is a modification of the bondprc() function to make it easier to calculate the YTM. For those familiar with the **internal rate of return** calculation, the YTM is basically the same calculation. In other words, given a bond price and a bond's cash flows, the YTM is the discount rate that would make the net present value of investing in the bond equal to zero.

**Step 1: Create Vector of Bond Cash Flows**  Continuing from our example above, we first construct a vector of the bond's cash flows. Make sure that there is at least one sign change in the cash flows. In our example below, we enter the purchase price of the bond of £1443.184 as a cash outflow and, thus, assign a negative sign to that cash flow. The interim coupon payments and the principal repayment at maturity are then entered as cash inflows (i.e., positive cash flows). Without the sign change, we will not be able to compute the YTM and the calculation returns an error.

```
1  > bond.cf <- c(-value,
2  +      rep(coupon.period * par,periods - 1),
3  +      par+coupon.period * par)
4  > bond.cf
5   [1] -1443.184    28.750    28.750    28.750    28.750    28.750    28.750
6   [8]    28.750    28.750    28.750    28.750    28.750    28.750    28.750
7  [15]    28.750    28.750    28.750    28.750    28.750    28.750    28.750
8  [22]    28.750  1028.750
```

**Step 2: Create Function to Find the Yield to Maturity** We then create the first of two functions. This function calculates the value of the bond. We use the character *i* as the variable that represents the per period yield.

```
1  > bondval <- function(i, bond.cf,
2  +      t = seq(along = bond.cf)) sum(bond.cf / (1 + i)^t)
```

The second function uses the uniroot() function to find the root of the cash flows (i.e., its YTM).

```
1  > bond.ytm <- function(bond.cf) {
2  +      uniroot(bondval,
3  +      c(0, 1), bond.cf = bond.cf)$root}
```

**Step 3: Annualize the Periodic YTM** We calculate the YTM of bond by multiplying the calculated per period yield by the coupon frequency. The result of this calculation is a YTM of 1.39%, which is identical to the YTM of 1.39% we were using in our prior calculations and what we expect. Note that the estimation procedure to calculate the YTM may cause minor differences from the actual YTM (especially differences due to rounding). We used digits = 4 option to reduce the number of decimal places shown to demonstrate that the YTM is effectively 1.391%.

```
1  > (YTM <- bond.ytm(bond.cf) * coupon.freq)
2  [1] 0.01390919
3  > options(digits=4)
4  > YTM
5  [1] 0.01391
6  > options(digits=7)
```

## 9.6 Duration and Convexity

Fixed income instruments are sensitive to changes in interest rates. As such, managing interest rate risk is an important consideration when dealing with fixed income portfolios. **Duration** and **convexity** are two techniques that help investors immunize against small changes in interest rates. The idea behind duration is that it estimates the impact of small changes in interest rates on the value of the fixed income instrument, and the idea behind convexity is to enhance this estimate by also considering the impact of the rate of the change in interest rates.

The Macaulay duration can loosely be interpreted as the weighted-average time to maturity of the bond. Technically, this definition only holds when coupons are reinvested at the YTM of the bond. The Macaulay duration is calculated as:

$$\text{Macaulay Duration} = \frac{\sum_{t=1}^{T} t_t PV(CF_t)}{Price}, \tag{9.2}$$

where $t_t$ is the cash flow period and $PV(CF_t)$ is the present value of the period $t$ cash flows, and $Price$ is the price of the bond.

Modified duration adjusts Macaulay duration by considering the yield of the bond, and gives the price sensitivity of the bond measured as the percentage change in price for a given change in the yield. Modified duration is calculated as:

$$\text{Modified Duration} = \frac{\text{Macaulay Duration}}{1 + YTM/freq}, \tag{9.3}$$

where $YTM$ is the yield to maturity and $freq$ is the coupon frequency.

### 9.6.1  Calculating Duration and Convexity

We now show an example of how to calculate the duration and convexity of the 5.75% Wal-Mart Bond due December 19, 2030. Recall that our estimated value for the bond as of December 19, 2019 was £1443.184. We use our estimated value to show how well duration and convexity adjustments estimate the change in the price of a bond by comparing the resulting estimate to the full valuation of the bond by changing the yield by the same basis point change.

**Step 1: Create Vector of Periods and Bond Cash Flows** We create a *period* variable, which equals the number of the bond's cash flow payment periods. Then, we construct a vector *cf* of the bond's cash flows. This equals the £28.75 on each semi-annual coupon payment period prior to the last coupon payment date. On the last coupon payment date, the bond pays £28.75 of coupons plus the £1000 of principal.

```
 1  > price <- value
 2  > period <- seq(1, maturity * coupon.freq, by = 1)
 3  > cf <- c(rep(par * coupon / coupon.freq,
 4  +      maturity * coupon.freq - 1),
 5  +      par * (1 + coupon / coupon.freq))
 6  > duration <- data.frame(period)
 7  > duration$cf <- cf
 8  > head.tail(duration)
 9     period    cf
10  1       1 28.75
11  2       2 28.75
12  3       3 28.75
```

```
13     period        cf
14  20      20    28.75
15  21      21    28.75
16  22      22  1028.75
```

**Step 2: Calculate Duration of Each Period's Cash Flows**  We need to create two variables before we can calculate the duration. First, in the tcf column, we multiply the period by the cash flow. Second, in the PV.factor column, we enter the present value factor that we apply to the values in tcf to calculate the duration in Line 3.

```
1  > duration$tcf <- duration$period * duration$cf
2  > duration$PV.factor <- (1 + yield / coupon.freq)^(-period)
3  > duration$dur <- (duration$tcf * duration$PV.factor) / price
4  > head.tail(duration)
5    period    cf   tcf PV.factor          dur
6  1      1 28.75 28.75 0.9930930 0.01978364
7  2      2 28.75 57.50 0.9862338 0.03929399
8  3      3 28.75 86.25 0.9794219 0.05853389
9    period    cf      tcf PV.factor          dur
10 20      20 28.75   575.00 0.8705601   0.3468527
11 21      21 28.75   603.75 0.8645471   0.3616798
12 22      22 1028.75 22632.50 0.8585758 13.4644800
```

**Step 3: Calculate Macaulay and Modified Duration**  Using the above equations, we can calculate the Macaulay and Modified durations. The output below shows the Macaulay duration is 8.82 and the Modified duration is 8.76.

```
1  > (mac.duration <- sum(duration$dur) / coupon.freq)
2  [1] 8.816838
3  > (mod.duration <- mac.duration / (1 + yield / coupon.freq))
4  [1] 8.755941
```

**Step 4: Calculate Convexity Adjustment**  We now turn to estimating the convexity of this bond. To calculate the convexity, we construct the variables $c1$ and $c2$, where $c1$ is equal to $(t^2 + t) * cf$ and $c2$ is equal to $price * (1 + YTM/2)^2$. Then, we calculate the per period convexity adjustment $conv$ as $(c1 \times \text{PV.factor})/c2$.

```
1  > convexity <- duration
2  > convexity$c1 <- (period^2 + period) * cf
3  > convexity$c2 <- price * (1 + yield / coupon.freq)^2
4  > convexity$conv <- (convexity$c1 * convexity$PV.factor) / convexity$c2
5  > head.tail(convexity)
6    period    cf   tcf PV.factor          dur    c1       c2       conv
7  1      1 28.75 28.75 0.9930930 0.01978364  57.5 1463.328 0.03902259
8  2      2 28.75 57.50 0.9862338 0.03929399 172.5 1463.328 0.11625919
9  3      3 28.75 86.25 0.9794219 0.05853389 345.0 1463.328 0.23091238
10   period    cf      tcf PV.factor          dur      c1       c2
    conv
11 20      20 28.75   575.00 0.8705601   0.3468527 12075.0 1463.328     7.183634
12 21      21 28.75   603.75 0.8645471   0.3616798 13282.5 1463.328     7.847419
13 22      22 1028.75 22632.50 0.8585758 13.4644800 520547.5 1463.328 305.419876
```

We can then find the convexity by taking the sum of the values in *conv* variable, which yields a convexity adjustment of 367.73.

```
1  > (convexity <- sum(convexity$conv))
2  [1] 367.7321
```

**Step 5: Estimate Price Change Based on Duration**  Duration is typically the measure for a 100bp change in yields. Suppose that yields go up by 100bp, what is the estimated price change due to duration? The answer is given by $\Delta P = -D * \Delta y * P$, where $\Delta P$ is the change in price, $D$ is the Modified Duration, $\Delta y$ is the change in yield (in decimals), and $P$ is the contemporaneous price of the bond. Note that in front of the Modified Duration is a minus sign, because there is a negative or inverse relationship between the change in yield and the change in the bond price. Put differently, an increase in yield will result in a lower bond price and, conversely, a decrease in yield will result in a higher bond price. To estimate the price change due to duration alone, we create a variable called *delta.yld*, which we assume is equal to an increase of 1% or 100 basis points. Using the above equation, we find the estimated price decline based on duration alone of £126. As such, given a price of £1443 and an increase in yield of 100 bp, the estimated new price based on the duration adjustment is £1317.

```
1  > delta.yld <- 0.01
2  > (delta.dur <- -mod.duration * delta.yld * price)
3  [1] -126.3643
4  > (px.dur <- price + delta.dur)
5  [1] 1316.819
```

**Step 6: Estimate Price Change Using Duration and Convexity**  We now turn to estimate the change in price based on duration and convexity. This is given by the formula $\Delta P = P * [(-D * \Delta y) + (0.5 * C * (\Delta y)^2))]$, where $C$ is the Convexity and the other terms are the same as defined above for duration. The estimate of the price decline with the duration and convexity adjustment is equal to approximately *negative* £100. We find that the price of the Wal-Mart bond making the adjustments for Duration and Convexity is equal to £1343.

```
1  > (delta.dur.conv <- ((-mod.duration * delta.yld) +
2  +     (0.5 * convexity * (delta.yld^2))) * price)
3  [1] -99.82905
4  > (px.dur.conv <- price + delta.dur.conv)
5  [1] 1343.354
```

**Step 7: Summarize Effects of Duration and Convexity**  We can summarize the results of the duration and convexity calculations into one output.

```
1  > (est.change <- cbind(price, delta.dur, px.dur,
2  +     delta.dur.conv, px.dur.conv))
3          price delta.dur    px.dur delta.dur.conv px.dur.conv
4  [1,] 1443.184 -126.3643 1316.819      -99.82905    1343.354
```

**Step 8: Compare Result to Full Valuation** We can then compare the estimated price with the bond price using full valuation. To get the full valuation price of the bond, we use the `bondprc()` function we constructed in the prior section but increasing the yield by 100bp. The full valuation estimates the bond price is £1323, which is in between our estimated price change based on duration only and duration plus convexity.

```
1  > bondprc(0.0575, 11, 0.01391 + delta.yld, 1000, 2)
2  [1] 1323.208
```

## 9.6.2 Duration and Convexity Functions

To simplify the above process, we can create a function for duration and convexity. The function to calculate duration based on the above methodology is as follows:

```
1  > duration <- function(price, yield, coupon.freq, par){
2  +     period <- seq(1, maturity * coupon.freq, by = 1)
3  +     cf <- c(rep(par * coupon / coupon.freq,
4  +         maturity * coupon.freq - 1),
5  +         par * (1 + coupon / coupon.freq))
6  +     tcf <- period * cf
7  +     dur <- tcf * (1 + yield / coupon.freq)^(-period) / price
8  +     dx <- sum(dur / coupon.freq) / (1 + yield / coupon.freq)
9  +     return(dx)
10 + }
11 > duration(price, yield, coupon.freq, par)
12 [1] 8.755941
```

The function to calculate convexity is as follows:

```
1  > convexity <- function(price, yield, coupon.freq, par) {
2  +     period <- seq(1, maturity * coupon.freq, by = 1)
3  +     cf <- c(rep(par * coupon / coupon.freq,
4  +         maturity * coupon.freq - 1),
5  +         par * (1 + coupon / coupon.freq))
6  +     num <- (period^2 + period) * cf * (1 + yield / coupon.freq)^(-period)
7  +     denom <- price * (1 + yield / coupon.freq)^2
8  +     convex <- sum(num / denom)
9  +     return(convex)
10 + }
11 > convexity(price, yield, coupon.freq, par)
12 [1] 367.7321
```

Note that some of the steps from the prior subsection have been combined when creating the function. But, as the results show, the duration and convexity calculated by the function are identical to the results we calculated earlier. Similarly, we can verify that we get the same estimates of price as we did previously.

```
1  > delta.yld <- 0.01
2  > (delta.dur <- -duration(price, yield, coupon.freq, par) *
```

```
3  +     delta.yld * price)
4  [1] −126.3643
5  > (px.dur <− price + delta.dur)
6  [1] 1316.819
7  > (delta.dur.conv <− ((−duration(price, yield, coupon.freq, par) *
8  +     delta.yld) + (0.5 * convexity(price, yield, coupon.freq, par) *
9  +     (delta.yld^2))) * price)
10 [1] −99.82905
11 > (px.dur.conv <− price + delta.dur.conv)
12 [1] 1343.354
```

As we can see, these duration and convexity functions give us the same estimated price as we previously calculated.

### 9.6.3  Comparing Estimates of Value to Full Valuation

It may be easier to understand how duration and convexity stack up compared to a full valuation of the bond at different yields. We know the correct value of the bond is given by the full valuation. We can use the bondprc() function we created above to calculate the bond value at different yield levels. Then, we would want to use the duration() and convexity() functions to estimate the price change. We can then compare these estimates with the correct bond value using full valuation.

For this example, we use a hypothetical bond with price of $1000, coupon of 5%, yield of 5%, a 10 year maturity, and pays coupons once per year.

```
1  > price <− 1000
2  > coupon <− 0.05
3  > yield <− 0.05
4  > maturity <− 10
5  > coupon.freq <− 1
```

We now calculate the bond value under full valuation, the estimate of the value based on duration alone, and the estimate of the value based on duration and convexity under different yields and changes in yields. We estimate each of these by adding and subtracting from the base yield of 5%. We do this by first creating a yield variable that runs from 0.1% to 20.0% in increments of 0.01%. Then, we create a variable *delta.yld* that gives us the difference between each of the yields in *yld* and the current yield of 5.0%. Recall that duration is an estimate of the change in price given a change in yield, so we need to identify the baseline for the yield. In Lines 3–5, we create three variables that we will use to store these value estimates. We then run the full valuation, duration, and duration plus convexity calculations in a loop (Lines 6–16).

```
1  > yld <− seq(0.001, 0.2, .001)
2  > delta.yld <− yld − 0.05
3  > px <− rep(0, 200)
4  > dur.mat <− rep(0, 200)
```

```
5   > dur.conv.mat <- rep(0, 200)
6   > for (i in 1:200) {
7   +      new.yld <- 0.05 + delta.yld[i]
8   +      px[i] <- bondprc(coupon, maturity, new.yld, par, coupon.freq)
9   +      delta.dur <- -duration(price, yield, coupon.freq, par) *
10  +          delta.yld[i] * price
11  +      dur.mat[i] <- price + delta.dur
12  +      delta.dur.conv <- ((-duration(price, yield, coupon.freq, par) *
13  +          delta.yld[i] + (0.5 * convexity(price, yield, coupon.freq, par) *
14  +          (delta.yld[i]^2))) * price)
15  +      dur.conv.mat[i] <- price + delta.dur.conv
16  + }
17  > combo <- data.frame(cbind(yld, delta.yld, px, dur.mat, dur.conv.mat))
18  > head.tail(combo)
19      yld delta.yld        px  dur.mat dur.conv.mat
20  1 0.001    -0.049 1487.316 1378.365     1468.400
21  2 0.002    -0.048 1474.762 1370.643     1457.041
22  3 0.003    -0.047 1462.337 1362.922     1445.756
23        yld delta.yld        px   dur.mat dur.conv.mat
24  198 0.198     0.148 375.2772 -142.8168     678.5578
25  199 0.199     0.149 373.1949 -150.5385     681.9733
26  200 0.200     0.150 371.1292 -158.2602     685.4637
```

The *combo* dataset contains the full valuation, value as estimated by duration only, and value as estimated by duration plus convexity. It is easier to compare the full valuation with the estimates when we plot the data. The code to plot the data is below and the chart it generates is shown in Fig. 9.13.

```
1   > plot(x = yld,
2   +      xlab = "Yield",
3   +      y = combo$px,
4   +      ylab = "Value",
5   +      type = "l",
6   +      lwd = 2,
7   +      main = "Full Valuation, Duration Estimate
8   +      Amd Duration Plus Convexity Estimate")
9   > lines(x = yld,
10  +      y = combo$dur.mat,
11  +      col = "red")
12  > lines(x = yld,
13  +      y = combo$dur.conv.mat,
14  +      col = "blue")
15  > abline(h = 1000, lty = 3)
16  > legend("topright",
17  +      c("Full Valuation", "Duration Estimate",
18  +        "Duration Plus Convexity Estimate",
19  +        "Price of $1,000"),
20  +      col = c("black", "red", "blue", "black"),
21  +      lwd = c(2, 1, 1, 1),
22  +      lty = c(1, 1, 1, 3))
```

As Fig. 9.13 shows, the full valuation (black line) is the actual value of the bond as we discounted the bond's cash flows at the different yields. This is our baseline. The duration estimate (red line) and duration plus convexity estimate (blue line) are

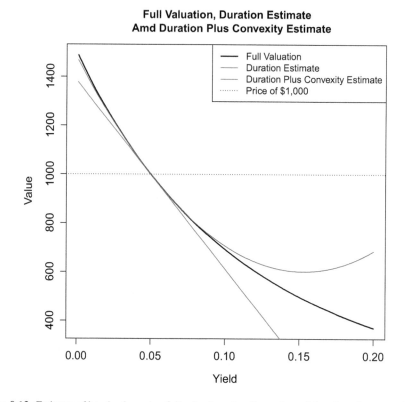

**Full Valuation, Duration Estimate Amd Duration Plus Convexity Estimate**

**Fig. 9.13** Estimate of bond value using full valuation, duration only, and duration plus convexity

equal to the full valuation of the bond when the yield is equal to the base yield of 5%. That is the point where all three lines intersect with the horizontal black line. We also see from the chart that, regardless of whether the yield changes are small or large, the duration estimate is always lower than the full valuation. The amount of understatement increases the farther away from 5% the yield becomes (i.e., larger yield changes). As such, duration is a good estimate for values close to the current yield of 5%, but the quality of the estimate deteriorates as yield changes become larger. We can also see from the chart that adding the convexity adjustment improves the estimates of bond value for larger yield changes relative to the base yield. The blue line is much closer to the black line for a much longer period. However, when the yield increase gets sufficiently large, we can see that the duration plus convexity estimate overestimates the bond's value.

One may wonder why then do we bother with duration and convexity when we can simply value the bond. The reason is that for a portfolio with many bonds, full valuation is a costly exercise in terms of time and resources. By contrast, portfolio duration is the weighted-average duration of the bonds in the portfolio with the weights equal to the market value of the bonds relative to all the other bonds.

Thus, using duration and convexity could be much more efficient to implement especially in portfolios that have many bonds and, as we have seen, using duration and convexity results in reasonable estimates of the bond's value for small changes in yield.

## 9.7   Short Rate Models

In this section, we will look at the valuation of zero coupon bonds using **short rate models** or models of short-term interest rates. Two of the most common short rate models are the **Vasicek** and **Cox, Ingersoll, and Ross** (CIR) models. These models generate a term structure of zero coupon prices. Both of these models assume a risk-neutral process for the short rate $r$ with one source of uncertainty. Specifically, the drift and volatility parameters depend only on the short rate $r$ and not on time. Another common feature of these two models is that they exhibit mean reversion of the short rate. The difference between the two models is in how they handle volatility.

### 9.7.1   Vasicek

The **Vasicek model** for changes in the short rate $r$ is stochastic and of the following form:

$$dr = a(b - r)dt + \sigma_r dz, \tag{9.4}$$

where a very small change in the short rate $(dr)$ in the time increment $dt$ is pulled back to the mean level $b$ at a rate of $a$. The second term is the volatility of the short rate $\sigma_r$ and it involves uncertainty $dz$. $dz$ is a normally distributed variable with zero mean and variance $dt$. We assume that the short rate $r$ is the instantaneous rate at time $t$ appropriate for continuous compounding.

Under the Vasicek model, the price of a zero coupon bond at time $t$, $P_t$, can be found as follows:

$$P_{t,s} = A_{t,s} \exp(-B_{t,s} r_t), \tag{9.5}$$

where $r_t$ is the value of the short rate at time $t$,

$$B_{t,s} = \frac{1 - \exp(-a(s - t))}{a} \tag{9.6}$$

and

$$A_{t,s} = \exp \left[ \frac{(B_{t,s} - (s-t)) \left(a^2 b - \frac{\sigma_r^2}{2}\right)}{a^2} - \frac{\sigma_r^2 B_{t,s}}{4a} \right] \qquad (9.7)$$

The Vasicek model requires the current value of the short rate $r$ and the value of the three parameters $a$, $b$, and $\sigma_r$ estimated using historical data. For purposes of our analysis, we use the parameters in [1]. In that paper, the authors estimate $a = 0.1779$, $b = 0.0866$, and $\sigma_r = 0.02$. Using these inputs and an assumed current short rate of 7%, we can implement the Vasicek model as follows.

**Step 1: Enter Model Inputs**  Since short rate modeling is a new type of analysis, we can consider clearing the R environment before starting using `rm(list = ls())`. We then input the assumed value of the parameters and the current short rate value.

```
1  > rm(list = ls())
2  > a <- 0.1779
3  > b <- 0.0866
4  > (X <- b * a)
5  [1] 0.01540614
6  > v <- 0.02
7  > r <- 0.07
```

**Step 2: Enter Assumption About Life of Zero Coupon Bond**  We assume a 10-year zero coupon bond. We assume the time today $t0$ is 0 and the maturity of the zero $s$ of 10 years. Thus, the time to maturity of the zero $T$ is 10 years.

```
1  > t0 <- 0
2  > s <- 10
3  > (T <- s - t0)
4  [1] 10
```

**Step 3: Calculate $B_{t,s}$ and $A_{t,s}$**  Using the equations above, we can calculate the value of $B_{t,s}$ and $A_{t,s}$ as follows:

```
1  > (B <- ifelse(a == 0, T, (1- exp(-a * T )) / a))
2  [1] 4.672249
3  > (A <- ifelse(a == 0, exp(v^2 * T^3 / 6),
4  +      exp(((B - T) * (a^2 * b - 0.5 * v^2)) / a^2
5  +          - (v^2 * B^2) / (4 * a))))
6  [1] 0.6440456
```

**Step 4: Calculate Vasicek Zero Price and Volatility**  Assuming a zero coupon bond has a par value of $1000. The Vasicek model estimates the price of a 10-year zero coupon bond is $464.38 and the volatility of that zero yield is 7.7%.

```
1  > (P <- A * exp(-B * r))
2  [1] 0.4643841
3  > (R <- ifelse(T > 0, -log(P) / T, r))
4  [1] 0.07670432
```

## 9.7.2 Cox, Ingersoll, and Ross

The **Cox, Ingersoll, and Ross** (CIR) model is another commonly used interest rate model. The difference between CIR and Vasicek is that the former assumes the volatility depends also on the square root of the short rate. The CIR model is as follows:

$$dr = a(b - r)dt + \sigma\sqrt{r}dz, \tag{9.8}$$

where all the terms are defined in the same manner as in the Vasicek model. Assuming $\sigma_r = \sigma\sqrt{r}$, the CIR model assumes a higher value of $\sigma_r$.

Under the CIR model, the price of a zero coupon bond at time $t$, $P_t$, can be found as follows:

$$P_{t,s} = A_{t,s}\exp(-B_{t,s}r_t), \tag{9.9}$$

where $r_t$ is the value of the short rate at time $t$,

$$\gamma = \sqrt{a^2 + a\sigma^2},$$

$$e_T = \exp(\gamma T - 1),$$

$$B_{t,s} = \frac{2e_T}{(\gamma + a)e_T + 2\gamma}, \quad \text{and}$$

$$A_{t,s} = \frac{2\gamma \cdot \exp(\gamma + a) \cdot 0.5T}{((\gamma + a)e_T + 2\gamma)^{2ab/\sigma^2}}.$$

Based on estimates by Chan et al. [1], $\sigma_r = 0.0854$, $a = 0.2339$, and $b = 0.0808$ under CIR.

**Step 1: Enter Model Inputs** We first enter the CIR parameter inputs. We assume the current value of the short rate is 7%.

```
1  > a <- 0.2339
2  > b <- 0.0808
3  > (X <- a * b)
4  [1] 0.01889912
5  > v <- 0.0854
6  > r <- 0.07
```

**Step 2: Enter Assumption About Life of Zero Coupon Bond** This is the same 10-year time to maturity assumption that we used in the Vasicek model implementation.

```
1  > t0 <- 0
2  > s <- 10
3  > (T <- s - t0)
4  [1] 10
```

**Step 3: Calculate** $B_{t,s}$ **and** $A_{t,s}$  Using the equations above, we calculate the values of $B_{t,s}$ and $A_{t,s}$.

```
1  > (gamma <- sqrt(a^2 + 2 * v^2))
2  [1] 0.2632404
3  > (exp_T <- exp(gamma * T) - 1)
4  [1] 12.90717
5  > (B <- 2 * exp_T / ((gamma + a) * exp_T + 2 * gamma))
6  [1] 3.717954
7  > (A <- ((2 * gamma * exp((gamma + a) * 0.5 * T)) /
8  +     ((gamma + a) * exp_T + 2 * gamma))^(2 * a * b / v^2))
9  [1] 0.6156529
```

**Step 4: Calculate CIR Zero Price and Volatility**  Assuming a zero coupon bond has a par value of $1000. The Vasicek model estimates the price of a 10-year zero coupon bond is $474.58 and the volatility of that zero yield is 7.5%.

```
1  > (P <- A * exp(-B * r))
2  [1] 0.4745782
3  > (R <- ifelse(T > 0, -log(P) / T, r))
4  [1] 0.07453288
```

## 9.8  Further Reading

An excellent overview of the bond markets can be found in [7]. More details on bond calculations can be found in [5] and [8]. Two excellent books on bond strategies are [3] and [6].

A wealth of information about bonds can be found on various regulatory websites. For example, the website of the Financial Industry Regulatory Authority (FINRA) (http://www.finra.org) contains, among other things, information on US corporate and agency data. Another regulator, the Municipal Securities Rulemaking Board (MSRB) provides information and market data through its Electronic Municipal Market Access (EMMA) website (http://emma.msrb.org/). Sterling (UK) corporate bond and bond index prices, as well as other information on Sterling bond markets can be obtained from Fixed Income Investor (http://www.fixedincomeinvestor.co.uk).

## References

1. Chan, K., Karolyi, G., Longstaff, F., & Sanders, A. (1992). An empirical comparison of alternative models of the short-term interest rates. *Journal of Finance, 47*, 1209–1227.
2. Cox, J., Ingersoll, J., & Ross, S. (1985). A theory of the term structure of interest rates. *Econometrica, 53*, 385–407.
3. Crescenzi, A. (2010). *The strategic bond investor: Strategies and tools to unlock the power of the bond market* (2nd ed.). New York: McGraw-Hill.

4. Estrella, A., & Trubin, M. (2006). The yield curve as a leading indicator: some practical issues. *Current Issues in Economics and Finance, 12,* 1–7.
5. Fabozzi, F. (2009). *Bond markets, analysis and strategies* (7th ed.). New York: Prentice-Hall.
6. Huggins, D., & Schaller, C. (2013). *Fixed income relative value analysis: a practitioner's guide to the theory, tools, and trades.* New York: Bloomberg Press/Wiley.
7. Thau, A. (2011). *The bond book* (3rd ed.). New York: McGraw-Hill.
8. Tuckman, B., & Serrat, A. (2012). *Fixed income securities: Tools for today's markets* (3rd ed.). New York: Wiley.
9. Vasicek, O. (1977). An equilibrium characterization of the term structure. *Journal of Financial Economics, 5,* 177–188.

# Chapter 10
# Options

One of the most important formulas in all of finance is the Options Pricing Model (OPM) developed by Black and Scholes [3] and Merton [8] (BSM). Many traders use the intuition behind the BSM OPM in their trading strategies. A complete discussion and derivation of the BSM OPM is beyond the scope of this book. However, we can go through a simplified example to show the underlying principle of how the options are valued under the BSM framework.

Consider a stock that sells for $70 per share today, and this stock could be worth either $100 per share or $50 per share tomorrow. How would we value an option to buy 100 shares of the stock for $80 per share tomorrow? This value is derived from the fact that we make the choice to purchase the stock tomorrow *after* we know what the stock price is. Put differently, we have to pay something for the right to buy the stock tomorrow only when its price is higher than $80 tomorrow. To do this, we can look at the value of this choice at the end of the day tomorrow. If the stock price is $100, we would make $2000 [= max($100 − $80 or 0) * 100 shares]. If the stock price is $50, we would not do anything because the intrinsic value of the option is zero (i.e., max($50 − $80, 0) = 0).

The BSM OPM is based on the **principle of no arbitrage**. Loosely speaking, if we have two assets that have the same cash flows in every state of the world, then these two assets must have the same price. In the context of the BSM OPM, if we can find a portfolio of stocks and bonds that replicate the payoff of the option (i.e., $2000 when the stock price is $100 and $0 if the stock price is $50), the price of the portfolio of stocks and bonds must equal the price of the option.

Let us assume that we purchased 40 shares of the stock today. You would pay a price of $2800 [= 40 shares * $70 per share]. If the stock price goes up to $100 tomorrow, then the 40 shares of stock we hold are now worth $4000 [= 40 shares * $100 per share]. On the other hand, if the stock price goes down to $50 tomorrow, then the 40 shares of stock we hold are now worth $2000 [= 40 shares * $50 per share]. This means that the payoff from holding 40 shares of the stock is always $2000 more than the payoff from holding an option to buy 100 shares of the

C.S. Ang, *Analyzing Financial Data and Implementing Financial Models Using R*, Springer Texts in Business and Economics, https://doi.org/10.1007/978-3-030-64155-9_10

stock, regardless of the state of the world tomorrow. Based on this, if no arbitrage opportunities exist, we know that the price of the option today must be $2000 less than the price of the 40 shares of stock today. That is, the option price is equal to $800 for 100 shares [= $2800 price for 40 shares of the stock − $2000] or $8 per share.

A **call** (**put**) option gives the holder the *right but not the obligation* to buy (sell) the stock for the **strike price** through the **time to maturity**. Therefore, the option holder will not exercise the option if the option has an intrinsic value of zero as the option is not profitable. It does not cost the option holder anything to let the option expire worthless. However, if the option has a positive intrinsic value, the option holder will exercise the option. In the case of a call option, the option holder will pay an amount equal to the strike price and receive the stock. In the case of a put option, the option holder will give up the stock and receive an amount equal to the strike price.

In Sect. 10.1, we first discuss how to use options chain data from the Chicago Board of Options Exchange (CBOE). We then show in Sect. 10.2 how to calculate call and put option values using the Black–Scholes–Merton (BSM) Options Pricing Model (OPM). Next, we discuss put–call parity, the Greeks, implied volatility, and the volatility smile. Then, in Sect. 10.3, we show how we can use implied volatility to gauge how much risk investors embed in market prices. We also show how to calculate option prices using the Binomial Model of [4] in Sect. 10.4. We conclude the chapter with American option pricing using the CRR model and the Bjerksund–Stensland approximation in Sect. 10.5.

## 10.1   Obtaining Options Chain Data

An options chain is a list of traded options for a particular company's stock at a given point in time. A publicly available source for options chain data is the Chicago Board of Options Exchange (CBOE). We can obtain the quote data from the CBOE website at http://www.cboe.com/delayedquote/quote-table. Then, we can select the stock's or index's options chain data that we are interested in. We can then select from a variety of filters, such as options of a particular maturity date and whether we want only options near the money or all available options.

The above CBOE page shows the current quotes for various options. To be able to reproduce the results we report in this book, the full AMZN options chain data as of the close of trading on December 31, 2019 is available on my website, http://www.cliffordang.com under the filename AMZN Options 12-31-2019.dat. This options chain data is provided as a courtesy by CBOE Exchange, Inc. (CBOE).

**Step 1: Import AMZN Options Chain Data**  We saved the raw options data we obtained from the CBOE as AMZN Options 12-31-2019.dat. Although this file has a .DAT extension, we can still use `read.csv()` to import the data. We can also see the data by opening it in Excel. Doing so, we can see that the data we

want starts in the third row, so we add the argument `skip` = 2. Then, to tell R
that the first row of data it is reading in (i.e., the third row of the raw data) are
column labels, we use `header` = TRUE. The values under Expiration.Date are
considered Factors by R, so we use `mdy()` in the `lubridate` package to convert
those into values R recognizes as dates.

```
1   > options <- read.csv("CBOE AMZN Options 12-31-2019.dat",
2   +     skip = 2, header = TRUE)
3   > names(options)
4    [1] "Expiration.Date" "Calls"         "Last.Sale"    "Net"
5    [5] "Bid"             "Ask"           "Vol"          "IV"
6    [9] "Delta"           "Gamma"         "Open.Int"     "Strike"
7   [13] "Puts"            "Last.Sale.1"   "Net.1"        "Bid.1"
8   [17] "Ask.1"           "Vol.1"         "IV.1"         "Delta.1"
9   [21] "Gamma.1"         "Open.Int.1"
10  > head.tail(options[, 1:5])
11      Expiration.Date              Calls Last.Sale Net    Bid
12  1         01/03/2020 AMZN200103C01040000   838.40   0 805.4
13  2         01/03/2020 AMZN200103C01060000   825.70   0 785.4
14  3         01/03/2020 AMZN200103C01080000   816.25   0 765.4
15      Expiration.Date              Calls Last.Sale   Net    Bid
16  2468      01/21/2022 AMZN220121C02600000    90.2 -0.30 85.05
17  2469      01/21/2022 AMZN220121C02700000    74.3 -3.95 70.30
18  2470      01/21/2022 AMZN220121C02800000     0.0  0.00 58.50
19  > class(options$Expiration.Date)
20  [1] "factor"
21  > library(lubridate)
22  > options$expiry <- mdy(options$Expiration.Date)
```

**Step 2: Create Separate Datasets for Calls and Puts** We want to subset the
*options* data to calls and puts that expire on March 20, 2020. Columns 2–12 and
Column 23 (i.e., the expiry column we created) are relevant to the calls, while
Columns 12–23 refer to the puts. But even within each set of columns, we only
need a few relevant columns, namely, the Last.Sale, Vol (volume), IV (implied
volatility), Open.Int (open interest), Strike, and Expiration. We identify those
relevant column numbers from the *options* dataset and use them to subset the data
in Lines 1 and 16. To ensure that the options we use in our analysis have traded
prices on that day, we only include options with positive volume (Lines 4 and 19).
In that same line of code, we also include strike prices that are close to the money,
rather than the full set of strike prices available. The closing stock price of AMZN
on December 31, 2019 is $1847.84. For the example below, we selected strike prices
between $1500 and $2500. Doing the above, we end up with 80 call options and 57
put options.

```
1   > calls <- subset(options[, c(3, 7, 8, 11, 12, 23)], expiry == "2020-03-20")
2   > names(calls) <-
3   +   c("Last.Sale", "Vol", "IV", "Open.Int", "Strike", "Expiration")
4   > calls <- subset(calls, Vol > 0 & (Strike >= 1500 & Strike <= 2500))
5   > rownames(calls) <- seq(1, nrow(calls), 1)
6   > head.tail(calls)
7       Last.Sale Vol    IV Open.Int Strike Expiration
```

```
8    1     274.30   3 0.2626        31    1590 2020-03-20
9    2     262.13   7 0.2602       260    1600 2020-03-20
10   3     240.55   2 0.2571        14    1630 2020-03-20
11      Last.Sale Vol      IV Open.Int Strike Expiration
12   78       1.71   1 0.2832        15    2460 2020-03-20
13   79       1.53   1 0.2881        28    2490 2020-03-20
14   80       1.54   2 0.2843       103    2500 2020-03-20
15   >
16   > puts <- subset(options[, c(12, 14, 18, 19, 22, 23)], expiry == "2020-03-20")
17   > names(puts) <-
18   +    c("Strike", "Last.Sale", "Vol", "IV", "Open.Int", "Expiration")
19   > puts <- subset(puts, Vol > 0 & (Strike >= 1500 & Strike <= 2500))
20   > rownames(puts) <- seq(1, nrow(puts), 1)
21   > head.tail(puts)
22      Strike Last.Sale Vol      IV Open.Int Expiration
23   1    1500      5.50  30 0.2874      526 2020-03-20
24   2    1520      6.60   1 0.2828      129 2020-03-20
25   3    1540      8.15  13 0.2784      175 2020-03-20
26      Strike Last.Sale Vol      IV Open.Int Expiration
27   55   2240    400.25   3 0.2832        1 2020-03-20
28   56   2250    405.85  18 0.2873        7 2020-03-20
29   57   2260    415.85   5 0.2913        6 2020-03-20
```

**Step 3: Determine Consistency of the Stock Price and the Intrinsic Value of the Calls** The **intrinsic value** of a call option is equal to $\max(S_t - K, 0)$, where $S_t$ is the stock price today and $K$ is the strike price of the option. However, on dates prior to their expiration, the price of an option should be higher than its intrinsic value because there is a chance that the option value will end up higher (i.e., the option contains what is called "time value"). As such, if the price of the option is less than the option's intrinsic value in our data, we have to consider the cause of such a result as well as whether we should discard such data. Below, we create a variable called diff, which is the difference between the AMZN's closing stock price on December 31, 2019 of $1847.84 and the strike price of the option. We then create a dummy variable that is set to 1 if diff is less than the option price (i.e., what we would expect) and 0 otherwise. As the output below shows, there are no such call options (Lines 15–18). The final step is to remove the extraneous columns, which are Expiration, diff, and dummy.

```
1   > stock.price <- 1847.84
2   >
3   > calls$diff <- stock.price - calls$Strike
4   > calls$dummy <- ifelse(calls$diff < calls$Last.Sale, 1, 0)
5   > head.tail(calls)
6      Last.Sale Vol      IV Open.Int Strike Expiration    diff dummy
7    1    274.30   3 0.2626        31   1590 2020-03-20 257.84     1
8    2    262.13   7 0.2602       260   1600 2020-03-20 247.84     1
9    3    240.55   2 0.2571        14   1630 2020-03-20 217.84     1
10     Last.Sale Vol      IV Open.Int Strike Expiration    diff dummy
11   78     1.71   1 0.2832        15   2460 2020-03-20 -612.16     1
12   79     1.53   1 0.2881        28   2490 2020-03-20 -642.16     1
13   80     1.54   2 0.2843       103   2500 2020-03-20 -652.16     1
```

```
14  >
15  > subset(calls, calls$dummy == 0)
16  [1] Last.Sale  Vol         IV          Open.Int   Strike      Expiration diff
17  [8] dummy
18  <0 rows> (or 0-length row.names)
19  >
20  > calls <- calls[, -c(6:8)]
21  > head.tail(calls)
22     Last.Sale Vol    IV Open.Int Strike
23  1    274.30   3 0.2626      31   1590
24  2    262.13   7 0.2602     260   1600
25  3    240.55   2 0.2571      14   1630
26     Last.Sale Vol    IV Open.Int Strike
27  78     1.71  1 0.2832      15   2460
28  79     1.53  1 0.2881      28   2490
29  80     1.54  2 0.2843     103   2500
```

**Step 4: Determine Consistency of the Stock Price and the Intrinsic Value of the Puts** Similar to what we did with the call options, we now do the same analysis for the put options. The difference is that the intrinsic value of the put option is calculated as $\max(K - S_t, 0)$. Despite the different formula, we still expect that the price of the put option should be greater than the intrinsic value of the put option. Thus, the dummy variable we use will still have a value of 1 if diff is less than the put option price and 0 otherwise. As the output below shows, there are such put options (Lines 13–16). The final step is to remove the extraneous columns, which are Expiration, diff, and dummy.

```
1   > puts$diff <- puts$Strike - stock.price
2   > puts$dummy <- ifelse(puts$diff < puts$Last.Sale, 1, 0)
3   > head.tail(puts)
4      Strike Last.Sale Vol    IV Open.Int Expiration    diff dummy
5   1    1500     5.50  30 0.2874      526 2020-03-20 -347.84     1
6   2    1520     6.60   1 0.2828      129 2020-03-20 -327.84     1
7   3    1540     8.15  13 0.2784      175 2020-03-20 -307.84     1
8      Strike Last.Sale Vol    IV Open.Int Expiration    diff dummy
9   55    2240   400.25   3 0.2832        1 2020-03-20 392.16     1
10  56    2250   405.85  18 0.2873        7 2020-03-20 402.16     1
11  57    2260   415.85   5 0.2913        6 2020-03-20 412.16     1
12  >
13  > subset(puts, puts$dummy == 0)
14  [1] Strike     Last.Sale  Vol        IV         Open.Int   Expiration diff
15  [8] dummy
16  <0 rows> (or 0-length row.names)
17  >
18  > puts <- puts[, -c(6:8)]
19  > head.tail(puts)
20     Strike Last.Sale Vol    IV Open.Int
21  1    1500     5.50  30 0.2874      526
22  2    1520     6.60   1 0.2828      129
23  3    1540     8.15  13 0.2784      175
24     Strike Last.Sale Vol    IV Open.Int
25  55    2240   400.25   3 0.2832        1
```

| 26 | 56 | 2250 | 405.85 | 18 0.2873 | 7 |
| 27 | 57 | 2260 | 415.85 | 5 0.2913 | 6 |

We will use the above call and put options datasets when we conduct our analysis below.

## 10.2   Black–Scholes–Merton Options Pricing Model

In the Black–Scholes–Merton (BSM) Options Pricing Model (OPM), the options are valued based on the value of a portfolio of stocks and bonds that replicate the cash flows of the option. The beauty of the BSM OPM is that we do not need to know stochastic calculus to implement the model, as the BSM OPM is a closed-form model. That is,

$$c = S_t \mathbb{N}(d_1) + K \exp(-r_f * T)\mathbb{N}(d_2) \tag{10.1}$$

and

$$p = K \exp(-r_f * T)\mathbb{N}(-d_2) - S\mathbb{N}(-d_1), \tag{10.2}$$

where

$$d_1 = \frac{ln(S_t/K) + (rf + 0.5 * \sigma^2) * T}{\sigma\sqrt{T}}$$

and

$$d_2 = d_1 - \sigma\sqrt{T}.$$

In the above equations, $S_t$ is the underlying stock price today, $K$ is the strike price of the option, $T$ is the time to maturity of the option, $r_f$ is the risk-free rate, $\sigma$ is the volatility of the underlying asset, and $\mathbb{N}()$ is the standard normal density.

In this section, we will estimate the value of March 2020 AMZN call and put options using the BSM OPM. Specifically, we will value the two call and put options that are the closest to being at the money as of December 31, 2020. Given that the price of AMZN stock as of December 31, 2013 was \$1847.84, we will choose options with strike prices of \$1845 and \$1850.

**Step 1: Enter Observable Inputs for the BSM OPM**   Aside from the strike price of the options, there are three other observable inputs to the BSM OPM. These are the stock price, risk-free rate, and time to maturity of the option. We first create variables for AMZN's close price as of December 31, 2019 and the risk-free rate as of December 31, 2019 as represented by the 3-Month Constant Maturity Treasury yield.

```
1  > price <- 1847.84
2  > (rf <- 0.0155)
3  [1] 0.0155
4  >
5  > expiry.date <- as.Date("2020-03-20")
6  > value.date <- as.Date("2019-12-31")
7  > (TTM <- as.numeric((expiry.date - value.date) / 365))
8  [1] 0.2191781
```

**Step 2: Estimate AMZN Historical Volatility as Proxy for Volatility Input in the BSM OPM**  Of the inputs to the BSM OPM, the volatility is the only unobservable input. As such, to calculate the option value, we need to have an input for the stock's volatility. A starting point could be the historical volatility of the stock. Below, we calculate the annualized daily volatility of AMZN returns over the last 3 years. Note that to annualize the daily volatility, we multiply the standard deviation of daily returns by the square root of 252. This yields a volatility estimate of 29.0%.

```
1  > amzn <- load.data("AMZN Yahoo.csv", "AMZN")
2  > head.tail(amzn)
3           AMZN.Open AMZN.High AMZN.Low AMZN.Close AMZN.Volume AMZN.Adjusted
4  2014-12-31  311.55   312.98   310.01    310.35     2048000
   310.35
5  2015-01-02  312.58   314.75   306.96    308.52     2783200
   308.52
6  2015-01-05  307.01   308.38   300.85    302.19     2774200
   302.19
7           AMZN.Open AMZN.High AMZN.Low AMZN.Close AMZN.Volume AMZN.Adjusted
8  2019-12-27 1882.92  1901.40  1866.01   1869.80    6186600      1869.80
9  2019-12-30 1874.00  1884.00  1840.62   1846.89    3674700      1846.89
10 2019-12-31 1842.00  1853.26  1832.23   1847.84    2506500      1847.84
11 >
12 > rets.amzn <- diff(log(amzn$AMZN.Adjusted))
13 > rets.amzn <- rets.amzn[-1, ]
14 > head.tail(rets.amzn)
15          AMZN.Adjusted
16 2015-01-02 -0.005914077
17 2015-01-05 -0.020730670
18 2015-01-06 -0.023098010
19          AMZN.Adjusted
20 2019-12-27  0.0005510283
21 2019-12-30 -0.0123283480
22 2019-12-31  0.0005142195
23 >
24 > (daily.vol <- sd(rets.amzn))
25 [1] 0.01825005
26 > (vol <- daily.vol * sqrt(252))
27 [1] 0.2897105
```

**Step 3: Identify Calls with the Closest Strike Price Around Stock Price**  Based on the AMZN's stock price of $1847.84, we will keep the options with the strike price of $1845 and $1850. For this analysis, we only keep the Last.Sale and Strike columns. These are Columns 1 and 5.

```
1  > bs.call <- subset(calls, Strike >= 1845 & Strike <= 1850)[, c(1, 5)]
2  > rownames(bs.call) <- seq(1, nrow(bs.call), 1)
3  > bs.call
4    Last.Sale Strike
5  1     89.91   1845
6  2     87.70   1850
```

**Step 4: Calculate the BSM Call Option Value**  The output below shows the steps to calculate d1, d2, and the BSM call value. The BSM value of the call with a strike of $1845 is $104.28 and a strike of $1850 is $101.87. Note that both these BSM OPM values are higher than the last price of the call options reported by the CBOE (Last.Sale) for these two options.

```
1  > bs.call$d1 <- (log(price / bs.call$Strike) +
2  +     (rf + 0.5 * (vol^2)) * TTM) / (vol * sqrt(TTM))
3  > bs.call$d2 <- bs.call$d1 - vol * sqrt(TTM)
4  > bs.call$optval <- price * pnorm(bs.call$d1, mean = 0, sd = 1) -
5  +     bs.call$Strike * exp(-rf * TTM) *
6  +     pnorm(bs.call$d2, mean = 0, sd = 1)
7  > bs.call
8    Last.Sale Strike         d1          d2    optval
9  1     89.91   1845 0.10420401 -0.03142820 104.2776
10 2     87.70   1850 0.08425033 -0.05138188 101.8683
```

**Step 5: Identify Puts with the Closest Strike Price Around Stock Price**  Based on AMZN's stock price of $1847.84, we will keep the options with the strike price of $1845 and $1850. For this analysis, we only keep the Last.Sale and Strike variables.

```
1  > bs.put <- subset(puts, Strike >= 1845 & Strike <= 1850)[, c(1, 2)]
2  > rownames(bs.put) <- seq(1, nrow(bs.put), 1)
3  > bs.put
4    Strike Last.Sale
5  1   1845     82.06
6  2   1850     84.70
```

**Step 6: Calculate the BSM Put Option Value**  The output below shows the steps to calculate d1, d2, and the BSM call value. The BSM value of the call with a strike of $1845 is $95.18 and strike of $1850 is $97.75. Note that both these BSM OPM values are higher than the last price of the put options reported by the CBOE for these two options.

```
1  > bs.put$d1 <- (log(price / bs.put$Strike) +
2  +     (rf + 0.5 * (vol^2)) * TTM) /
3  +     (vol * sqrt(TTM))
4  > bs.put$nd1 = -bs.put$d1
5  > bs.put$d2 <- bs.put$d1 - vol * sqrt(TTM)
6  > bs.put$nd2 <- -bs.put$d2
7  > bs.put$optval <- bs.put$Strike *
```

```
8   +       exp(-rf * TTM) * pnorm(bs.put$nd2, mean = 0, sd = 1) -
9   +       price * pnorm(bs.put$nd1, mean = 0, sd = 1)
10  > bs.put
11    Strike Last.Sale        d1          nd1         d2          nd2   optval
12  1   1845      82.06 0.10420401 -0.10420401 -0.03142820 0.03142820 95.18025
13  2   1850      84.70 0.08425033 -0.08425033 -0.05138188 0.05138188 97.75407
```

As we can see from the output above, there is a large difference from the last traded price for the calls and puts and the BSM value of those same options based on the inputs that we used. The time to maturity of this option is short and the annualized risk-free rate is low, so the risk-free rate will not be a primary determinant of that difference. Thus, the primary input that causes the difference is the asset volatility input. We recognized earlier that volatility is the one input that is unobservable, so in our example we estimated it using historical volatility. As shown above, the historical volatility of 29% was too low. Thus, to the extent we believe the traded price is reflective to the true value of the option, then the volatility should be higher. How much higher? In Sect. 10.3, we show how to extract the volatility embedded in option prices.

### 10.2.1 BSM Function

If we were to repeatedly calculate option values using the Black–Scholes–Merton model, we would benefit from converting the above calculation into a function. We allow the function to be flexible enough that identifying whether the option type is a call or a put will allow us to calculate the appropriate option value. We accomplish this using an if() statement in the code. Note that to denote equality we have to use two equal signs (i.e., ==) in the code.

```
1   > bs.opm <- function(S, K, T, riskfree, sigma, type){
2   +   d1 <- (log(S / K) + (riskfree + 0.5 * sigma^2) * T) /
3   +       (sigma * sqrt(T))
4   +   d2 <- d1 - sigma * sqrt(T)
5   +   if(type == "Call"){
6   +       opt.val <- S * pnorm(d1) - K * exp(-riskfree * T) * pnorm(d2)
7   +       }
8   +   if(type=="Put"){
9   +       opt.val <- K * exp(-riskfree * T) * pnorm(-d2) - S * pnorm(-d1)
10  +       }
11  +   opt.val
12  + }
```

We can now calculate the value of the call and put options we calculated in the previous section. As we can see, our BSM function calculates identical values to the manual approach we used in the previous section.

```
1   > (call.1845 <- bs.opm(price, 1845, TTM, rf, vol, "Call"))
2   [1] 104.2776
3   > (call.1850 <- bs.opm(price, 1850, TTM, rf, vol, "Call"))
```

```
4   [1] 101.8683
5   > (put.1845 ← bs.opm(price, 1845, TTM, rf, vol, "Put"))
6   [1] 95.18025
7   > (put.1850 ← bs.opm(price, 1850, TTM, rf, vol, "Put"))
8   [1] 97.75407
```

### 10.2.2  Put–Call Parity

If we have the value of the call (put) option under BSM, we can find the value of
a put (call) option on the same underlying with the same strike price and maturity
using a relationship known as **put–call parity**. Put–call parity states that

$$p + S = c + K * \exp(-r_f * T), \tag{10.3}$$

where all the terms are defined in the same manner as Eqs. (10.1) and (10.2). To find
the value of a put option, we manipulate Eq. (10.3) to get

$$p = c + K * \exp(-r_f * T) - S. \tag{10.4}$$

Similarly, to get the value of the call option, we use the following formula:

$$c = p + S - K * \exp(-r_f * T). \tag{10.5}$$

Below, we show that put–call parity holds when we use the BSM option values we
calculated above. Lines 1, 6, 11, and 16 use put–call parity, and Lines 3, 8, 13, and
18 show the BSM calculation.

```
1   > call.1845 + 1845 * exp(−rf * TTM) − price
2   [1] 95.18025
3   > put.1845
4   [1] 95.18025
5   >
6   > call.1850 + 1850 * exp(−rf * TTM) − price
7   [1] 97.75407
8   > put.1850
9   [1] 97.75407
10  >
11  > put.1845 + price − 1845  * exp(−rf * TTM)
12  [1] 104.2776
13  > call.1845
14  [1] 104.2776
15  >
16  > put.1850 + price − 1850  * exp(−rf * TTM)
17  [1] 101.8683
18  > call.1850
19  [1] 101.8683
```

Since the market price of the options is different from the BSM option value, the above analysis implies that put–call parity will not hold if we use the market price of options.

## 10.2.3  The Greeks

When estimating the BSM value of the option, we may be interested in the sensitivity of the option price to changes in the inputs. These sensitivities are known as the **Greeks**. Specifically, the Greeks are **delta** (change in option price given a change in the underlying price), **gamma** (change in option price given the rate of change in the underlying price), **vega** (change in option price given a change in volatility), **theta** (change in option price given a change in the time to maturity), and **rho** (change in option price given a change in the risk-free rate).

The formulas for the Greeks for European calls are as follows:

$$\text{Delta:} \quad \mathbb{N}(d_1)$$

$$\text{Gamma:} \quad \frac{\mathbb{N}'(d_1)}{S_0 \sigma \sqrt{T}}$$

$$\text{Vega:} \quad S_0 \mathbb{N}'(d_1)\sqrt{T}$$

$$\text{Theta:} \quad \frac{S_0 \mathbb{N}'(d_1)\sigma}{2\sqrt{T}} - r_f K \exp(-r_f * T)\mathbb{N}(d_2)$$

$$\text{Rho:} \quad K T \exp(-r_f * T) + \mathbb{N}(d_2)$$

and the Greeks for European puts are as follows:

$$\text{Delta:} \quad \mathbb{N}(d_1) - 1$$

$$\text{Gamma:} \quad \frac{\mathbb{N}'(d_1)}{S_0 \sigma \sqrt{T}}$$

$$\text{Vega:} \quad \mathbb{N}'(d_1)\sqrt{T}$$

$$\text{Theta:} \quad -\frac{S_0 \mathbb{N}(d_1)\sigma}{2\sqrt{T}} + r_f K \exp(-r_f * T)\mathbb{N}(-d_2)$$

$$\text{Rho:} \quad K T \exp(-r_f * T) + \mathbb{N}(d_2),$$

where $\mathbb{N}'(\cdot)$ is equal to $(1/\sqrt{2\pi})\exp(-x^2/2)$, or in R, we can use `dnorm()`. For $\mathbb{N}(\cdot)$, we can use `pnorm()` in R.

We continue using the two call options with strike prices that are the closest to being at the money. We calculate the Greeks for each of those options. Since we only need the strike and last close price for these options, we can subset the *bs.call* data object and only keep the first two columns.

```
 1   > (greeks.call <- bs.call[, 1:2])
 2     Last.Sale Strike           ,
 3   1    89.91   1845
 4   2    87.70   1850
```

Using the equations laid out above for the Greeks of a European call option, we calculate the following:

```
 1   > (greeks.call <- bs.call[, 1:2])
 2     Last.Sale Strike
 3   1    89.91   1845
 4   2    87.70   1850
 5   > greeks.call$delta <- pnorm(bs.call$d1, mean = 0, sd = 1)
 6   > greeks.call$gamma <- dnorm(bs.call$d1, mean = 0, sd = 1) /
 7   +     (price * vol * sqrt(TTM))
 8   > greeks.call$vega <- price *
 9   +     dnorm(bs.call$d1, mean = 0, sd = 1) * sqrt(TTM)
10   > greeks.call$theta <- -((price * vol *
11   +     dnorm(bs.call$d1, mean = 0, sd = 1)) /
12   +     (2 * sqrt(TTM))) - (rf * greeks.call$Strike * exp(-rf * TTM) *
13   +     pnorm(bs.call$d2, mean = 0, sd = 1))
14   > greeks.call$rho <- greeks.call$Strike * TTM * exp(-rf * TTM) *
15   +     pnorm(bs.call$d2, mean = 0, sd = 1)
16   > greeks.call
17     Last.Sale Strike    delta       gamma      vega     theta       rho
18   1    89.91   1845 0.5414963 0.001583161 343.2536 -240.7500 196.4539
19   2    87.70   1850 0.5335713 0.001586140 343.8996 -240.9873 193.7723
```

Similarly, we calculate the Greeks for the two put options that are the closest to being in the money. For the puts, we subset the *bs.put* data object and only keep the first two columns. Using the equations laid out above for the Greeks of a European put option, we calculate the following:

```
 1   > (greeks.put <- bs.put[, 1:2])
 2     Strike Last.Sale
 3   1   1845     82.06
 4   2   1850     84.70
 5   > greeks.put$delta <- pnorm(bs.put$d1, mean = 0, sd = 1) - 1
 6   > greeks.put$gamma <- dnorm(bs.put$d1, mean = 0, sd = 1) /
 7   +     (price * vol * sqrt(TTM))
 8   > greeks.put$vega <- price *
 9   +     dnorm(bs.put$d1, mean = 0, sd = 1) * sqrt(TTM)
10   > greeks.put$theta <- -((price * vol *
11   +     dnorm(bs.put$d1, mean = 0, sd = 1)) /
12   +     (2 * sqrt(TTM))) + (rf * greeks.put$Strike * exp(-rf * TTM) *
13   +     pnorm(bs.put$nd2, mean = 0, sd = 1))
14   > greeks.put$rho <- -greeks.put$Strike * TTM * exp(-rf * TTM) *
15   +     pnorm(bs.put$nd2, mean = 0, sd = 1)
16   > greeks.put
17     Strike Last.Sale     delta       gamma      vega     theta       rho
18   1   1845     82.06 -0.4585037 0.001583161 343.2536 -212.2495 -206.5582
19   2   1850     84.70 -0.4664287 0.001586140 343.8996 -212.4096 -210.3320
```

## 10.3  Implied Volatility

Assuming we believe that the BSM OPM is the appropriate model to value options, the asset volatility is one of the five required inputs of that model that is not directly observable. If the options are publicly traded, we can use the price of the options to infer what investors think about the volatility of the underlying asset. The process of doing so is straight-forward. We find the volatility that when used in the BSM OPM gives us the market price of the option. The volatility calculated in this manner is called the option's **implied volatility**.

### 10.3.1  Implied Volatility Function

The implied volatility is essentially found using trial-and-error. However, we do not need to do this manually. There are several numerical methods that can be used to automate the trial-and-error process. We will use the **bisection method**. The bisection method first uses some value for the volatility and compares that value to the market price of the option (*price*). If the difference between the two values is greater in absolute value terms than the acceptable error (*epsilon*), another volatility number is used which is half of the original volatility and a volatility that is either higher ( *sigma.up*) or lower (*sigma.down*). That is where the name bisection comes from. The process repeats until we reach an acceptable level of difference between the estimated price and the market price of the option (i.e., the absolute value of the difference must be less than 0.00001 in our example) or if we get to 1000 iterations and no solution is found. The latter is needed because issues with the data may cause the iterations to go on an infinite loop. Setting the limit to 1000 iterations is usually sufficient to reach a reasonable answer, as those situations in which the program does not reach a solution after 1000 iterations typically result in odd-looking outputs.

```
1   > iv.opt <- function(S, K, T, riskfree, price, type){
2   +     sigma <- vol
3   +     sigma.up <- 1
4   +     sigma.down <- 0.001
5   +     count <- 0
6   +     epsilon <- bs.opm(S, K, T, riskfree, sigma, type) - price
7   +     while(abs(epsilon) > 0.00001 & count < 1000){
8   +         if(epsilon < 0){
9   +             sigma.down < -sigma
10  +             sigma <- (sigma.up + sigma) / 2
11  +             } else{
12  +                 sigma.up <- sigma
13  +                 sigma <- (sigma.down + sigma) / 2
14  +             }
15  +         epsilon <- bs.opm(S, K, T, riskfree, sigma, type) - price
16  +         count <- count + 1
17  +         }
```

```
18  +          if(count == 1000){
19  +              return(NA)
20  +              } else{
21  +              return(sigma)
22  +          }
23  + }
```

To test that this function works, we can reverse the calculations for the call option with a strike of $1845. We calculated a BSM OPM value for that option of $104.2776 (*call.1845*). As the output shows, we get a volatility of 29.0%, which is equal to the historical volatility we used when we calculated the BSM OPM value of this particular call option.

```
1  > iv.opt(price, 1845, TTM, rf, call.1845, "Call")
2  [1] 0.2897105
3  > vol
4  [1] 0.2897105
```

Earlier, we saw that the BSM option values for the calls and puts with strike price of $1845 and $1850 were very different from the last traded price for those options. We can now use our implied volatility function to see how far the implied volatility is from the historical volatility of 29% we used. Before we run the calculations, we know that the traded prices were *lower* than the BSM option values. Because volatility and BSM values are directly related, we would expect that the implied volatility would be *lower* than the 29% volatility that we used. We confirm this with the output below:

```
1  > iv.opt(price, 1845, TTM, rf, 89.91, "Call")
2  [1] 0.2478621
3  > iv.opt(price, 1850, TTM, rf, 87.70, "Call")
4  [1] 0.2485234
5  > iv.opt(price, 1845, TTM, rf, 87.70, "Put")
6  [1] 0.2679208
7  > iv.opt(price, 1850, TTM, rf, 97.78, "Put")
8  [1] 0.2897859
```

Under BSM, the assumption is that the volatility is constant. However, as we see from the above output, this is not the case. This is known as the volatility smile, which we discuss next.

### 10.3.2   *Volatility Smile*

If the assumptions of the BSM were true, we would expect a constant implied volatility across all options regardless of the strike price for the same firm with the same time to maturity. However, in practice, it is common to see implied volatilities of options that are further out of the money to be higher than options that are closer to the money. This is known as the **volatility smile**. Using the CBOE options data in the *calls* and *puts* dataset, we can see this characteristic of traded options.

**Step 1: Subset Calls and Puts Data**  The stock price of AMZN when we pulled the options data was approximately $1850. The IVs of deep out of the money options can behave oddly due to the lack of trading, so we limit our analysis to options that are at or close to the money. For our illustration, we will set the range as plus or minus $550, so we subset the call and put options data to those with strike prices from $1300 to $2400. To limit less reliable implied volatility numbers, we also restrict the data to those with IV greater than zero.

```
1   > iv.calls <- subset(calls[, c(1, 3, 5)],
2   +      Strike >= 1300 &
3   +      Strike <= 2400 &
4   +      IV > 0)
5   > head.tail(iv.calls)
6     Last.Sale      IV Strike
7   1    274.30 0.2626   1590
8   2    262.13 0.2602   1600
9   3    240.55 0.2571   1630
10    Last.Sale      IV Strike
11  74     4.09 0.2581   2290
12  75     3.75 0.2592   2300
13  76     2.55 0.2741   2400
14  >
15  > iv.puts <- subset(puts[, c(1, 2, 4)],
16  +      Strike >= 1300 &
17  +      Strike <= 2400 &
18  +      IV > 0)
19  > head.tail(iv.puts)
20    Strike Last.Sale     IV
21  1   1500     5.50 0.2874
22  2   1520     6.60 0.2828
23  3   1540     8.15 0.2784
24    Strike Last.Sale     IV
25  55   2240   400.25 0.2832
26  56   2250   405.85 0.2873
27  57   2260   415.85 0.2913
```

**Step 2: Plot the Implied Volatility**  We plot the chart using the default hollow markers for the call implied volatilities. We then use pch = 16 to plot the put implied volatilities using a solid marker. Figure 10.1 shows the output of this code and demonstrates the volatility smile based on traded option prices for AMZN March 2020 options as of January 1, 2020.

```
1   > (y.range <- range(calls$IV, puts$IV))
2   [1] 0.2380 0.2913
3   > plot(x = calls$Strike,
4   +      y = calls$IV,
5   +      ylim = y.range,
6   +      col = "blue",
7   +      ylab = "Implied Volatility",
8   +      xlab = "Strike Price",
9   +      main = "Implied Volatility of AMZN March 2020 Options
10  +      As of December 31, 2019")
```

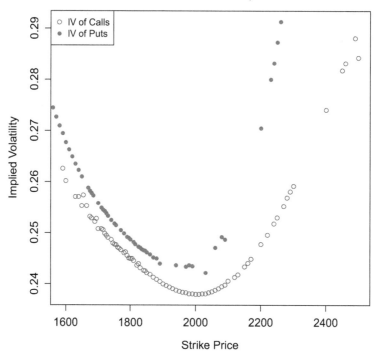

**Fig. 10.1** Volatility smile of AMZN March 2020 options as of January 1, 2020. Data Source: CBOE data is provided by Chicago Board Options Exchange, Incorporated (CBOE), and CBOE makes no warranties of any kind with respect to this data

```
11   > points(x = puts$Strike,
12   +       y = puts$IV, col = "red", pch = 16)
13   > legend("topleft",
14   +       c("IV of Calls", "IV of Puts"),
15   +       pch = c(1, 16),
16   +       col = c("blue", "red"))
```

### 10.3.3   Gauging Market Risk

In this section, we plot the CBOE Volatility Index (VIX) over the last 5 years. The VIX was introduced in 1993. The VIX is intended to measure the market's expectation of 30-day volatility as it calculates the volatility implied in the prices of S&P 500 Index options with approximately 1 month to maturity. The VIX is often referred to as the "investor fear gauge," because it reflects the investors' view of expected future stock market volatility.

**Step 1: Import VIX Data** We can obtain the CBOE Volatility Index (VIXCLS) from FRED. The data we downloaded starts in 1990. When we downloaded the data, we saved it under the filename VIXCLS.csv. We now import that data into R using the `load.fred()` function we created. If not already loaded into the R environment, we can use the code in Appendix B to load the function.

```
1  > vix <- load.fred("VIXCLS.csv", "VIXCLS")
2  > head.tail(vix)
3               VIXCLS
4  1990-01-02  17.24
5  1990-01-03  18.19
6  1990-01-04  19.22
7               VIXCLS
8  2019-12-30  14.82
9  2019-12-31  13.78
10 2020-01-02  12.47
```

**Step 2: Subset VIX Data to 2015–2019** We want to plot 5 years of data, so we need to subset the data to the period 2015–2019. As the output shows, we have 1258 observations over this period. We will use this number to check against the historical volatility we calculate below to make sure that we have the same number of observations.

```
1  > vix <- vix["2015-01-01/2019-12-31"]
2  > names(vix) <- "VIX"
3  > head.tail(vix)
4                VIX
5  2015-01-02 17.79
6  2015-01-05 19.92
7  2015-01-06 21.12
8                VIX
9  2019-12-27 13.43
10 2019-12-30 14.82
11 2019-12-31 13.78
12 > dim(vix)
13 [1] 1258    1
```

**Step 3: Import S&P 500 Index Data** We can obtain the S&P 500 Index (SP500) data from FRED. The data we downloaded starts in 2010. When we downloaded the data, we saved it under the filename SP500.csv. We now import that data into R using the `load.fred()` function we created.

```
1  > spx <- load.fred("SP500.csv", "SP500")
2  > head.tail(spx)
3                 SP500
4  2010-01-04 1132.99
5  2010-01-05 1136.52
6  2010-01-06 1137.14
7                 SP500
```

```
 8    2019–12–31 3230.78
 9    2020–01–02 3257.85
10    2020–01–03 3234.85
```

**Step 4: Calculate Historical Volatility from 2015 to 2019**  We want to plot 5 years of volatility data, so we would need volatility data for the period 2015–2019. Since the S&P 500 Index data is the level of the S&P 500, we first have to convert this to returns and then calculate the 30-day standard deviation of those returns. To do this, we first subset the data to a period longer than 5 years. We arbitrarily picked November 1, 2014 as the start date, but we will subset the data again later, so the excess data at the beginning will be removed anyway.

```
 1    > spx <- spx["2014–11–01/2019–12–31"]
 2    > head.tail(spx)
 3                    SP500
 4    2014–11–03 2017.81
 5    2014–11–04 2012.10
 6    2014–11–05 2023.57
 7                    SP500
 8    2019–12–27 3240.02
 9    2019–12–30 3221.29
10    2019–12–31 3230.78
```

We then calculate the log returns of the S&P 500 Index. This is calculated by taking the difference of the log prices. Then, we clean up the data by keeping only the returns data and then removing the first observation, which has a return value of NA.

```
 1    > spx$SPX <- diff(log(spx$SP500))
 2    > head.tail(spx)
 3                    SP500           SPX
 4    2014–11–03 2017.81             NA
 5    2014–11–04 2012.10 −0.002833812
 6    2014–11–05 2023.57  0.005684325
 7                    SP500           SPX
 8    2019–12–27 3240.02  3.395098e−05
 9    2019–12–30 3221.29 −5.797602e−03
10    2019–12–31 3230.78  2.941694e−03
11    >
12    > spx <- spx[−1, 2]
13    > head.tail(spx)
14                           SPX
15    2014–11–04 −0.002833812
16    2014–11–05  0.005684325
17    2014–11–06  0.003768396
18                           SPX
19    2019–12–27  3.395098e−05
20    2019–12–30 −5.797602e−03
21    2019–12–31  2.941694e−03
```

Next, we use rollapply() to recursively apply the standard deviation formula to the S&P 500 Index returns data. The second argument in rollapply() tells R to calculate standard deviation over 30-day rolling windows. The na.rm = TRUE is not necessary in this dataset but is useful if there are any NAs in the data. We then annualize the daily volatility to an annual volatility by multiplying it by $\sqrt{252}$. We turn the volatility into percentage points by multiplying it by 100.

```
1  > spx.sd <- rollapply(spx, 30, sd, na.rm = TRUE) * sqrt(252) * 100
2  > head.tail(spx.sd)
3             SPX
4  2014-11-04  NA
5  2014-11-05  NA
6  2014-11-06  NA
7                 SPX
8  2019-12-27 6.999198
9  2019-12-30 7.313896
10 2019-12-31 7.081939
```

Finally, we subset the standard deviation data to the period 2015–2019. We rename the column to Hist.Vol because that makes more sense as a volatility measure than SPX. We then use dim() to show that there are 1258 observations in this dataset, which matches the number of observations that we have in the *vix* dataset.

```
1  > spx.sd <- spx.sd["2015-01-01/2019-12-31"]
2  > names(spx.sd) <- "Hist.Vol"
3  > head.tail(spx.sd)
4             Hist.Vol
5  2015-01-02 13.31100
6  2015-01-05 14.35380
7  2015-01-06 14.54006
8             Hist.Vol
9  2019-12-27 6.999198
10 2019-12-30 7.313896
11 2019-12-31 7.081939
12 > dim(spx.sd)
13 [1] 1258    1
```

**Step 5: Combine Implied and Historical Volatility Data**  We then combine the two data series together using merge().

```
1  > vol.idx <- merge(VIX, spx.sd)
2  > head.tail(vol.idx)
3             VIXCLS Hist.Vol
4  2015-01-02  17.79 13.31100
5  2015-01-05  19.92 14.35380
6  2015-01-06  21.12 14.54006
7             VIXCLS Hist.Vol
8  2019-12-27  13.43 6.999198
9  2019-12-30  14.82 7.313896
10 2019-12-31  13.78 7.081939
```

**Step 9: Plot the Volatility Data**  Before we plot the data, we need to check what the range of volatilities is in the data. The output below shows that the range is from 3.56% to 40.74%. Based on this, we select a y-axis range of zero to 45%. We then use plot() to chart the data. Figure 10.2 shows the output of the code below. The

**Implied vs. Historical Market Volatility, 2015 − 2019**

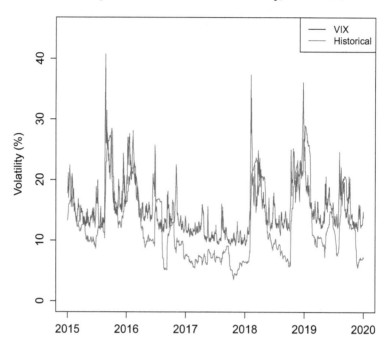

**Fig. 10.2** CBOE Volatility Index and 30-Day Historical Volatility of the S&P 500 Index, 2015–2019. Data Source: VIX data is provided as a courtesy by CBOE Exchange, Inc. (Cboe). S&P 500 Index data obtained from Federal Reserve Electronic Database. S&P® and S&P 500® are registered trademarks of Standard & Poor's Financial Services LLC, and Dow Jones® is a registered trademark of Dow Jones Trademark Holdings LLC. © 2020 S&P Dow Jones Indices LLC, it affiliates and/or its licensors. All rights reserved

figure shows that the implied volatility (VIX) is generally higher than the historical volatility from 2015 to 2019. However, the shape of the two volatility estimates is generally the same.

```
 1  > range(vol.idx)
 2  [1]  3.563562 40.740000
 3  > (y.range <- c(0, 45))
 4  [1]  0 45
 5  > plot(y = vol.idx$VIX,
 6  +     x = index(vol.idx),
 7  +     xlab = "",
 8  +     ylab = "Volatility (%)",
 9  +     ylim = y.range,
10  +     type = "l",
11  +     col = "blue",
12  +     main = "Implied vs. Historical Market Volatility, 2015 − 2019")
13  > lines(y = vol.idx$Hist.Vol, x = index(vol.idx), col = "red")
14  > legend("topright",
```

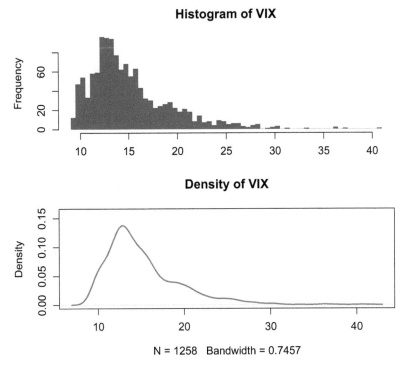

**Fig. 10.3** CBOE Volatility Index Histogram and Density, 2015–2019. Data Source: VIX data provided as a courtesy by CBOE Exchange, Inc. (Cboe)

```
15   +    c("VIX", "Historical"),
16   +    lwd = 1,
17   +    col = c("black", "red"))
```

Another interesting feature of the VIX is that its distribution is positively skewed. As we can see from Fig. 10.3, a histogram and a plot of the density of the VIX appear to show that volatility is better measured using a logarithmic scale (i.e., the log of the VIX would result in a much more symmetric distribution). To plot the histogram, we use hist(), and to plot the density, we use plot() on density(vix). Line 1 is to stack the histogram on top of the density chart, and Line 13 is to revert subsequent charts back to one per page.

```
1   > par(mfrow = c(2, 1))
2   > hist(vix,
3   +    xlab = "",
4   +    breaks = 100,
5   +    col = "blue",
6   +    border = 0,
7   +    main = "Histogram of VIX")
8   > plot(density(vix),
9   +    ylim = c(0, 0.16),
```

```
10   +      lwd = 2,
11   +      col = "red",
12   +      main = "Density of VIX")
13   > par(mfrow = c(1, 1))
```

## 10.4   The Cox, Ross, and Rubinstein Binomial OPM

Most options problems in practice cannot be solved using a closed-form solution (e.g., Black–Scholes–Merton). Numerical methods have to be implemented to solve such issues. These models can become extremely complex. In this section, we will discuss the Binomial Options Pricing Model, from which we can learn the basic intuition of how these types of models work. Moreover, when certain criteria are met, the option value calculated from the binomial model with converge to the result of the Black–Scholes–Merton model. In this section, we use the Binomial Options Pricing Model (OPM) based on [4] (CRR).

The CRR Binomial OPM is a lattice-based or tree-based model that allows us to model the path of the underlying asset's price in discrete time steps. To describe how the binomial model works, consider a stock with value $V$ today. In a binomial model, the value in 6 months can either go up by $u$ or go down by $d$. That is, the value of the stock can be either $Vu$ or $Vd$ at the end of 6 months. At the end of the year, the value of the stock depends on whether you start at node $Vu$ or node $Vd$. Starting in node $Vu$, we can either go up to node $Vuu$ or go down to node $Vud$. Starting in node $Vd$, we can either go up to node $Vdu$ or go down to node $Vdd$. Note that one feature of the binomial tree is that since $u$ and $d$ are fixed increments, the tree recombines, that is, $Vud = Vdu$. Below is a graphical representation of how the underlying stock price can move as time passes.

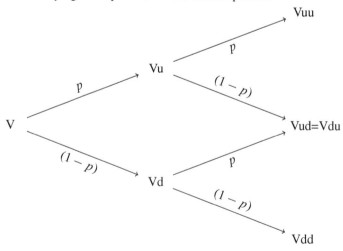

The above shows that there is a *risk-neutral probability p* that denotes the likelihood that the stock would go up. Therefore, the likelihood that the stock would go down is $(1 - p)$. This will be useful in determining the value of the option.

The risk-neutral probability is not the same as **real world probabilities** (also called subjective probabilities), which is what we would normally think of when speaking of probabilities. However, there is a specific set of risk-neutral probabilities that correspond to a set of real world probabilities. In pricing derivatives, it is almost always easier to use risk-neutral probabilities because we can use the risk-free rate to discount cash flows instead of having to deal with a risk premium in the discount rate.

We can now put some numbers to show how to calculate the option value using the CRR Binomial Model. Consider the call option on AMZN above with a strike price of $1845 expiring in March 2020 (TTM = 0.2192). Amazon's stock price on December 31, 2019 was $1847.84 and the volatility of the underlying stock was 28.97%. Using the risk-free rate of 1.55%, what is the value of the call option using the Binomial Model?

First, we have to calculate the up move factor, $u$, which is calculated as

$$u = \exp(\sigma * \sqrt{dt}) = 1.100656, \qquad (10.6)$$

where $\sigma$ is the volatility of the underlying asset and $dt$ is the time increment (i.e., TTM / number of periods in tree = 0.22 / 2 = 0.1096). We can then calculate the down move factor, $d$, as

$$d = 1/u = 1/1.100656 = 0.908549. \qquad (10.7)$$

Below, we show how the stock price evolves based on the $u$ and $d$ we calculated above. The stock price today is $1847.84. At the end of the first period, our stock price can go from $1847.84 to either $2033.84 [=$1847.84 * 1.101] or $1678.85 [=$1847.84 * 0.909]. If the stock price was at $2033.84 at the end of the first period, it can end up at either $2238.55 [=$2033.84 * 1.101] or $1847.84 [=$2033.84 * 0.909] by the end of the second period. If the stock price was at $1678.85 at the end of the second period, it can end up at $1847.84 [=$1678.85 * 1.101] or $1525.32 [=$1678.85 * 0.909] by the end of the second period.

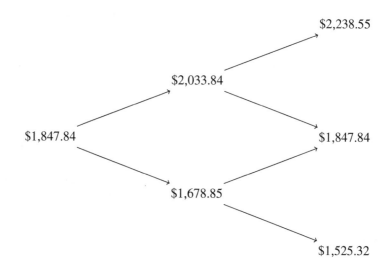

Then, we calculate the risk-neutral probability of an up move $p$ as

$$p = ((1 + r_f * dt) - d)/(u - d)$$
$$= ((1 + 0.0155 * 0.110) - 0.909)/(1.101 - 0.909)$$
$$= 0.4849, \tag{10.8}$$

where $r_f$ is the risk-free rate. Therefore, for the down move, we have $1 - p = 0.5151$. We then add the risk-neutral probabilities to our binomial tree.

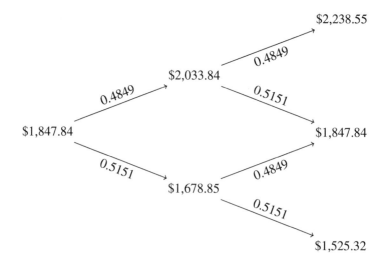

To calculate the value of the call option with a strike price of $1845, we have to start at the time to maturity of the option. That is, we start at the end of the second period and work our way to the present using backward substitution. Taking the $Vuu = \$2238.55$ node, the intrinsic value of the call option at that node is $393.55[=max(\$2238.55 - \$1845.00, 0)]$. Similarly, at the $Vud = Vdu = \$1847.84$ node, the intrinsic value of the call option at that node is \$2.84 $[=max(\$1847.84 - \$1845.00, 0)]$. Lastly, at the $Vdd = \$1525.32$ node, the intrinsic value of the call option at that node is zero because the strike price exceeds the stock price at that node.

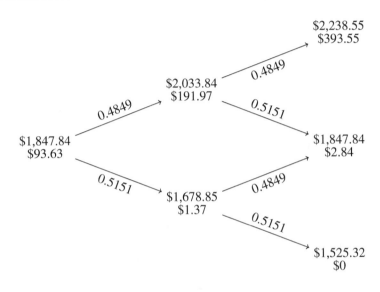

Continuing to work backward through time, we now calculate the first period option values for the up-node and down-node. Recall that the time increment we have is 0.110 years (i.e., half the time from valuation date of December 31, 2019 to the option expiration date of March 20, 2019). Therefore, the call option value at node $Vu$ is equal to $191.97 [=(0.4849 * \$393.55 + 0.5151 * \$2.85)/(1 + (0.0155 * 0.110))]$ and at node $Vd$ is equal to $1.37 [=(0.4849 * \$2.84 + 0.5151 * \$0) / (1 + (0.0155 * 0.110))]$. We then work backward to get to the price today for the call option using backward substitution, which is $93.63 [= (0.4849 * \$191.97 + 0.5151 * \$1.37) / (1 + (0.0155 * 0.110))]$.

## 10.4.1 CRR: The Long Way

Now that we understand the mechanics of the CRR Binomial Model, we can then program this specific calculation. We start with the same example of only having two time steps ($n = 2$).

**Step 1: Enter Inputs for Options**  An option would have a strike price and time to maturity. The option would be based on the underlying asset's price, volatility, and risk-free rate. The last choice we have to make is to determine how many time steps we would want in our calculation.

```
1   > S <- 1847.84
2   > K <- 1845
3   > TTM
4   [1] 0.2191781
5   > (r <- rf)
6   [1] 0.0155
7   > (sigma <- vol)
8   [1] 0.2897105
9   > n <- 2
```

**Step 2: Calculate Time Increment**  We divide the *TTM* by the number of time increments *n*.

```
1   > (dt <- TTM / n)
2   [1] 0.109589
```

**Step 3: Calculate Periodic Discount Factor**  Since there are *dt* periods, we have to scale down the risk-free rate, which is in annual terms, into the appropriate periodic risk-free rate.

```
1   > (disc <- (1 + r * dt))
2   [1] 1.001699
```

**Step 4: Calculate Up and Down Move Sizes**  The up move and down move are calculated based on the volatility of the underlying asset.

```
1   > (u <- exp(sigma * sqrt(dt)))
2   [1] 1.100656
3   > (d <-1 / u)
4   [1] 0.908549
```

**Step 5: Calculate Risk-Neutral Probability of an Up Move**  Using Eq. (10.8), we calculate the risk-neutral probability.

```
1   > (p <- ((1 + r * dt) - d) / (u - d))
2   [1] 0.4848838
```

**Step 6: Generate Values at All Nodes at Maturity**  The shortcut to this procedure is to calculate an upstate multiplicative factor and downstate multiplicative factor and multiply the two together.

```
1   > (UP <- u^(0:n))
2   [1] 1.000000 1.100656 1.211444
3   >
4   > (DOWN <- d^(n:0))
5   [1] 0.8254613 0.9085490 1.0000000
6   >
7   > (terminal <- S * UP * DOWN)
8   [1] 1525.320 1847.840 2238.554
```

**Step 7: Calculate Intrinsic Value of Call Option at Maturity**  Similar to our example above, we only do this at the terminal nodes.

```
1  > (terminal.optval <- ifelse(terminal - K < 0, 0, terminal - K))
2  [1]    0.0000   2.8400 393.5544
```

**Step 8: Calculate Option Value Today Using Backward Substitution**  The most efficient way to use this is that we need to use two loops. We should go through this step-by-step as the four lines of code may look pretty daunting. The first line of code is the loop that we want to start from one time increment prior to maturity (`from = n - 1`) and walk backward in time to the present (`to = 0`) one time increment at a time (`by = -1`). The second line of code is the loop that calculates the value at each node. The third line of code calculates the value of the option at each node.

```
1  > for (j in seq(from = n - 1, to = 0, by = -1)) {
2  +   for (i in 0:j) {
3  +     terminal.optval[i + 1] =
4  +        (p * terminal.optval[i + 2] +
5  +           (1 - p) * terminal.optval[i + 1]) / disc
6  +   }
7  + }
8  > terminal.optval
9  [1]   93.62984 191.96502 393.55439
```

**Step 9: Extract Call Option Value**  The call option value is the first element of the list in `terminal.optval`, which is $93.63. This equals the value we calculated previously above.

```
1  > (call.optval <- terminal.optval[1])
2  [1] 93.62984
```

However, the value using $n = 2$ of $93.63 is very far off from $104.28 calculated using the BSM approach above. Below, we create a function that calculates Binomial OPM, so we can calculate the option value with larger time steps $n$. We show that the larger the time steps, the closer our result becomes to the BSM value.

### 10.4.2   CRR Function

We can also create a function to calculate the call and put option values using a CRR Binomial Model. Note that we add a variable called type, which takes the value of call or put, so we can calculate either option type with one function. We create a variable $x=1$ for a call option and $x=-1$ for a put option. We then set up a warning if the option type is not equal to call or put. This warning will be in the form of an error message. The rest of the calculations should look very similar to the calculations above.

```
1   > EuroCRR <- function(S, K, T, r, sigma, n, type){
2   +   x <- NA
3   +   if (type == "Call") x = 1
4   +   if (type == "Put") x = -1
5   +   if (is.na(x)) stop("Option Type can only be call or put")
6   +   dt <- T / n
7   +   u <- exp(sigma * sqrt(dt))
8   +   d <- 1 / u
9   +   p <- ((1 + r * dt) - d) / (u - d)
10  +   disc <- (1 + r * dt)
11  +   OptVal <- x * (S * u^(0:n) * d^(n:0) - K)
12  +   OptVal <- ifelse(OptVal < 0, 0, OptVal)
13  +   for (j in seq(from = n - 1, to = 0, by = -1)) {
14  +     for (i in 0:j) {
15  +     OptVal[i + 1] <- (p * OptVal[i + 2] +
16  +         (1 - p) * OptVal[i + 1]) / disc
17  +     }
18  +   }
19  +   value <- OptVal[1]
20  +   results <- rbind(u, d, p, value)
21  +   results
22  + }
```

We now calculate the value of a European call using two time steps. We see that the value generated by the Binomial Model Function is equal to $93.63, which is identical to the two calculations we performed above.

```
1   > EuroCRR(1847.84, 1845, 0.2191781, 0.0155, 0.2897105, 2, "Call")
2                   [,1]
3   u       1.1006561
4   d       0.9085490
5   p       0.4848838
6   value 93.6298404
```

The Binomial OPM is said to converge to the Black–Scholes–Merton OPM when the time increments approach infinity. Using the Binomial Model Function, we can see the effect of increasing the time increments. We begin by increasing the time increment from two to 252 and then to 504 [= 252 * 2]. We see that the value at 504 time increments is equal to $104.27, which is very close to the value we calculated above using the BSM OPM of $104.28.

```
1   > EuroCRR(1847.84, 1845, 0.2191781, 0.0155, 0.2897105, 252, "Call")
2                   [,1]
3   u       1.0085806
4   d       0.9914924
5   p       0.4986529
6   value 104.2432526
7   > EuroCRR(1847.84, 1845, 0.2191781, 0.0155, 0.2897105, 252 * 2, "Call")
8                   [,1]
9   u       1.0060598
10  d       0.9939767
11  p       0.4990475
12  value 104.2719222
13  > call.1845
14  [1] 104.2776
```

Similarly, we can calculate the value of the put options using the Binomial OPM. At 2 time increments, the put option value is \$84.54, which is substantially lower than the BSM OPM put value of \$95.18. However, at 504 time increments, the Binomial OPM value of the put option is equal to \$95.17, which is much closer to the BSM OPM put value for the option.

```
1   > EuroCRR(1847.84, 1845, 0.2191781, 0.0155, 0.2897105, 2, "Put")
2                 [,1]
3   u        1.1006561
4   d        0.9085490
5   p        0.4848838
6   value 84.5378290
7   > EuroCRR(1847.84, 1845, 0.2191781, 0.0155, 0.2897105, 252, "Put")
8                 [,1]
9   u        1.0085806
10  d        0.9914924
11  p        0.4986529
12  value 95.1459839
13  > EuroCRR(1847.84, 1845, 0.2191781, 0.0155, 0.2897105, 252 * 2, "Put")
14                [,1]
15  u        1.0060598
16  d        0.9939767
17  p        0.4990475
18  value 95.1746324
19  > put.1845
20  [1] 95.18025
```

## 10.5  American Option Pricing

American options differ from European options in that the former can be exercised at any time through maturity. This means that at any point during the life of an American option, the holder can exercise the option at time $t$ and receive $S_t - K$, where $S_t$ is the underlying stock price at time $t$ and $K$ is the option strike price. By contrast, a European option can only be exercised at maturity. The ability to exercise prior to maturity makes it difficult to arrive at exact closed-form solutions (e.g., Black–Scholes–Merton for European options) to pricing American options in many cases. Thus, numerical methods have to be used. In this section, we discuss the use of binomial trees using the approach laid out by Cox, Ross, and Rubinstein. Unfortunately, numerical methods can be computationally intensive. As such, approximations are sometimes used when pricing American options. A common approximation method used follows [1].

## 10.5.1  CRR Binomial Tree

The CRR Binomial Tree we discussed above can be applied to American options as well. We show below how to implement this. We follow the methodology of the CRRBinomialTreeOption function in the fOptions package.

**Step 1: Enter Inputs for Option**  We enter the inputs needed to calculate the American option price. From our prior discussion, we are familiar with the following inputs: price of the stock $S$, the strike price of the option $K$, annualized risk-free rate $r$, asset volatility $v$, time to maturity of the option $T$, and number of time steps $n$. The additional variable that sets this model apart from its European option pricing counterpart is the cost of carry $b$. The cost of carry is an annualized percentage, but we input that number as a decimal (i.e., 1% is 0.01). This reflects the cost of holding the stock (i.e., not exercising) through the expiration date and is equal to risk-free rate less the dividend yield, which is assumed equal to 0.55%.

```
1  > S <- 50
2  > K <- 30
3  > r <- 0.0155
4  > v <- 0.30
5  > T <- 2
6  > b <- 0.01
7  > n <- 100
```

**Step 2: Determine Whether the Option Is a Call or Put**  We include a variable cp that allows us to choose whether the calculation is for a call option or a put option. In this example, we are valuing a call option.

```
1  > cp <- "Call"
2  > if (cp == "Call") {
3  +    z = +1
4  + } else {z = -1}
5  > z
6  [1] 1
```

**Step 3: Calculate Time Steps and Movements**  We calculate the size of the time steps $dt$. We then calculate the size of the up move ($u$) and down move ($d$).

```
1  > (dt = T / n)
2  [1] 0.02
3  > (u  = exp(v * sqrt(dt)))
4  [1] 1.043339
5  > (d  = 1 / u)
6  [1] 0.958461
```

**Step 4: Calculate Risk-Neutral Probability of an Up Move**  Based on the features of the option and up/down movements, we can calculate the risk-neutral probability of an up move. Note that we use the cost of carry $b$ here, which embeds the risk-free rate.

```
1   > (p  =  (exp(b * dt) − d)/(u − d))
2   [1] 0.4917515
```

**Step 5: Calculate Discount Factor**  Since we use continuous discounting in this case, we use $\exp(-r * dt)$.

```
1   > (Df = exp(−r * dt))
2   [1] 0.99969
```

**Step 6: Calculate Intrinsic Values**  In this step, we do a shortcut on the tree building by multiplying the various combinations of the up and down state through the various steps (Line 1). We then calculate the intrinsic value of the various notes in Line 6.

```
1   > opt.val <− z * (S * u^(0:n) * d^(n:0) − K)
2   > head.tail(opt.val)
3   [1] −29.28152 −29.21789 −29.14863
4   [1] 2906.454 3166.497 3449.569
5   >
6   > opt.val <− ifelse(opt.val < 0 , 0, opt.val)
7   > head.tail(opt.val)
8   [1] 0 0 0
9   [1] 2906.454 3166.497 3449.569
```

**Step 7: Build the Tree**  We use a for-loop to calculate the present value through backward substitution.

```
1   for (j in seq(n − 1, 0, −1)) {
2     for (i in 0:j) {
3       opt.val[i + 1] = max((z * (S * u^i * d^(abs(i − j))) − K)),
4           p * opt.val[i + 2] + (1 − p) * opt.val[i + 1]) * Df
5     }
6   }
```

**Step 8: Extract Option Values**  When the prior step is finished, we would have calculated the present value of the various nodes and walked that value back to the present. Thus, the first element of *opt.val* is the value of the option.

```
1   > print(opt.val[1])
2   [1] 21.16155
```

We can use CRRBinomialTreeOption() in the fOptions package to compare the results of our calculation. In Line 24, we see that the option price is \$21.16, which is consistent with the calculation we performed above.

```
1   > library(fOptions)
2   > start_time <− Sys.time()
3   > CRRBinomialTreeOption("ca", 50, 30, 2, 0.0155, 0.01, 0.30, 100)
4
5   Title:
6     CRR Binomial Tree Option
7
```

```
 8  Call:
 9  CRRBinomialTreeOption(TypeFlag = "ca", S = 50, X = 30, Time = 2,
10      r = 0.0155, b = 0.01, sigma = 0.3, n = 100)
11
12  Parameters:
13              Value:
14  TypeFlag ca
15  S        50
16  X        30
17  Time     2
18  r        0.0155
19  b        0.01
20  sigma    0.3
21  n        100
22
23  Option Price:
24  21.16312
25
26  Description:
27  Tue Mar 10 22:17:38 2020
28
29  > end_time <- Sys.time()
30  > end_time - start_time
31  Time difference of 0.02528286 secs
```

We also use the code in Lines 2 and 29–30 to calculate how long it took for R to calculate the results of the binomial model. As we can see, it took us 0.025 s to run the Binomial model. We did this to compare with the results of the Bjerksund–Stensland approximation we discuss next.

## 10.5.2  Bjerksund–Stensland Approximation

In [1], Bjerksund and Stensland wrote a paper in which they laid out an approximation for pricing American style options with continuous dividends and a constant dividend yield. The Bjerksund–Stensland approximation assumes optimal exercise when the underlying asset price is greater than or equal to a trigger price. At that time, the value of the option is equal to $S - K$. Given this strategy, the value of an American call equals: (1) a European up-and-out call with knock-out barrier $X$, strike $K$, and maturity date $T$ and (2) a rebate $X - K$ the holder receives at the knock-out date if the option is knocked out prior to the maturity date.

We can use the BSAmericanApproxOption() function in the fOptions package to implement the Bjerksund–Stensland approximation. The output below shows that the option price is $21.17, which is consistent with the option price we calculated using the CRR model.

```
1  > library(fOptions)
2  > start_time <- Sys.time()
3  > BSAmericanApproxOption("c", 50, 30, 2, 0.0155, 0.01, 0.30)
4
5  Title:
```

```
6    BS American Approximated Option
7
8    Call:
9    BSAmericanApproxOption(TypeFlag = "c", S = 50, X = 30, Time = 2,
10        r = 0.0155, b = 0.01, sigma = 0.3)
11
12   Parameters:
13                  Value:
14   TypeFlag    c
15   S           50
16   X           30
17   Time        2
18   r           0.0155
19   b           0.01
20   sigma       0.3
21   TrigerPrice 148.78333142886
22
23   Option Price:
24     21.16919
25
26   Description:
27     Tue Mar 10 22:18:59 2020
28
29   > end_time <- Sys.time()
30   > end_time - start_time
31   Time difference of 0.004168034 secs
```

Similar to what we did for the CRR Binomial Tree above, we calculated the time it took to run the Bjerksund–Stensland code. As we can see, it took 0.004 s to run this code. This appears considerably faster than the CRR approach, which clocked in at 0.025 s. Although it appears that the Bjerksund–Stensland approach is 6× faster than the CRR approach, from a practical perspective the difference at least in this case is not meaningful.

## 10.6 Further Reading

Discussion of options and derivatives can get very complicated very quickly. As such, depending on your level of training, different texts may be appropriate. Excellent introductory texts include [5] and [10]. A slightly more advanced discussion can be found in [6] and [2]. For the more technically inclined, an excellent formal treatment of options pricing can be found in [9] and [7].

# References

1. Bjerksund, P., & Stensland, G. (1993). Closed-form approximation of American options. *Scandinavian Journal of Management, 9*, 87–99.
2. Bjork, T. (2009). *Arbitrage theory in continuous time* (3rd ed.). Oxford: Oxford University Press.
3. Black, F., & Scholes, M. (1973). Pricing of options and corporate liabilities. *Journal of Political Economy, 81*, 637–654.
4. Cox, J., Ross, S., & Rubinstein, M. (1979). Options pricing: a simplified approach. *Journal of Financial Economics, 7*, 229–263.
5. Hull, J. (2011). *Options, futures, and other derivatives* (8th ed.). New York: Prentice Hall.
6. Joshi, M. (2008). *The concepts and practice of mathematical finance* (2nd ed.). Cambridge: Cambridge University Press.
7. Karatzas, I., & Shreve, S. (1998). Methods of mathematical finance. New York: Springer.
8. Merton, R. (1973). Theory of rational option pricing. *Bell Journal of Economics and Management Science, 4*, 141–183.
9. Oksendal, B. (2003). *Stochastic differential equations: an introduction with applications* (6th ed.). New York: Springer.
10. Wilmott, P. (2007). *Paul Wilmott introduces quantitative finance* (2nd ed.). New York: Wiley.

# Chapter 11
# Simulation

If the value of the inputs to various levers that drive the value of an asset could be determined with relative precision and the relationship between those inputs and the asset value can be analytically derived, then we could use closed-form formulas (e.g., like Black–Scholes–Merton for European options) to value the asset. However, in many cases, there is a higher degree of uncertainty that affects the value of the asset or the inputs that drive the value of those assets. In those situations, we can turn to **numerical methods** to help solve the problem. An example of that is the Cox, Ross, and Rubinstein model we discussed in the previous chapter. In many cases, we would rely on a general method called **simulations**, in which we generate many possible outcomes to reflect the different ways things could turn out in the future based on the uncertainty of each input. In this chapter, we demonstrate how to apply simulations when modeling stocks and options.

In Sect. 11.1, we first show how to generate a stock price path using a Geometric Brownian Motion (GBM). GBM is a common process used to simulate stock prices. We next show in Sect. 11.2 how to simulate stock prices when the stock pays a dividend. In some instances, we would need to model securities that move together (i.e., correlated with one another). We show how to perform this in Sect. 11.3.

We then show how to apply simulations in the context of Value-at-Risk (VaR) in Sect. 11.4. Then, we discuss how to use Monte Carlo simulation in the pricing of European options in Sect. 11.5 and an approach to improve the precision of the Monte Carlo estimate using antithetic variables in Sect. 11.6. Finally, in Sect. 11.7, we show how to use simulations to value several exotic options. In particular, we implement models that value Asian, lookback, and barrier options.

C.S. Ang, *Analyzing Financial Data and Implementing Financial Models Using R*,
Springer Texts in Business and Economics,
https://doi.org/10.1007/978-3-030-64155-9_11

## 11.1   Simulating Stock Prices Using Geometric Brownian Motion

Before we simulate stock prices, we would have to first think about what are some of the properties of stock prices. This will help us think about what we should be expecting, as well as how we are going to run our simulation. If we look at Fig. 11.1, we can see that the closing price of the S&P 500 ETF looks like a squiggly line with what appears to be random up and down movements. The line is also continuous, which from a practical perspective means that we can draw the line without having to lift the pen off the paper.

```
1   > plot(y = data.spy$SPY.Close,
2   +     x = index(data.spy),
3   +     xlab = "",
4   +     ylab = "Close Price",
5   +     type = "l",
6   +     col = "blue",
7   +     main = "SPY Close Prices
8   +     December 31, 2014 — December 31, 2019")
```

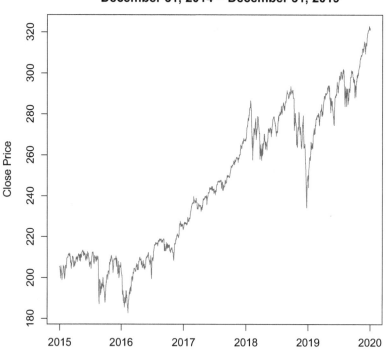

**Fig. 11.1**   S&P 500 ETF close prices, December 31, 2014–December 31, 2019. Data source: Price data reproduced with permission of CSI ©2020. www.csidata.com

In addition, stock prices can never fall below zero. Since stocks represent equity in the company and companies have limited liability, the shareholders of the company are not required to fork out more money in case the firm's assets fall short of the firm's liabilities. In other words, shareholders can *at most* lose only what they invested in the company. Thus, from an economic perspective, the lowest value for a company's stock price is zero.

One of the most common processes to simulate stock price paths is a Geometric Brownian Motion (GBM) because it is consistent with the properties of stock prices we discussed above. In a GBM, the stock price tomorrow $S_{t+1}$ is

$$S_{t+1} = S_t * \exp((r_f - 0.5\sigma^2)dt + \sigma z\sqrt{(dt)}), \tag{11.1}$$

where $r_f$ is the risk-free rate, $\sigma$ is the volatility of the stock, and $z \sim \mathbb{N}(0, 1)$ (i.e., a random normal variable). We now look at how to generate simulated stock prices in R using the following example.

**Step 1: Layout the Inputs** In this example, we will assume that the starting stock price $S_0$ is \$100. We will simulate daily stock prices over a 1-year period (i.e., $T = 1$). The risk-free return $r_f$ is assumed to be 1.55% and the stock's standard deviation $\sigma$ is assumed to be equal to 40%. Since our increments are daily and there are approximately 252 trading days in a year, we create a time increment variable $dt$ equal to 1/252 or 0.00397.

```
1  > S_0 <- 100
2  > T <- 1
3  > rf <- 0.0155
4  > sigma <- 0.40
5  > (dt <- 1 / 252)
6  [1] 0.003968254
7  > (n <- T / dt)
8  [1] 252
```

Lines 7 and 8 give us the number of prices we need to generate. In this case, we need to simulate prices for one calendar year, which means we need 252 simulated prices because we only count the number of trading days in a year. If we were simulating 2 years of prices, then we would need to simulate 504 prices.

**Step 2: Create Dataset to Store Simulated Stock Prices** Before we run the simulation, we need to create a dataset with temporary values. This will be a placeholder that we will later replace with prices created in our simulation. In Line 1, we create a data frame by joining together the starting stock price $S_0$ and the 252 zeroes. We create this using rep(), which tells us to repeat zero $n$ times and where $n$ is 252. Line 2 then renames the column header to Price. As we can see, the first observation is equal to the price of the stock today of \$100, and then it is followed by 252 zeroes.

```
1  > s <- data.frame(c(S_0, rep(0,n)))
2  > names(s) <- "Price"
3  > head.tail(s)
```

```
4       Price
5   1    100
6   2      0
7   3      0
8       Price
9   251     0
10  252     0
11  253     0
```

**Step 3: Simulate 252 Stock Prices**  We use a for-loop to fill the other 252 elements of *s*. Because we will use a random number generator, we add the set.seed(1) in Line 1 for reproducibility. Specifically, we should be able to see the output below in our R console. Otherwise, if we do not set the seed, a different set of random prices will be generated. In practice, however, we would typically not include set.seed().

```
1  > set.seed(1)
2  > for (i in 1:n){
3  +    s[i + 1, ] <- s[i, ] * exp (((rf - 0.5 * sigma^2) * dt +
4  +       sigma * rnorm(1, 0, 1) * sqrt(dt)))
5  + }
6  > head.tail(s)
7         Price
8  1 100.00000
9  2  98.40869
10 3  98.83981
11         Price
12  251 107.8602
13  252 108.2034
14  253 109.2912
```

Line 2 loops through a variable *i* that takes on a value from 1 to 252 (remember, $n = 252$). In Line 3, we see that the value of the stock price at time $t + 1$ is equal to the value of the stock price at time *t* growing at a return equal to what is described in the equation above for a Geometric Brownian Motion. The *z* in Eq. (11.1) is represented by rnorm(), which generates a standard normal random variable. Recall that a **standard normal** variable has mean zero and variance of unity. The first argument in rnorm() tells R how many random variables it should generate, which in this case is one random variable for each loop. The second argument is the mean (i.e., 0), and the third argument is the standard deviation (i.e., 1). Recall that the standard deviation is the positive square root of the variance. So, the square root of 1 is also equal to 1. As we can see in the output, we start the year at a price of \$100 and end the year at a price of \$109.29.

**Step 4: Plot Simulated Prices**  Although the prior listing can show some of the simulated stock prices, we can visually see what we have generated by creating a simple plot of stock price dataset *S*. The graph of the simulated stock price is shown in Fig. 11.2.

```
1  > plot(x = s$Price,
2  +    xlab = "Trading Days",
```

**Simulated Stock Price**

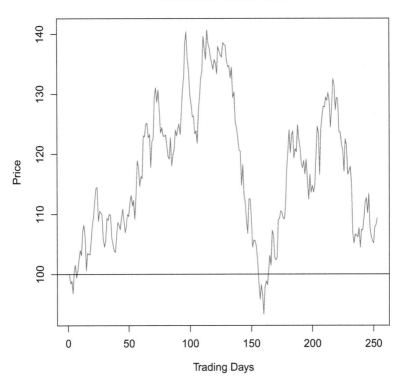

**Fig. 11.2** Simulated price path using Geometric Brownian Motion. Simulating 252 trading days of stock prices based on a starting stock price of $100, mean return of 10%, and standard deviation of 40%

```
3  +    ylab = "Price",
4  +    type = "l",
5  +    col = "blue",
6  +    main = "Simulated Stock Price")
7  > abline(h = 100)
```

Here, we use `plot()` to create a line chart (i.e., `type = "l"`). To make it easier to follow, we also add simple axis labels. The `abline(h = 100)` creates a horizontal line and gives us a point of reference for the starting point of the stock price (i.e., $S_0 = \$100$). We can then see whether the stock price is higher or lower than the starting price of $100 much more easily. As we can see from the chart, our simulated stock price only falls briefly below the initial price of $100 at the beginning and then again after 150 trading days.

### 11.1.1  Simulating Multiple Ending Stock Price Paths

We simulated a single stock price path above. We can repeat the process multiple times to simulate multiple stock price paths. Sometimes, we do not need to see the daily evolution of stock prices but only care about the price at the end of the simulation period. To the extent, we only need the ending date (e.g., valuing European options), then this can lead to faster calculations.

**Step 1: Create Dataset to Store Simulated Ending Prices**  In this example, we will simulate 10,000 ending prices. We then create a temporary ending price dataset $p$ that is a vector of 10,000 zeroes. When we run the simulation, we will replace the elements in this vector by the simulated ending prices. We also create a temporary price path dataset of $s$, which starts with the starting price of $100 and 252 zeroes.

```
1  > sim <- 10000
2  > p <- c(rep(0, sim))
3  > s <- c(100, rep(0, n))
```

**Step 2: Simulate Stock Prices**  Lines 1 through 4 are the inputs and assumptions we use in our simulation. Our starting stock price is $100, the risk-free rate is 1.55%, we assume 252 trading days in a year, and we use a time increment of 0.00397. Next, we run the simulation similar to the above, except that we generate 252 standard normal random variables. We run two for-loops. The second for-loop generates one simulated stock price path, and then the ending price is saved as an element of $p$. The first for-loop then runs this 10,000 times, so we get 10,000 ending prices. The average ending price from our simulation is $101.40.

```
1   > S_0 <- 100
2   > rf <- 0.0155
3   > n <- 252
4   > (dt <- 1 / n)
5   [1] 0.003968254
6   >
7   > set.seed(1)
8   > for (i in 1:sim){
9   +   for (j in 1:n) {
10  +     s[j + 1] <- s[j] * exp((rf - 0.5 * sigma^2) * dt +
11  +       sigma * rnorm(1, 0, 1) * sqrt(dt))
12  +   }
13  +   p[i] <- s[length(s)]
14  + }
15  > mean(p)
16  [1] 101.4021
17  > min(p)
18  [1] 20.0268
19  > max(p)
20  [1] 559.653
```

**Step 3: Plot Histogram of End of Year Prices**  We can use hist() to plot the frequency of the ending prices. We can specify the number of breakpoints we would

**10,000 Simulated Ending Prices**

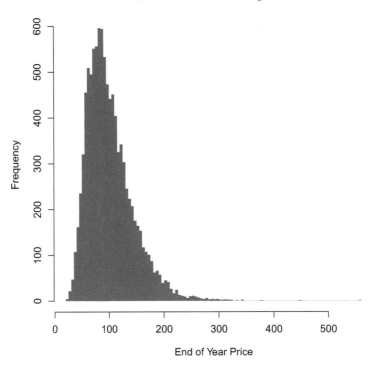

**Fig. 11.3** Histogram of 10,000 simulated ending stock prices based on a Geometric Brownian Motion

like. In this example, we set the breakpoints equal to 60. The graph of the histogram is shown in Fig. 11.3.

```
1  > hist(p,
2  +      breaks = 100,
3  +      col = "blue",
4  +      border = 0,
5  +      xlab = "End of Year Price",
6  +      main = "10,000 Simulated Ending Prices")
```

## 11.1.2 Comparing Theoretical to Empirical Distributions

Although the empirical distribution shown above is based on some theoretical model of the stock price, it is possible that the distribution may not yield identical results to the theoretical distribution. We can consider the example of calculating the probability of loss using the following:

$$d = -\frac{(\mu - 0.5\sigma^2)T}{\sigma\sqrt{T}}, \tag{11.2}$$

and the probability of loss is calculated as pnorm(d). In our example above, a loss would be an ending stock price that is less than $100, which is our starting value. As the below shows, the theoretical probability of loss is 56.4%. In Line 6, we calculate the empirical probability of loss by counting the number of ending prices from our simulation that is less than $100. We do this using sum() but imposing a condition in what we are summing. If we simply wrote sum(p), R would have summed all 10,000 prices. However, by imposing a condition that we only sum if $p < 100$, R counts the number of observations that satisfies that condition. Applying this condition results in 5657 observations that are less than $100. We then divide that amount by the number of ending prices we simulated, which is 10,000. Thus, the empirical probability of a loss is 5657 out of 10,000 or 56.6%. This indicates that our simulation generated a very similar probability of loss to what is predicted by theory.

```
1   > (d <- -(rf - 0.5 * sigma^2) * T / (sigma * sqrt(T)))
2   [1] 0.16125
3   > (pr.loss <- pnorm(d))
4   [1] 0.5640517
5   >
6   > sum(p < 100) / sim
7   [1] 0.5657
```

## 11.2   Simulating Stock Prices with and Without Dividends

When a stock pays dividends, its price falls by the amount of dividend it expects to pay on the ex-dividend date or ex-date. This is because the day prior to the ex-date is the last day the stock trades with the dividend. An intuitive way to think about why this happens is to think in terms of a market value balance sheet. For simplicity, assume that a company holds cash of $10 per share and has debt of $1 per share. The equity of the company would be $9 per share [= $10 cash–$1 debt]. Recall that shareholders have a claim on the firm's equity, so the stock price is also $9 per share. Now, suppose this same firm decides to pay $0.50 per share of dividends in cash to shareholders as of day $T$. Thus, shareholders that bought the stock on day $T + 1$ no longer have a claim on $9 per share in cash, but only to $8.50 of equity as $0.50 will be paid out as dividends. Thus, the stock price drops to $8.50 because those shareholders only have a claim on $8.50 per share worth of equity (i.e., $10 cash minus $0.50 dividend minus $1 debt).

The above implies that if we have two identical firms, but one pays a dividend while the other retains those cash flows, we should expect the stock price of the firm that pays dividends to be lower than the stock price of an identical firm that does not pay dividends. This may suggest that investors will lose money when holding a stock that pays dividends, but this is not the case in reality. Although the stock price falls, the cash from the dividend payment goes to the investor and offsets the decline in stock price. The way to see this is if we calculate the total returns for both stocks. Recall that the total return is $R_t = (P_t + D_t)/P_{t-1} - 1$, where $P$ is the price of the stock and $D$ is the dividend payment. Continuing from our example above, the total return on the stock that did not pay dividends is 0% [= ($9 after + $0 dividend)/$9 before - 1]. On the other hand, after the dividend is paid out, the total return on the stock that paid dividends is also 0% [= ($8.50 after + $0.50 dividend)/$9 before - 1]. Because investors receive the dividends, they are neither better off nor worse off. Intuitively, that should be the case because you are simply trading off capital gains and dividends. Thus, all else equal, the return on a stock on the ex-dividend date should be the same for an investor in a stock that pays dividends and a stock that does not.

Aside from potentially having a differential tax effect to different investors, there could also be an impact of dividends on how certain derivatives are valued. For example, if you are valuing a security that has a trigger based on the actual stock price (e.g., average of last 5 days), then we would need to model stock prices to account for the fact that they would go down based on the amount of dividends they are expected to pay. In this section, we show an example of how dividend payments could be incorporated in simulating stock prices.

In the following example, we consider a stock that has a price of $100 today and pays $2.50 per share of dividends each quarter. We want to model daily prices for 2 years, and the stock has a mean return of 10% and standard deviation of 40%. We assume a risk-free rate of 2%.

**Step 1: Enter Inputs** We create variables for the key inputs in our example. We also add two additional variables. First is $dt$, which is the time increment. In this example, we are doing daily time increments. However, we only count the number of trading days in a year. Typically, it is assumed that there are approximately 252 trading days in a year. So, $dt$ is equal to 1/252. In the final step, we also create the number of time steps we would be simulating. Since we are doing this over a 2-year period and there are 252 trading days each year, we create a variable $n$ that is equal to 504.

```
1   > div <- 2.50
2   > S_0 <- 100
3   > T <- 2
4   > rf <- 0.02
5   > sigma <- 0.40
6   > (dt <- 1 / 252)
7   [1] 0.003968254
8   > (n <- T / dt)
9   [1] 504
```

**Step 2: Create Vectors to Store Simulated Prices and Returns** We create the vector $s$ for stock prices and $r$ for returns. Each vector has 505 elements. The first element of *s* is the starting stock price of $100, and then it is followed by 504 zeroes. For *r*, all 505 elements are zeroes.

```
1  > s <- c(S_0, rep(0, n))
2  > head.tail(s)
3  [1] 100    0   0
4  [1] 0 0 0
5  > length(s)
6  [1] 505
7  >
8  > r <- c(0, rep(0, n))
9  > head.tail(r)
10 [1] 0 0 0
11 [1] 0 0 0
12 > length(r)
13 [1] 505
```

**Step 3: Simulate Returns Over the 2-Year Period** The first element of *r* will be zero. Elements 2 through 505 are then populated with returns based on a Geometric Brownian Motion.

```
1  > set.seed(1)
2  > for (i in 1:n) {
3  +     r[i + 1] <- (rf - 0.5 * sigma^2) * dt +
4  +         sigma * rnorm(1, 0, 1) * sqrt(dt)
5  + }
6  > head.tail(r)
7  [1]   0.000000000 -0.016023248  0.004389282
8  [1] -7.718482e-03 -3.005300e-02  4.645376e-05
```

**Step 4: Add Dividends to the Price Dataset** The dataset *s* currently has $100 in the first observation and 504 zeroes. Dividends are typically paid quarterly, so we would want to create a series that has the value of $2.50 once every quarter and zero otherwise. Given we are assuming 252 trading days in a year, the end of each quarter is approximately 63 trading days apart from each other. To do this, we first create a sequence of trading days that we can use as an index. We do this in Line 1. We start at 0, so when the value of *t* is 63, it is actually the 64th trading day and there are 63 trading days from the starting price. Then, in Line 6, we use ifelse() to tell R that on every 63rd trading day, we want there to be a dividend payment *div* or else the value of div is zero. As the output shows, the last observation, which is the 504th trading day, has a dividend payment of $2.50 per share. To more efficiently implement this, we used seq() and tell R to start at number 63 and go to 63 * 8 or 504 using increments of 63.

```
1  > t <- seq(0, n, 1)
2  > head.tail(t)
3  [1] 0 1 2
4  [1] 502 503 504
5  >
6  > div <- ifelse(t %in% seq(63, 63*8, 63), div, 0)
7  > head.tail(div)
```

```
8    [1] 0 0 0
9    [1] 0.0 0.0 2.5
10   >
11   > s <- cbind(s, t, div)
12   > head.tail(s)
13           s t div
14   [1,] 100 0   0
15   [2,]   0 1   0
16   [3,]   0 2   0
17          s   t div
18   [503,] 0 502 0.0
19   [504,] 0 503 0.0
20   [505,] 0 504 2.5
```

**Step 5: Create a Second Price Series**  To determine what the price series would look like with and without dividends, we create a second stock price series that assumes no dividend payments. We label this s2. The starting stock price will also be $100.

```
1    > s2 <- c(S_0, rep(0, n))
2    > s <- cbind(s, s2)
3    > head.tail(s)
4           s t div  s2
5    [1,] 100 0   0 100
6    [2,]   0 1   0   0
7    [3,]   0 2   0   0
8          s   t div s2
9    [503,] 0 502 0.0  0
10   [504,] 0 503 0.0  0
11   [505,] 0 504 2.5  0
```

**Step 6: Simulate Price Series**  We first convert s into a data frame object using data.frame(). Then, we use a for-loop to calculate the stock price. We take the exponential of the log returns generated in by our simulation in Step 3. For s, we subtract the value in the div column. Thus, on those ex-dividend dates, the stock price goes down, and then the subsequent return is generated off the lower price. Thus, by the end of the 2 years, we would see the stock price in column s is lower at $95.44. By contrast, the stock price for the non-dividend paying stock in column s2 ends at $113.91.

```
1    > s <- data.frame(s)
2    > for (i in 1:n) {
3    +    s[i + 1, 1] <- (s[i, 1] * exp(r[i + 1])) - s[i + 1, 3]
4    +    s[i + 1, 4] <- (s[i, 4] * exp(r[i + 1]))
5    + }
6    > head.tail(s)
7              s t div        s2
8    1 100.00000 0   0 100.00000
9    2  98.41044 1   0  98.41044
10   3  98.84334 2   0  98.84334
11            s   t div        s2
12   503 100.92774 502 0.0 117.3784
```

```
13   504   97.93969 503 0.0 113.9033
14   505   95.44424 504 2.5 113.9086
```

**Step 7: Plot the Two Price Series** It may be easier to visualize the difference in the two price series using a simple line chart. Figure 11.4 shows the output of the code below. The red line denotes the price of the stock that pays dividends, which is lower than the blue line. The blue line denotes the price of the stock that does not pay dividends. As we can see, the red and blue lines begin to diverge near the middle of trading days 0 and 100, because the first dividend payment is on trading day 63. Then, the gap widens on each ex-dividend date, which we assume to occur every 63rd trading day. Looking simply at the price return of the dividend paying stock, we may think that the investor lost money as the red line ends below the horizontal line denoting $100. The above shows that the difference between the ending price

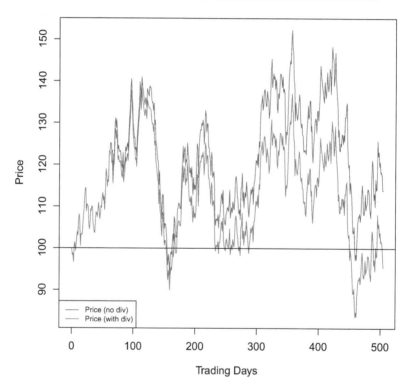

**Fig. 11.4** Comparison of simulated stock price with and without dividends. The starting stock price is $100 for both dividend and non-dividend paying stocks. The daily returns applied to both stocks are the same. The dividend paying stock receives quarterly dividends of $2.50 over the 2-year period

of the blue line ($113.91) and red line ($95.44) is $18.46. However, the investor received nominally $20 in dividends over the 2-year period in this example.

```
1   > (y.range <- range(s$s, s$s2))
2   [1]   83.61193 152.30753
3   > plot(y = s$s2,
4   +      x = s$t,
5   +      xlab = "Trading Days",
6   +      ylim = y.range,
7   +      ylab = "Price",
8   +      type = "l",
9   +      col = "blue",
10  +      main = "Simulated Stock Price With and Without Dividends")
11  > lines(y = s$s, x = s$t, col = "red")
12  > abline(h = 100)
13  > legend("bottomleft",
14  +      legend = c("Price (no div)", "Price (with div)"),
15  +      col = c("blue", "red"),
16  +      lty = 1,
17  +      cex = 0.75)
```

## 11.3  Simulating Stocks with Correlated Prices

Because of the potential for similar factors affecting all stocks, we may want to consider simulating two or more asset prices the extent that the prices of those assets move together. Stocks within say, the same industry are affected by some similar forces. For example, the supply and demand for light vehicles may impact car sales at Ford and General Motors (GM). Therefore, if we want to model the stock prices of those two companies, it may make more sense to conduct a simulation in which we consider the correlation between Ford stock and GM stock rather than modeling the prices of the two stock as if they were independent of one another. In this section, we will go through an example of how to model the stock prices of two correlated assets.

Consider two stocks that were not correlated in the first year but are expected to be much more highly correlated in the second year. This scenario is highly stylized but will make the point we are making much clearer. Standing at the end of Year 1, we want to know how we are going to model the stock prices of the two firms going forward if we assume that the two stocks are now correlated by 0.7. We first simulate the prices for both stocks in Year 1 with the following assumptions. The first stock has a price of $10, mean return of 10%, and standard deviation of 40%. The second stock has a price of $10, mean return of 12%, and standard deviation of 50%. We assume a risk-free rate of 1.55%. We then assume that the two stocks are correlated by 0.7 and simulate the prices for Year 2 accordingly.

**Step 1: Enter Inputs for First Year of Simulation**  The simulation requires the risk-free rate (1.55% as of December 31, 2019), starting stock prices for the two securities ($100), and volatility of 40% for security 1 and 50% for security 2. We assume that there are a number of trading days $n$ of 252 in a year, so the time increment $dt$ is equal to 1/252 or 0.00397 because we are simulating daily prices.

```
1   > rf <- 0.0155
2   > s1 <- 10
3   > sigma1 <- 0.40
4   > s2 <- 10
5   > sigma2 <- 0.50
6   > T <- 1
7   > n <- 252
8   > (dt <- 1 / n)
9   [1] 0.003968254
```

**Step 2: Create Two Vectors to Store Stock Prices**  We create two vectors that have 253 elements. The first element is equal to the stock price, which is equal to $10 for both. Then, we temporarily use a value of 0 for the remaining 252 observations. Since we would have generated *path2* in the same manner, we can simply assign *path1* to *path2*.

```
1   > path1 <- c(s1, rep(0, n))
2   > head.tail(path1)
3   [1] 10  0  0
4   [1] 0 0 0
5   > length(path1)
6   [1] 253
7   > path2 <- path1
```

**Step 3: Simulate Stock Prices**  Note here that we use `set.seed(12345)` instead of `set.seed(1)`. That number is not chosen wholly at random. We selected this value by trial-and-error with the goal of generating two assets that have low correlation for our example.

```
1   > set.seed(12345)
2   > for (i in 1:n){
3   +     path1[i + 1] <- path1[i] * exp((rf - 0.5 * sigma1^2) * dt +
4   +         sigma1^2 * rnorm(1, 0, 1) * sqrt(dt))
5   +     path2[i + 1] <- path2[i] * exp((rf - 0.5 * sigma2^2) * dt +
6   +         sigma2^2 * rnorm(1, 0, 1) * sqrt(dt))
7   + }
8   > head.tail(path1)
9   [1] 10.00000 10.05662 10.04297
10  [1] 12.17861 12.00244 12.05814
11  > head.tail(path2)
12  [1] 10.00000 10.10796 10.03167
13  [1] 11.41955 10.97963 10.81394
```

**Step 4: Calculate Stock Returns**  We calculate the log returns by taking the difference of the logarithm of the prices.

```
1   > return1 <- diff(log(path1))
2   > head.tail(return1)
```

```
3    [1]  0.005645623 −0.001357626  0.005850819
4    [1] −0.001379352 −0.014571471  0.004629524
5    >
6    > return2 <- diff(log(path2))
7    > head.tail(return2)
8    [1]  0.010738516 −0.007576433 −0.029064639
9    [1] −0.005341789 −0.039285156 −0.015206199
```

**Step 5: Calculate Correlation of Returns** Using `cor()`, we calculate the correlation of the returns of the two stocks in *return1* and *return2*. We find that the correlation between our two stocks is 0.007.

```
1    > cor(return1, return2)
2    [1] 0.007064377
```

**Step 6: Calculate the Mean Return, Volatility, and Price Vectors** We create a vector *mu.port* for the average return of the two assets and another vector *sigma.port* for the standard deviation of the two assets. We also create a vector for the last price of the two assets *price.port*.

```
1    > (mu.port <- as.matrix(c(mean(return1), mean(return2))))
2                     [,1]
3    [1,] 0.0007426764
4    [2,] 0.0003105187
5    > (sigma.port <- as.matrix(c(sd(return1), sd(return2))))
6                     [,1]
7    [1,] 0.009995675
8    [2,] 0.015777745
9    > (price.port <- as.matrix(c(path1[length(path1)],
10   +       path2[length(path2)])))
11                    [,1]
12   [1,] 12.05814
13   [2,] 10.81394
```

**Step 7: Create a Matrix of Returns** Using `as.matrix()`, we create a matrix of returns from the return series of the two assets.

```
1    > return.mat <- as.matrix(cbind(return1, return2))
2    > head.tail(return.mat)
3                  return1        return2
4    [1,]   0.005645623  0.010738516
5    [2,]  −0.001357626 −0.007576433
6    [3,]   0.005850819 −0.029064639
7                  return1        return2
8    [250,] −0.001379352 −0.005341789
9    [251,] −0.014571471 −0.039285156
10   [252,]  0.004629524 −0.015206199
```

**Step 8: Calculate Actual Correlation of Portfolio** Using `cor()`, we calculate the correlation of the returns of the two assets. As the output shows, the correlation between the two assets (i.e., the values in the off-diagonal) is low at 0.007.

```
1  > (corr.port <- cor(return.mat))
2              return1      return2
3  return1 1.000000000 0.007064377
4  return2 0.007064377 1.000000000
```

**Step 9: Assume Correlation is 0.7 Instead** Although the historical correlation is low, let us assume that we expect the correlation of the two assets to increase to 0.7 in the subsequent year. We replace the two off-diagonal values by 0.7. We will later use this correlation matrix in Step 12.

```
1  > corr.port[1, 2] <- 0.7
2  > corr.port[2, 1] <- 0.7
3  > corr.port
4            return1 return2
5  return1      1.0      0.7
6  return2      0.7      1.0
```

**Step 10: Create a Matrix to Store the Simulated Prices** We create a matrix with 252 rows and 2 columns.

```
1   > paths <- matrix(nrow = n, ncol = nrow(mu.port))
2   > head.tail(paths)
3          [,1] [,2]
4   [1,]    NA   NA
5   [2,]    NA   NA
6   [3,]    NA   NA
7          [,1] [,2]
8   [250,]  NA   NA
9   [251,]  NA   NA
10  [252,]  NA   NA
```

**Step 11: Generate Random Numbers** We create a matrix that has two columns and 252 rows. Each element in this matrix will be a random normal variable with mean 0 and variance 1. We then add each day's figure to the prior day's cumulative figure.

```
1   > set.seed(12345)
2   > p <- matrix(rnorm(n * nrow(mu.port), 0, 1), ncol = nrow(mu.port))
3   > p <- apply(p, 2, cumsum)
4   > head.tail(p)
5               [,1]        [,2]
6   [1,] 0.5855288  0.4560525
7   [2,] 1.2949948 -0.9781978
8   [3,] 1.1856915 -1.2435026
9               [,1]        [,2]
10  [250,] 31.84898 5.301701
11  [251,] 31.04030 5.786417
12  [252,] 32.04142 4.848444
```

**Step 12: Multiply _p_ with Cholesky Matrix** We implement Cholesky decomposition to decompose the correlation matrix. This is a highly technical step, but for our current purposes the only thing to note is that this step is required to generate the correlated assets. Essentially, we use Cholesky to generate correlated random variables by multiplying it to the matrix of standard normal variables.

```
1   > (chol_mat <- chol(corr.port))
2               return1    return2
3   return1        1 0.7000000
4   return2        0 0.7141428
5   >
6   > p <- p %*% chol_mat
7   > head.tail(p)
8               return1      return2
9   [1,] 0.5855288   0.73555680
10  [2,] 1.2949948   0.20792342
11  [3,] 1.1856915  -0.05805443
12              return1  return2
13  [250,] 31.84898 26.08046
14  [251,] 31.04030 25.86054
15  [252,] 32.04142 25.89148
```

**Step 13: Simulate Prices** We simulate prices using two steps. First, we create a function gbm(), so we can simulate prices based on a Geometric Brownian Motion. Then, we use a for-loop to replace with simulated prices the temporary matrix of prices we created in Step 10 called *paths*.

```
1   > gbm <- function(N, sigma, rf, S_0, Wt) {
2   +     t <- (1:N) / n
3   +     S_t = S_0 * exp((rf - 0.5 * sigma^2) * t + sigma * Wt)
4   +     return(S_t)
5   + }
6   >
7   > for (i in 1:nrow(mu.port)) {
8   +     paths[, i] <- gbm(n, sigma.port[i], mu.port[i],
9   +        price.port[i], p[, i])
10  + }
11  > head.tail(paths)
12              [,1]       [,2]
13  [1,] 12.12895 10.94018
14  [2,] 12.21530 10.84949
15  [3,] 12.20200 10.80406
16              [,1]       [,2]
17  [250,] 16.58965 16.32196
18  [251,] 16.45613 16.26543
19  [252,] 16.62168 16.27339
```

**Step 14: Check the Correlation of Returns of the Simulated Prices** In this case, we used ROC(), which is an alternative way of calculating log returns in R. This is the same as taking the difference of the log prices. We then remove the first element of the vector, which will be an NA. As we can see that the correlation of this series is 0.74, which is much higher than the almost zero correlation in the historical data and closer to our assumed 0.70 correlation.

```
1   > rets1 <- ROC(paths[, 1])
2   > rets1 <- rets1[-1]
```

```
 3  > head.tail(rets1)
 4  [1]   0.007094340 −0.001089812 −0.004530261
 5  [1]   0.007528105 −0.008080495  0.010009617
 6  >
 7  > rets2 <− ROC(paths[, 2])
 8  > rets2 <− rets2[−1]
 9  > head.tail(rets2)
10  [1] −0.008324127 −0.004195793  0.002223299
11  [1] −0.0194807336 −0.0034690331  0.0004888651
12  >
13  > cor(rets1, rets2)
14  [1] 0.7424496
```

**Step 15: Plot the Simulated Prices**   We can plot the correlated prices using a line chart. Figure 11.5 shows the output of the code below. The starting point of the two assets is different because the Year 1 close price of the assets is different. One asset had a price of $12.06, while the other had a price of $10.81.

```
 1  > (y.range <− c(min(paths), max(paths)))
 2  [1] 10.38606 19.31994
```

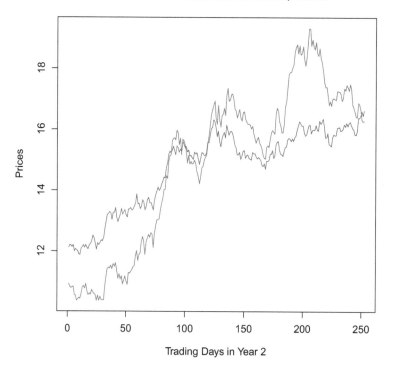

**Prices of Two Correlated Assets, Year 2**

**Fig. 11.5**  Simulated prices of two correlated assets with assumed correlation of 0.7

```
3  > days <- seq(1, n, 1)
4  > plot(x = days,
5  +    xlab = "Trading Days in Year 2",
6  +    y = paths[, 1],
7  +    ylab = "Prices",
8  +    ylim = y.range,
9  +    type = "l",
10 +    col = "blue",
11 +    main = "Prices of Two Correlated Assets, Year 2")
12 > lines(x = days,
13 +    y = paths[, 2],
14 +    col = "red")
```

## 11.4  Value-at-Risk Using Simulation

In this section, we compare the results of a historical Value-at-Risk (VaR) calculation with those of a simulation based on the features of the security. For our analysis, we will use Apple stock, so we import into R the data for AAPL and use daily returns from 2016 to 2019. Our final step is to convert the object from xts to a data frame.

```
1  > aapl <- load.data("AAPL Yahoo.csv.", "AAPL")
2  > rets <- diff(log(aapl$AAPL.Adjusted))
3  > rets <- rets[-1, ]
4  > names(rets) <- "return"
5  > rets <- rets["2016-01-01/2019-12-31"]
6  > rets <- data.frame(rets)
7  > head.tail(rets)
8                      return
9  2016-01-04  0.000854462
10 2016-01-05 -0.025378378
11 2016-01-06 -0.019763664
12                      return
13 2019-12-27 -0.0003795553
14 2019-12-30  0.0059175879
15 2019-12-31  0.0072799852
```

For the historical VaR, we want to sort the returns data from smallest (the most negative) to largest (the biggest positive). We do this using order().

```
1  > rets.sorted <- rets[order(rets$return), ]
2  > head.tail(rets.sorted)
3  [1] -0.10492433 -0.06863306 -0.06796461
4  [1] 0.06294017 0.06610106 0.06805252
```

Next, we calculate the cut-off based on a 95% VaR (i.e., $\alpha = 0.05$). Applying it to the number of returns in our dataset (i.e., 1006), results in the cut-off being equal to the 50th smallest return, which in this case is $-2.39\%$.

```
1  > length(rets.sorted)
2  [1] 1006
3  > alpha <- 0.05
```

```
4  > (cut.off <- length(rets.sorted) * alpha)
5  [1] 50.3
6  > (cut.off <- round(cut.off))
7  [1] 50
8  > (critical.value <- (rets.sorted[cut.off]))
9  [1] -0.02394501
```

Therefore, applying this critical value to the December 31, 2019 AAPL close price of \$293.65 yields a 5% historical VaR of \$7.03. Note that we follow the convention of reporting VaR as a positive number.

```
1  > (last.price <- as.numeric(aapl$AAPL.Close[nrow(aapl)]))
2  [1] 293.65
3  > (VaR.historical <- abs(last.price * critical.value))
4  [1] 7.031451
```

Now, we turn to simulating VaR using Monte Carlo. First, we calculate the average return and standard deviation of the historical returns as proxy for our expected return and expected standard deviation. This gives us a daily average return of 0.11% and daily standard deviation of 1.54%.

```
1  > (mean.rets <- mean(rets$return))
2  [1] 0.001087205
3  > (sd.rets <- sd(rets$return))
4  [1] 0.015357
```

Next, we need two more variables before we can run the simulation. First, because we are simulating prices on a daily basis, we need to define the time increment, which is equal to 1/252. Second, the risk-free rate is assumed to be equal to 1.55%.

```
1  > dt <- 1 / 252
2  > r <- 0.0155
```

Then, we run the 10,000 simulations of the next day's price. We use set.seed(1) for reproducibility of the output. In practice, we can drop that line of code.

```
1   > sim <- 10000
2   > prc <- data.frame(c(rep(0, sim)))
3   > names(prc) <- "End.Price"
4   > set.seed(1)
5   > for (t in 1 : sim) {
6   +     prc[t, ] <- last.price * exp(((r - 0.5 * sigma^2) * dt +
7   +         sigma * sqrt(dt) * rnorm(1, mean = 0, sd = 1)))
8   + }
9   > simprc.sorted <- prc[order(prc$End.Price), ]
10  > head.tail(simprc.sorted)
11  [1] 277.5364 278.0983 278.4790
12  [1] 310.4403 310.5128 311.3279
```

The cut-off for 10,000 observations assuming a 5% significance level is the 500th smallest price, which in this case is \$286.13. Subtracting that price from the December 31, 2019 price of \$293.65 yields a 1-day VaR of \$7.52. In other words, the maximum 1-day loss with 95% confidence is \$7.52 under the simulated VaR approach. This result is close to the maximum 1-day loss with 95% confidence of \$7.03 under the historical VaR approach.

```
1  > length(simprc.sorted)
2  [1] 10000
3  > (sim.cut.off <- length(simprc.sorted) * alpha)
4  [1] 500
5  > simprc.sorted[sim.cut.off]
6  [1] 286.1348
7  > (VaR.simulation <- abs(simprc.sorted[sim.cut.off] - last.price))
8  [1] 7.515213
```

## 11.5  Monte Carlo Pricing of European Options

In Chap. 10, we value European options using the Black–Scholes–Merton model
and the Binomial model. In this section, we will value European options using
Monte Carlo simulation. Simulations can also be used for more complex options
as we will see later in this chapter when we value exotic options.

**Step 1: Enter Inputs and Assumptions**  We set up the basic characteristics of the
underlying stock. For this example, we assume that the underlying asset is a non-
dividend paying stock ($q = 0$) with a price today of $100 with a volatility of 40%.
The option has a strike price of $110 and a time to maturity of 1 year. The risk-free
rate is 2%. For the simulation, we assume 100,000 runs. We also create *opt.val*,
which is a vector that takes on the value of beginning stock price 100,001 times.

```
1  > S <- 100
2  > K <- 110
3  > v <- 0.40
4  > T <- 1
5  > r <- 0.02
6  > q <- 0.00
7  >
8  > sim <- 100000
9  > opt.val <- c(S, rep(S, sim))
```

**Step 2: Run the Simulation**  We use set.seed(1) for reproducibility of the
output. In practice, we do not need to enter Line 1 of the code. The for-loop tells R
to calculate the value of the option, which is the larger of zero or the intrinsic value
of the option value at the time to maturity. The last term in Line 5 that contains the
exponential discounts the intrinsic value at maturity to the present.

```
1  > set.seed(1)
2  > for (i in 1 : sim) {
3  +    opt.val[i] <- max(0,
4  +        (S * exp((rf - q - 0.5 * v^2) * T +
5  +            v * sqrt(T) * rnorm(1,0,1))) - K) * exp(-rf * T)
6  + }
7  > head.tail(opt.val)
8  [1] 0 0 0
9  [1]   21.91808  14.89727 100.00000
```

**Step 3: Calculate Option Value** When running a Monte Carlo simulation, the resulting option value is obtained by calculating the average of the present value of the intrinsic values we have calculated. To do this, we use mean () on the elements of the dataset *opt.val*. Doing so yields a call option value of $12.85.

```
1  > mean(opt.val)
2  [1] 12.85225
3  > sd(opt.val) / sqrt(sim)
4  [1] 0.08665133
```

## 11.6  Monte Carlo Option Pricing Using Antithetic Variables

In Step 3 above, Line 4 of the output shows the standard error of the Monte Carlo option value of $0.09. One way of improving the precision of the Monte Carlo estimate of the option value is to use **antithetic variables**. The idea behind this approach is to use pairs of negatively correlated random numbers to generate pairs of negatively correlated simulation results. We then average each pair of results and take the average of those values to arrive at our option value. Let us go through the following example to see how this approach is implemented and whether it does indeed lower the variability of our option value.

**Step 1: Create a Temporary Matrix** This matrix will later house the 100,000 pairs of simulated option values.

```
1   > opt.val <- c(S, rep(S, sim))
2   > opt.val2 <- cbind(opt.val, opt.val)
3   > head.tail(opt.val2)
4         opt.val opt.val
5   [1,]      100     100
6   [2,]      100     100
7   [3,]      100     100
8             opt.val opt.val
9   [99999,]     100     100
10  [100000,]    100     100
11  [100001,]    100     100
12  > class(opt.val2)
13  [1] "matrix"
```

**Step 2: Run the Simulation** For each row in *opt.val2*, we generate a pair of option values. These option values are generated by running two separate simulations for each run. We then calculate the intrinsic value of the call option at maturity and then discount that value back to the present.

```
1  > set.seed(1)
2  > for (i in 1 : sim) {
3  +     opt.val2[i, 1] <- max(0,
4  +         (S * exp((r - q - 0.5 * v^2) * T +
5  +             v * sqrt(T) * rnorm(1,0,1))) - K) * exp(-r * T)
```

```
 6  +     opt.val2[i, 2] <- max(0,
 7  +         (S * exp((r - q - 0.5 * v^2) * T -
 8  +             v * sqrt(T) * rnorm(1,0,1))) - K) * exp(-r * T)
 9  + }
10  > head.tail(opt.val2)
11        opt.val  opt.val
12  [1,]        0  0.00000
13  [2,]        0  0.00000
14  [3,]        0 20.34815
15              opt.val    opt.val
16  [99999,]   46.85377   45.11314
17  [100000,]   0.00000    0.00000
18  [100001,] 100.00000 100.00000
```

**Step 3: Calculate Option Value** We calculate the option value in two steps. First, we take the average of the pair of option values using `rowMeans()`. We have 100,000 pairs. Second, we take the average of those resulting values. The resulting option value is $12.82, which is close to $12.85 we estimate in the prior section using Monte Carlo simulation. This calculation yields a standard error of $0.06, which is one-third lower than $0.09 generated without using antithetic variables.

```
 1  > avg <- rowMeans(opt.val2[-1, ])
 2  > head.tail(avg)
 3  [1]   0.000000 10.174075  2.181153
 4  [1]   45.98345   0.00000 100.00000
 5  > length(avg)
 6  [1] 100000
 7  >
 8  > mean(avg)
 9  [1] 12.82193
10  > sd(avg) / sqrt(sim)
11  [1] 0.06104794
```

## 11.7  Exotic Option Valuation

In the prior section and earlier chapter on options, we dealt with European and American options. These types of options are known as plain vanilla options. The reason is that they have well-defined characteristics. The prices and implied volatilities of some plain vanilla options can be obtained by looking at market quotes for these instruments, so modeling those options may not be necessary if one wants to get an indication of the value of those options. However, there is an entire class of options that are not standardized and are transacted in over-the-counter (OTC) markets rather than on an exchange. These products are called **exotic options**. In this section, we will look at three examples of exotic options and implement models to value these exotic options.

## 11.7.1  Asian Options

The payoff of an **Asian option** depends on the arithmetic average of the price of the underlying asset during the life of the option. In other words, the intrinsic value of an Asian call option is $\max(S_{avg} - K, 0)$ and an Asian put option is $max(K - S_{avg}, 0)$. We can think of valuing these options by having to simulate the underlying stock price and calculate the average stock price value during the life of the option. We then compare that to the strike price and apply to determine the intrinsic value of the Asian call or put option.

Before delving specifically into the Asian option valuation, we first set up the simulated stock price. We will use this simulated stock price not only for the Asian option valuation but also for the valuation of the lookback and barrier options.

**Step 1: Enter Inputs and Assumptions** We set up the basic characteristics of the underlying stock and the simulation for use in the rest of the simulation and valuation exercise. For this example, we assume that the underlying asset is a stock with a price today of $100, expected return of 10%, and standard deviation of 40%. The option has a strike price $K$ of $110 and has 1-year time to maturity. The risk-free rate is assumed to be 2%. We also assume that there are 252 trading days in a year, so the time increment $dt$ is equal to 1/252. For the simulations, we will use 100,000 runs.

```
1   > S_0 <- 100
2   > sigma <- 0.40
3   > K <- 110
4   > T <- 1
5   > r <- 0.02
6   > (dt <- 1 / 252)
7   [1] 0.003968254
8   > sim <- 100000
```

**Step 2: Create a Matrix of Temporary Stock Prices** We first determine the number of the columns of the matrix of stock prices. This is a function of the time to maturity of the option (1 year) and the time increment (252 trading days). We then add 1 to this number because we need an extra observation for the starting stock price $S\_0$ of $100. Thus, this matrix has 253 columns. This will house the simulated price path of the underlying asset. The number of rows is equal to the number of simulations we will run, which in this illustration is 100,000. Thus, each row will be one price path for the underlying asset.

```
1   > (t <- T / dt + 1)
2   [1] 253
3   > s <- matrix(rep(0, t * sim), nrow = sim)
4   > dim(s)
5   [1] 100000    253
```

**Step 3: Generate Prices** We first change the values in the first column of the stock price matrix to equal the price of the stock today of $100. The first for-loop simulates the stock price for each of the subsequent 252 trading days, and the second for-loop simulates the stock prices for all 100,000 runs. Like in previous runs, we simulate the stock price using a GBM.

```
1   > s[, 1] <- S_0
2   > set.seed(1)
3   > for (i in 2:t){
4   +    for (j in 1:sim){
5   +      s[j, i] <- s[j, i - 1] * exp (((rf - 0.5 * sigma^2) * dt +
6   +         sigma * rnorm(1, 0, 1) * sqrt(dt)))
7   +    }
8   + }
9   > s[1, 1:10]
10   [1] 100.00000  98.40869 100.36520  99.77477 101.25174  99.89696  97.19071
11   [8]  97.42005  96.37471  95.14623
12   > s[1, 243:253]
13   [1] 87.02607 89.41397 87.77481 87.08010 86.27813 88.03872 88.57250 89.54893
14   [9] 88.94516 89.32014 90.38505
```

Now that we have the underlying stock prices for all 100,000 runs, and we are ready to value the Asian option.

**Step 4: Create Temporary Matrices**  We set up three temporary matrices to hold the average price, call option value, and put option value. These are very large matrices that initially would have zeroes for all their elements. The matrix will have $t = 253$ columns and $sim = 100,000$ rows.

```
1   > asian.avg <- matrix(0, ncol = t, nrow = sim)
2   > asian.call <- matrix(0, ncol = t, nrow = sim)
3   > asian.put <- matrix(0, ncol = t, nrow = sim)
```

**Step 5: Calculate Values to Fill In Arrays**  We run a for-loop to fill in each of the three arrays we created above. The *asian.avg* is equal to the mean of each of the 100,000 price paths. In other words, it has 100,000 values with each value denoting the average price for each simulation. The value of the Asian call option is equal to the maximum of the difference between the average stock price and the strike price or zero. The value of the Asian put option is equal to the maximum of the difference between the strike price and the average stock price or zero.

```
1   > for (i in 1:sim) {
2   +    asian.avg[i] <- mean(s[i, ])
3   +    asian.call[i] <- exp(-rf * T) * (max((asian.avg[i] - K), 0))
4   +    asian.put[i] <- exp(-rf * T) * (max((K - asian.avg[i]), 0))
5   + }
```

**Step 6: Calculate the Value of the Asian Option**  To calculate the value of the Asian option, we take the average of the American option values today (i.e., first column) calculated under each of the 100,000 simulations. As the output shows, our example yields an Asian call option value of $5.77 and an Asian put option value of $14.81.

```
1  > (asian.call.value ← mean(asian.call[, 1]))
2  [1] 5.766807
3  > (asian.put.value ← mean(asian.put[, 1]))
4  [1] 14.80564
```

### 11.7.2  Lookback Options

The payoff of a **lookback option** depends on the maximum or minimum price of the underlying asset during the life of the option. Specifically, the payoff of a lookback call is the amount that the final asset price exceeds the minimum asset price, while the payoff of a lookback put is the amount that the maximum asset price during the life of the option exceeds the final asset price. We now continue from our previous example and implement a model to value lookback options.

**Step 1: Create Temporary Matrices**  We set up four temporary matrices to hold the minimum price, maximum price, call option value, and put option value.

```
1  > lookback.min ← matrix(0, ncol = t, nrow = sim)
2  > lookback.max ← matrix(0, ncol = t, nrow = sim)
3  > lookback.call ← matrix(0, ncol = t, nrow = sim)
4  > lookback.put ← matrix(0, ncol = t, nrow = sim)
```

**Step 2: Calculate Values to Fill In Arrays**  We run a for-loop to fill in each of the four arrays we created above. The *lookback.min* equals the minimum price of the stock for each price path. The *lookback.max* equals the maximum price of the stock for each price path. The *lookback.call* holds the value of the call option, and the *lookback.put* holds the value of the put option.

```
1  > for (i in 1:sim) {
2  +     lookback.min[i] ← min(s[i, ])
3  +     lookback.max[i] ← max(s[i, ])
4  +     lookback.call[i] ← exp(−r * T) * (max((s[i, t] − lookback.min[i]), 0))
5  +     lookback.put[i] ← exp(−r * T) * (max(lookback.max[i] − s[i, t], 0))
6  + }
```

**Step 3: Calculate the Value of the Lookback Option**  To calculate the value of the lookback option, we take the average of the lookback option values today (i.e., first column) calculated under each of the 100,000 simulations. As the output shows, our example yields a lookback call option value of $27.67 and a lookback put option value of $33.18.

```
1  > (lookback.call.value ← mean(lookback.call[, 1]))
2  [1] 27.66831
3  > (lookback.put.value ← mean(lookback.put[, 1]))
4  [1] 33.17716
```

### *11.7.3  Barrier Options*

The payoff of a **barrier option** depends on whether the underlying asset's price reaches a certain level during a specific time period. There are different types of barrier options. Below, we will value two types: an up-and-in call and a down-and-in call. These both are considered knock-in options, which come into existence only when the underlying asset price reaches a barrier. An up-and-in call is a regular call option that comes into existence only if the underlying asset price reached a barrier, when the barrier is set above the initial asset price. A down-and-in call is a regular call option that comes into existence only if the underlying asset price reaches a barrier, when the barrier is set at or below the initial asset price. We first start with an up-and-in call.

**Step 1: Create a Matrix to Store Payoffs**  We set up one temporary matrix to hold the payoff values for the up-and-in call. Here, we only need one column for the matrix because in the next step we calculate the intrinsic value for each simulation and store only that intrinsic value in this payoff matrix.

```
1   > payoff1 <- matrix(0, ncol = 1, nrow = sim)
```

**Step 2: Calculate Values to Fill In Array**  The starting stock price is $100. We assume in this example that the barrier before the call comes into existence is $150. We then use some of the inputs from the earlier exotic option calculations. We use a for-loop that runs a simulation for each of the 100,000 runs. For each run, we check if the maximum stock price exceeds $150, the barrier. If it does, then we know the up-and-in call came into existence. Then, we determine the call option intrinsic value at the end of the life of the option.

```
1   > barrier1 <- 150
2   > for (i in 1:sim) {
3   +     if(lookback.max[i] > barrier1){
4   +         payoff1[i] <- max(s[i, t] - K, 0)
5   +     } else {
6   +         payoff1[i] <- 0
7   +     }
8   + }
```

**Step 3: Calculate the Up-and-in Call Value**  To calculate the value of the up-and-in call, we take the average of the 100,000 up-and-in call intrinsic values at the maturity of the option. In our example, that average value is $11.22. We then discount that average value to the present, to give us the value of the up-and-in call of $11.05.

```
1   > (payoff1.avg <- mean(payoff1))
2   [1] 11.21817
3   > (barrier.upin.value <- exp(-rf * T) * payoff1.avg)
4   [1] 11.04563
```

**Step 4: Calculate the Down-and-in Call Value**   The value of a down-and-in call is calculated similarly, so we will not go into the details of the code as much. We focus only on the key difference. In Line 3, as we can see the barrier is set at $70, which is below the initial stock price of $100. Then, in Line 5, we evaluate whether, for each stock price path, the simulated stock price falls below $70. If that is the case, the call option comes into existence. We can then calculate the call option value at maturity for each of those price paths. If the simulated stock price never falls below $70, the value of the down-and-in call for that price path is zero because the option never came into existence. As the output shows, the down-and-in call yields a present value of $0.28.

```
1   > payoff2 <- matrix(0, ncol = 1, nrow = sim)
2   >
3   > barrier2 <- 70
4   > for (i in 1:sim) {
5   +    if(lookback.min[i] <= barrier2){
6   +       payoff2[i] <- max(s[i, t] - K, 0)
7   +    } else {
8   +       payoff2[i] <- 0
9   +    }
10  + }
11  >
12  > (payoff2.avg <- mean(payoff2))
13  [1] 0.2865829
14  > (barrier.downin.value <- exp(-rf * T) * payoff2.avg)
15  [1] 0.2821751
```

## 11.8   Further Reading

The simulations we constructed above should give you a good foundation to learn more advanced simulation methods. There are a couple of interesting books that apply Monte Carlo to financial applications that may be worth considering. These are [1] and [4]. In addition, Christian Robert and George Casella have also written two excellent books on Monte Carlo simulation. The first focuses on Monte Carlo methods [2], while the other focuses on Monte Carlo using R [3]. These books go into the theory, as well as more advanced simulation methods.

## References

1. Glasserman, P. (2003). *Monte Carlo methods in financial engineering*. Berlin: Springer.
2. Robert, C., & Casella, G. (2004). *Monte Carlo statistical methods* (2nd ed.). Berlin: Springer.
3. Robert, C., & Casella, G. (2010). *Introducing Monte Carlo methods with R*. Berlin: Springer.
4. Shonkwiler, R. (2010). *Finance with Monte Carlo*. Berlin: Springer.

# Chapter 12
# Trading Strategies

Our goal in this chapter is to discuss some fundamental concepts related to trading strategies. Trading strategies and building trading systems are both very complex and sophisticated subjects. Entire books and even volumes of books have been devoted to these subjects. Thus, we do not attempt to provide a comprehensive discussion of trading strategies in this single chapter. In addition, the discussion in this chapter is not meant for us to make money off the examples or the implementation of the models. If I knew of such a profitable trading strategy, I would exploit it first until I no longer can profit from the strategy before writing this book. Having set the expectations, we can now proceed with our discussion of trading strategies.

We begin this chapter with a discussion of the efficient markets hypothesis (EMH) in Sect. 12.1. This helps setup the hurdle that many active investors face. The bottom-line implication of the EMH is that it is extremely difficult to develop a trading strategy that consistently generates positive risk-adjusted profits. We end our discussion with three common tests of the EMH: autocorrelation, variance ratio, and runs tests.

In Sect. 12.2, we discuss technical indicators. The profitability of a trading strategy ultimately relies on being able to time buys and sells. Technical indicators are used by some investors to signal when they should buy or sell a security. Next, we show how to build a basic trading strategy using simple moving average signals in Sect. 12.3.

Lastly, we explore some elementary machine learning techniques in Sect. 12.4. We provide a high-level discussion of ML and specifically show the implementation of three techniques: k-nearest neighbors, regressions with cross validation, and artificial neural networks.

C.S. Ang, *Analyzing Financial Data and Implementing Financial Models Using R*,
Springer Texts in Business and Economics,
https://doi.org/10.1007/978-3-030-64155-9_12

## 12.1   Efficient Markets Hypothesis

A relevant implication of the **efficient markets hypothesis** (EMH) is that it is very difficult to develop a consistently profitable trading strategy. The EMH says that market prices fully reflect available information and prices should be unpredictable. To be clear, the EMH does not mean that prices can never be predictable or we cannot develop a profitable trading strategy. What the EMH means is that any predictability will be competed away quickly by traders and any actionable mispricing will disappear. The term actionable indicates that there could still be some mispricing that remains but the costs of trading that mispricing away (e.g., commission and/or bid-ask spread) would outweigh the profits from taking advantage of the mispricing. In practice, virtually all profitable mispricing opportunities will not reach retail (i.e., individual) investors because such mispricing would have been competed away quickly by high frequency traders and large institutional investors.

In the academic literature, the EMH is usually described in three forms with the difference being what is included in the information set that is reflected in prices. First, the **weak form** of the EMH suggests that information from historical prices are fully reflected in the market price. This implies that we should not be able to consistently make **risk-adjusted profits** from trading strategies based on past market prices, such as patterns of those past prices. In Sect. 12.2, we discuss **technical analysis**, which are trading strategies that use patterns in past prices to make a profit. Thus, if the weak form of the EMH holds, then technical analysis should not consistently lead to risk-adjusted profits.

Second, the **semi-strong form** of the EMH implies that public information, such as historical prices, analyst reports, company filings, and news, are fully reflected in the market price. This means that, if the semi-strong form of the EMH holds, trading strategies that look at fundamentals (e.g., low versus high price-to-earnings or price-to-book stocks, income statement metrics, balance sheet metrics, discounted cash flows, etc.) should not lead to consistent risk-adjusted profits.

Finally, the **strong form** of the EMH says that all information is fully reflected in the market price. This implies that we should not be able to consistently make risk-adjusted profits from trading strategies based on private and public information. Thus, if the strong form of the EMH holds, insiders (e.g., company directors and officers) should not be able to make money off their private knowledge about the prospects of the company.

In developed financial markets like the USA, the empirical evidence is generally consistent with the semi-strong form of the EMH. This implies that it is very difficult to consistently make risk-adjusted profits using historical prices or publicly available information.

If we were to believe the empirical evidence that it is very difficult to develop a trading strategy that consistently makes money, we may wonder how certain investors we often hear about in the media beat the market. Such a result would appear to be a violation of the EMH. One way to think about it is as follows. First, we know the identity of these successful investors only *after* they have made their wildly profitable trades or the consecutive years of profitable trading. For us to benefit from the incredible returns such investors have made, we would have to identify these investors *before* they made their money. We also rarely hear about the flip side, which are the investors that have lost huge sums of money.

Second, past success is not an indicator of future success. Thus, if we invested in a fund because it beat the market 10 years in a row, we are not guaranteed that the fund would beat the market in the 11th year.

Third, although it is rare to beat the market, say, 10 years in a row, such a result cannot be distinguished from just being lucky. To see why that is the case, let us walk through a simpler version of the **infinite monkey theorem**. This theorem states that if you give an infinite number of monkeys each a typewriter, at least one of them will be able to come up with the complete works of William Shakespeare. In our simplified version, we assume that there are 10,000 fund managers in the market and we simulate 10 annual returns for each fund manager. We simulate the annual returns by using `runif()` to generate uniform random variables. The default uniform random number generator in R generates numbers between 0 and 1 with a mean of 0.5. We verify this also across all 100,000 random numbers we generated using `mean()`. Each of these randomly generated numbers will then represent a normalized annual return relative to the market return with 0.5 being equal to the market return.

```
1   > set.seed(1)
2   > mgr.ret <- matrix(runif(100000),
3   +     nrow = 10000, ncol = 10, byrow = TRUE)
4   > options(digits = 2)
5   > head.tail(mgr.ret)
6          [,1] [,2] [,3] [,4] [,5] [,6]   [,7] [,8] [,9] [,10]
7   [1,] 0.27 0.37 0.57 0.91 0.20 0.90 0.945 0.66 0.63 0.062
8   [2,] 0.21 0.18 0.69 0.38 0.77 0.50 0.718 0.99 0.38 0.777
9   [3,] 0.93 0.21 0.65 0.13 0.27 0.39 0.013 0.38 0.87 0.340
10             [,1] [,2] [,3] [,4] [,5]   [,6]   [,7] [,8]   [,9] [,10]
11     [9998,] 0.236 0.99 0.33 0.44 0.14 0.035 0.017 0.75 0.333  0.69
12     [9999,] 0.051 0.52 0.90 0.18 0.39 0.158 0.597 0.64 0.302  0.12
13    [10000,] 0.443 0.76 0.14 0.40 0.39 0.031 0.225 0.98 0.092  0.12
14  > options(digits = 7)
15  > mean(mgr.ret)
16  [1] 0.4996247
```

Now, assume that the fund manager is considered to have beat the market if the value generated in each year is greater than 0.5. We can then use `ifelse()` to identify which values are greater than 0.5. We denote those cases with a 1. Otherwise, the value will equal 0.

```
1   > beat <- ifelse(mgr.ret > 0.5, 1, 0)
2   > beat <- data.frame(beat)
3   > head.tail(beat)
4     X1 X2 X3 X4 X5 X6 X7 X8 X9 X10
5   1  0  0  1  1  0  1  1  1  1   0
6   2  0  0  1  0  1  0  1  1  0   1
```

```
7    3  1  0  1  0  0  0  0  0  1   0
8           X1 X2 X3 X4 X5 X6 X7 X8 X9 X10
9    9998   0  1  0  0  0  0  0  1  0   1
10   9999   0  1  1  0  0  0  1  1  0   0
11   10000  0  1  0  0  0  0  0  1  0   0
```

This then allows us to sum the 1s within each row. We do this using `rowSums()`. This lets us to count the number of years when a particular fund manager beat the market.

```
1    > beat$tot <- rowSums(beat)
2    > head.tail(beat)
3        X1 X2 X3 X4 X5 X6 X7 X8 X9 X10 tot
4    1   0  0  1  1  0  1  1  1  1   0   6
5    2   0  0  1  0  1  0  1  1  0   1   5
6    3   1  0  1  0  0  0  0  0  1   0   3
7           X1 X2 X3 X4 X5 X6 X7 X8 X9 X10 tot
8    9998   0  1  0  0  0  0  0  1  0   1   3
9    9999   0  1  1  0  0  0  1  1  0   0   4
10   10000  0  1  0  0  0  0  0  1  0   0   2
```

Next, we then count the number of times 10 appears in the column tot. We use `sum()` to do this.

```
1    > sum(beat$tot == 10)
2    [1] 9
```

As the above shows, just by random chance alone, we have nine fund managers out of 10,000 that beat the market 10 years in a row. That comes out to 0.09%, so the probability of beating the market 10 years in a row is rare enough that when such an accomplishment is achieved the feat is newsworthy. Of course, our investment portfolio does not care whether the fund manager beats the market 10 years in a row due to skill or luck. The increase in our portfolio values is still real and the proceeds from the sale of our investment in the fund will generate real money that we can spend.

Having said that, the market reflects the average performance of all investors. This implies that half the investors beat the market and half the investors do not beat the market. Moreover, since passive investors invest in index funds, which have low fees, they will perform like the market. This implies that, collectively, all active investors will also have to perform like the market. However, because of the higher fee structure of active investing, the net return to active investing would be less than the market return. This does not mean that there should only be passive investors. If every investor believes that the market price is correct, then no trading will occur and those mispricings will persist and make market prices less efficient. Therefore, for markets to be efficient, we need investors that are putting their money on the line making bets with, on average, half of them winning and half of them losing.

There are several tests that indicate or support market efficiency. Below, we perform three tests that help determine whether security prices exhibit return predictability.

### 12.1.1 Autocorrelation Test

If there are patterns in the residuals of factor models, this would imply that we may be able to profit from recognizing such patterns and, thus, the EMH is violated. Formally, a pattern in the error term of the regression is called **autocorrelation**. In the example below, we test the residuals of a market model for AMZN using returns from 2015 to 2019 using the **Durbin–Watson** test.

**Step 1: Import Data** We first import the price data for AMZN and SPY data using `load.data()`. Then, we calculate the log returns based off the adjusted close prices using `ROC()`. `ROC()` yields the same result as `diff(log())`. We then clean up the data by removing the first observation, which is for December 31, 2014. We then create a dataset that combines these two returns using `data.frame()`.

```
 1  > data.amzn <- load.data("AMZN Yahoo.csv", "AMZN")
 2  > r.amzn <- ROC(data.amzn$AMZN.Adjusted)
 3  > r.amzn <- r.amzn[-1, ]
 4  > head.tail(r.amzn)
 5               AMZN.Adjusted
 6  2015-01-02  -0.005914077
 7  2015-01-05  -0.020730670
 8  2015-01-06  -0.023098010
 9               AMZN.Adjusted
10  2019-12-27   0.0005510283
11  2019-12-30  -0.0123283480
12  2019-12-31   0.0005142195
13  >
14  > data.spy <- load.data("SPY Yahoo.csv", "SPY")
15  > r.spy <- ROC(data.spy$SPY.Adjusted)
16  > r.spy <- r.spy[-1, ]
17  > head.tail(r.spy)
18                SPY.Adjusted
19  2015-01-02  -0.0005354247
20  2015-01-05  -0.0182246098
21  2015-01-06  -0.0094636586
22                SPY.Adjusted
23  2019-12-27  -0.0002478074
24  2019-12-30  -0.0055284735
25  2019-12-31   0.0024263490
26  >
27  > returns <- data.frame(r.amzn, r.spy)
28  > names(returns) <- c("AMZN", "MKT")
29  > head.tail(returns)
30                AMZN            MKT
31  2015-01-02  -0.005914077  -0.0005354247
32  2015-01-05  -0.020730670  -0.0182246098
33  2015-01-06  -0.023098010  -0.0094636586
34                AMZN            MKT
35  2019-12-27   0.0005510283  -0.0002478074
36  2019-12-30  -0.0123283480  -0.0055284735
37  2019-12-31   0.0005142195   0.0024263490
```

**Step 2: Run Durbin–Watson Test** A common test for autocorrelation is the **Durbin–Watson** (DW) test. The DW test analyzes the residual of the market model regression to see whether the residuals are serially correlated. If the residuals are serially correlated, then this suggests that there is some element of predictability from residuals in time $t$ given residuals in time $t - 1$. To perform this test, we use dwtest() in the lmtest package. The two-sided for the alternative argument tells R that the null hypothesis is that autocorrelation is zero and that the alternative hypothesis is that the autocorrelation is positive or negative. As the output of our example shows, the DW statistic is 1.91 with a p-value of 9.2% (Line 8). If we use a 5% significance level, the above result suggests that we cannot reject the null hypothesis that there is no autocorrelation.

```
1   > library(lmtest)
2   > dwtest(AMZN ~ MKT, data = returns,
3   +          alternative = "two.sided")
4
5     Durbin-Watson test
6
7   data:  AMZN ~ MKT
8   DW = 1.9051, p-value = 0.0921
9   alternative hypothesis: true autocorrelation is not 0
```

### 12.1.2   Variance Ratio Test

Another test of the EMH is whether the ratio of variances of returns is the same across different samples. If the EMH holds, then the ratio of the variance between two periods should be not significantly different from each other (i.e., not significantly different from 1).

The argument for this test is as follows. Assume the stock price $P$ follows the following random walk process:

$$\ln P_{t+1} = \mu + \ln P_t + \epsilon_{t+1}, \tag{12.1}$$

where $\epsilon$ is assumed to be independent and identically distributed $\mathbb{N}(0, \sigma^2)$ over the interval $t$ to $t + 1$. This random walk model implies

$$var(r(1)) = \sigma^2, var(r(2)) = 2\sigma^2, \ldots, var(r(n)) = n\sigma^2. \tag{12.2}$$

Thus, we can generalize the above to $var(r(n)) = n * var(r(1))$. With a little algebra, we get

$$\frac{var(r(n))/n}{var(r(1))} = 1. \tag{12.3}$$

The above equation says that the average daily variance for $n$ days should be equal to the 1 day variance. The above formulation allows us to empirically test this relationship.

**Step 1: Split the Data into Two Periods** Continuing with our AMZN example, we first separate the AMZN returns data into 2015 to 2017 (period A) and 2018 to 2019 (period B).

```
1   > names(r.amzn) <- "AMZN"
2   >
3   > rets.A <- r.amzn["/2017-12-31"]
4   > head.tail(rets.A)
5                     AMZN
6   2015-01-02 -0.005914077
7   2015-01-05 -0.020730670
8   2015-01-06 -0.023098010
9                     AMZN
10  2017-12-27  0.004662962
11  2017-12-28  0.003242724
12  2017-12-29 -0.014119964
13  >
14  > rets.B <- r.amzn["2018-01-01/"]
15  > head.tail(rets.B)
16                    AMZN
17  2018-01-02 0.016570407
18  2018-01-03 0.012694369
19  2018-01-04 0.004466026
20                    AMZN
21  2019-12-27  0.0005510283
22  2019-12-30 -0.0123283480
23  2019-12-31  0.0005142195
```

**Step 2: Calculate the Variance of Period A and Period B** We use `var()` to calculate the variance of the returns of AMZN.

```
1   > (var.A <- var(rets.A))
2               AMZN
3   AMZN 0.0003136365
4   > (var.B <- var(rets.B))
5               AMZN
6   AMZN 0.0003624759
```

**Step 3: Calculate the Degrees of Freedom of Period A Data and Period B Data** Since we are estimating one parameter, we lose one degree of freedom. Thus, the degrees of freedom in this case is equal to the number of observations less one, i.e., $n - 1$.

```
1   > (df.A <- length(rets.A) - 1)
2   [1] 754
3   > (df.B <- length(rets.B) - 1)
4   [1] 502
```

**Step 4: Calculate Variance Ratio**  The variance ratio is calculated by dividing the larger variance by the smaller variance. In this case we divide the variance of period B by the variance of period A.

```
1  > (var.ratio <- var.B / var.A)
2            AMZN
3  AMZN 1.15572
```

**Step 4: Calculate the p-Value of the Variance Ratio**  We calculate the p-value using the code below. Note that the degrees of freedom for the period with the higher variance goes first. This shows a p-value of 7.4%. Thus, if our significance level is 5%, we cannot reject the null hypothesis that the ratio of the variances in the two periods is not equal to one.

```
1  > (vr.pval <- 2 * (1 - pf(var.ratio, df.B, df.A)))
2                AMZN
3  AMZN 0.07359709
```

We can compare the result of the above to that using `var.test()`. As we can see, we get the same result using `var.test()` as what we calculated above.

```
1  > var.test(rets.B, rets.A)
2
3    F test to compare two variances
4
5  data:  rets.B and rets.A
6  F = 1.1557, num df = 502, denom df = 754, p-value = 0.0736
7  alternative hypothesis: true ratio of variances is not equal to 1
8  95 percent confidence interval:
9   0.986292 1.357773
10 sample estimates:
11         AMZN
12 AMZN 1.15572
13 attr(,"names")
14 [1] "ratio of variances"
```

### 12.1.3  Runs Test

Another test of predictability of stock returns is there are runs in the returns series. That is, is there a pattern of successive positive returns or successive negative returns? Having runs violate the assumption that returns are independent and identically distributed or i.i.d. and implies some level of return predictability.

**Step 1: Convert Returns Data into a Factor**  Factors are categorical variables, like TRUE or FALSE. Thus, we convert returns into either positive returns (TRUE) or negative return (FALSE) (Line 1). We can then convert these into Factors using `factor()` in Line 8.

```
1   > rets.factor <- returns[, 1] > 0
2   > head.tail(rets.factor)
3   [1] FALSE FALSE FALSE
4   [1]  TRUE FALSE  TRUE
5   > class(rets.factor)
6   [1] "logical"
7   >
8   > rets.factor <- factor(rets.factor)
9   > class(rets.factor)
10  [1] "factor"
```

**Step 2: Perform Runs Test**  Using the `runs.test()` in the `tseries` package, we can perform the runs test on *rets.factor*. The null hypothesis of the runs test is that returns are independent and identically distributed or i.i.d. As the output below shows, the p-value of our runs test is 36%, which means that we cannot reject the null hypothesis that the returns are i.i.d.

```
1   > library(tseries)
2   > runs.test(rets.factor)
3
4     Runs Test
5
6   data:  rets.factor
7   Standard Normal = -0.91595, p-value = 0.3597
8   alternative hypothesis: two.sided
```

## 12.2  Technical Analysis

Technical analysis is the use of charts to study stock price and volume data for the purpose of forecasting future trends. Those who follow technical analysis are using stock price and volume data as an indication of the supply and demand for the stock. For example, a rising stock price may indicate that demand exceeds supply while a falling stock price may indicate that supply exceeds demand. As a trading strategy, profiting from technical analysis relies on being able to identify trends and the ability to catch on during the early stages of the trend. Moreover, the same technical indicators may be interpreted differently by different chartists depending on their investment philosophies (e.g., trend follower or contrarian).

There are three broad groups of technical indicators: (1) trend indicators, (2) volatility indicators, and (3) momentum indicators. Within each group, there are many possible technical indicators to choose from. In this chapter, we go through one example for each type of indicator starting with simple moving average cross over (trend), Bollinger Bands (volatility), and relative strength index (momentum). Note that the above examples, as well as many more technical indicators, can be implemented using `chartSeries()`, which we used earlier to create the Open-High-Low-Close (OHLC) chart. In constructing trading strategies, combining indicators is often used. For example, we can use the relative strength index to confirm signals identified by the simple moving average. Therefore, we can view these examples as building blocks to more complex strategies.

## 12.2.1  Trend: Simple Moving Average Crossover

A common technical analysis trend indicator is the **Simple Moving Average** (SMA) crossover. In an SMA, the moving average is calculated by taking the average of a firm's stock price over a certain number of days. The term "simple" comes from the fact that this type of average treats all days equally, regardless of how near or far those days are from the present. This is called an SMA "crossover" because we will use two SMA lines, a shorter-term average and a longer-term average, and make trading decisions when the lines cross.

In our example, we will use an average over 50 (200) days for the shorter (longer) term. Since we are interested in making decisions during the present, an extremely long time series of data may not be necessary. As such, we will demonstrate how to implement an SMA crossover for AMZN from 2018 to 2019.

**Step 1: Obtain Close Prices for AMZN**  We import the AMZN data from Yahoo Finance using the `load.data()` function. We then

```
 1  > data.amzn <- load.data("AMZN Yahoo.csv", "AMZN")
 2  > sma <- data.amzn$AMZN.Close
 3  > head.tail(sma)
 4             AMZN.Close
 5  2014-12-31    310.35
 6  2015-01-02    308.52
 7  2015-01-05    302.19
 8             AMZN.Close
 9  2019-12-27   1869.80
10  2019-12-30   1846.89
11  2019-12-31   1847.84
```

**Step 2: Calculate Rolling 50-Day and 200-Day Average Price**  To calculate the rolling or moving average, we use `rollmeanr()` and choose the window length as $k = 50$ for the 50-day moving average and $k = 200$ for the 200-day moving average. Note that the first three observations below under *sma50* and *sma200* are NA as R only reports data beginning the 50th and 200th observation, respectively.

```
 1  > sma$sma50 <- rollmeanr(sma$AMZN.Close, k = 50)
 2  > sma$sma200 <- rollmeanr(sma$AMZN.Close, k = 200)
 3  > head.tail(sma)
 4             AMZN.Close sma50 sma200
 5  2014-12-31    310.35    NA     NA
 6  2015-01-02    308.52    NA     NA
 7  2015-01-05    302.19    NA     NA
 8             AMZN.Close    sma50    sma200
 9  2019-12-27   1869.80 1775.790 1825.658
10  2019-12-30   1846.89 1776.978 1826.331
```

```
11   2019–12–31      1847.84 1778.784 1826.859
12   >
13   > sma[48:52, ]
14             AMZN.Close     sma50 sma200
15   2015–03–11      366.37        NA     NA
16   2015–03–12      374.24        NA     NA
17   2015–03–13      370.58 345.6926      NA
18   2015–03–16      373.35 346.9526      NA
19   2015–03–17      371.92 348.2206      NA
20   >
21   > sma[198:202, ]
22             AMZN.Close     sma50    sma200
23   2015–10–13      548.90 522.8242       NA
24   2015–10–14      544.83 523.0828       NA
25   2015–10–15      562.44 523.5914 432.0543
26   2015–10–16      570.76 524.4174 433.3563
27   2015–10–19      573.15 525.4280 434.6795
```

**Step 3: Subset to 2018 and 2019 Data**  We subset the data from January 1, 2018 through December 31, 2019 using xts-style date subsetting.

```
1   > sma <- sma["2018–01–01/2019–12–31"]
2   > head.tail(sma)
3             AMZN.Close     sma50    sma200
4   2018–01–02    1189.01 1133.787 1005.517
5   2018–01–03    1204.20 1138.213 1007.253
6   2018–01–04    1209.59 1143.078 1009.085
7             AMZN.Close     sma50    sma200
8   2019–12–27    1869.80 1775.790 1825.658
9   2019–12–30    1846.89 1776.978 1826.331
10  2019–12–31    1847.84 1778.784 1826.859
```

**Step 4: Plot the SMA**  To make sure we have the full range in the y-axis, we use range to find the minimum and maximum values in  *sma*. The output of the chart is shown as Fig. 12.1. A bullish crossover happens when the 50-day moving average crosses above the 200-day moving average. When this occurs, some traders may consider this as an indicator to buy the stock. Conversely, a bearish crossover happens when the 50-day moving average crosses below the 200-day moving average. When this occurs, some traders may consider this as an indication to sell the stock.

```
1   > (y.range <- range(sma))
2   [1] 1005.517 2039.510
3   > plot(y = sma$AMZN.Close,
4   +      x = index(sma),
5   +      xlab = "",
6   +      ylab = "Price / Moving Average",
7   +      type = "l",
8   +      lwd = 2,
9   +      col = "blue",
10  +      ylim = y.range,
11  +      main = "AMZN Simple Moving Average")
```

## AMZN Simple Moving Average

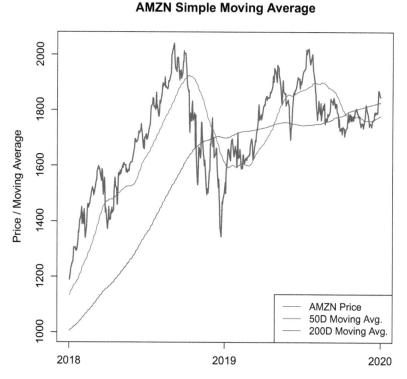

**Fig. 12.1**  50-day and 200-day simple moving average of AMZN, 2018–2019. Data source: Price data reproduced with permission of CSI ©2020. www.csidata.com

```
12  > lines(y = sma$sma50, x = index(sma), col = "red")
13  > lines(y = sma$sma200, x = index(sma), col = "darkgreen")
14  > legend("bottomright",
15  +     c("AMZN Price", "50D Moving Avg.", "200D Moving Avg."),
16  +     lty = 1,
17  +     col = c("blue", "red", "darkgreen"))
```

### 12.2.2  Volatility: Bollinger Bands

**Bollinger Bands** are frequently used volatility indicators .The Bollinger Bands have three components. The first component is a 20-day simple moving average (SMA). The second component is an upper band, which is two standard deviations above the 20-day SMA. The third component is a lower band, which is two standard deviations below the 20-day SMA. Bollinger Bands are considered volatility indicators because the Bollinger Bands widen (narrow) with more (less) volatility in the stock. When the bands narrow, it may be used as an indication that volatility is about to rise. Below we demonstrate how to apply Bollinger Bands to AMZN stock in 2019.

**Step 1: Obtain Closing Prices for AMZN Stock**  We create a new dataset *bb* that extracts the AMZN close price from the *data.amzn* dataset.

```
1   > bb <- data.amzn$AMZN.Close
2   > head.tail(bb)
3                 AMZN.Close
4    2014-12-31     310.35
5    2015-01-02     308.52
6    2015-01-05     302.19
7                 AMZN.Close
8    2019-12-27    1869.80
9    2019-12-30    1846.89
10   2019-12-31    1847.84
```

**Step 2: Calculate Rolling 20-Day Mean and Standard Deviation**  Similar to our calculation in the SMA, we use `rollmeanr()` with `k = 20` to calculate the 20-day moving average. For the standard deviation, we use `rollapply()`, which allows us to apply a function on a rolling basis. Here, the `FUN = sd` tells R that the function we want to apply is the standard deviation function and the `width = 20` tells R that it is a 20-day standard deviation that we want to calculate.

```
1   > bb$avg <- rollmeanr(bb$AMZN.Close, k = 20)
2   > bb$sd <- rollapply(bb$AMZN.Close, width = 20, FUN = sd, fill = NA)
3   > head.tail(bb)
4                 AMZN.Close avg sd
5    2014-12-31     310.35   NA NA
6    2015-01-02     308.52   NA NA
7    2015-01-05     302.19   NA NA
8                 AMZN.Close      avg        sd
9    2019-12-27    1869.80 1780.365 35.71382
10   2019-12-30    1846.89 1782.669 38.48159
11   2019-12-31    1847.84 1785.981 41.14320
```

Since we are using a 20-day moving average and rolling standard deviation, we will have NAs under the average and standard deviation columns for the first 19 observations.

```
1   > bb[18:22, ]
2                 AMZN.Close      avg        sd
3    2015-01-27     306.75        NA        NA
4    2015-01-28     303.91        NA        NA
5    2015-01-29     311.78 300.5385  8.232959
6    2015-01-30     354.53 302.7475 14.525972
7    2015-02-02     364.47 305.5450 20.037988
```

**Step 3: Subset to 2019 Data**  We now subset the data to only data in 2019 using xts-style date subsetting.

```
1   > bb <- bb["2019-01-01/2019-12-31"]
2   > head.tail(bb)
```

```
3                       AMZN.Close      avg         sd
4      2019-01-02       1539.13 1558.427 112.30781
5      2019-01-03       1500.28 1544.823 100.93262
6      2019-01-04       1575.39 1540.173  97.00545
7                       AMZN.Close      avg         sd
8      2019-12-27       1869.80 1780.365 35.71382
9      2019-12-30       1846.89 1782.669 38.48159
10     2019-12-31       1847.84 1785.981 41.14320
```

**Step 4: Calculate the Bollinger Bands** We calculate the bands that are two standard deviations around the average. Using January 2, 2019 as an example, this means that the upper Bollinger Band is equal to $1783.04 (= 1558.43 + (2 \times 112.31))$ and the lower Bollinger Band is equal to $1333.81 (= 1558.43 - (2 \times 112.31))$. Note that the difference between the upper and lower bands is whether we add or subtract from the mean term *2 * bb$sd*.

```
1  > bb$sd2up <- bb$avg + 2 * bb$sd
2  > bb$sd2down <- bb$avg - 2 * bb$sd
3  > head.tail(bb)
4                     AMZN.Close      avg         sd    sd2up  sd2down
5      2019-01-02     1539.13 1558.427 112.30781 1783.043 1333.812
6      2019-01-03     1500.28 1544.823 100.93262 1746.689 1342.958
7      2019-01-04     1575.39 1540.173  97.00545 1734.184 1346.162
8                     AMZN.Close      avg         sd    sd2up  sd2down
9      2019-12-27     1869.80 1780.365 35.71382 1851.793 1708.937
10     2019-12-30     1846.89 1782.669 38.48159 1859.633 1705.706
11     2019-12-31     1847.84 1785.981 41.14320 1868.268 1703.695
```

**Step 5: Plot the Bollinger Bands** As typical when we have multiple lines to plot, we would need to determine the range of possible values for the $y$-axis. We create a chart with four lines. We start with the AMZN close price. We then draw a dashed line for the 20-day moving average. Then, we draw two solid lines for the upper and lower bands.

```
1   > (y.range <- range(bb[, c(2, 4, 5)]))
2   [1] 1333.812 2101.166
3   > plot(y = bb$AMZN.Close,
4   +     x = index(bb),
5   +     xlab = "",
6   +     ylab = "",
7   +     type = "l",
8   +     lwd = 2,
9   +     col = "blue",
10  +     ylim = y.range,
11  +     main = "AMZN Bollinger Bands, 2019")
12  > lines(y = bb$avg, x = index(bb))
13  > lines(y = bb$sd2up, x = index(bb), col = "red")
14  > lines(y = bb$sd2down, x = index(bb), col = "darkgreen")
15  > legend("bottomright",
16  +     c("AMZN Price", "20D Moving Avg.",
17  +        "Upper Band", "Lower Band"),
18  +     lty = 1,
```

```
19  +      cex = 0.75,
20  +      col = c("blue", "black", "red", "darkgreen"))
```

The resulting plot is shown in Fig. 12.2. Assuming a normal distribution, two standard deviations in either direction from the mean should cover the vast majority of the data. As the chart shows, for 2019, most of the AMZN close prices fell within the Bollinger Bands. For a trend follower, when AMZN's stock price was right around the upper band as in May, July and October 2019, this may be taken as an indication that the stock is "overbought." Conversely, when AMZN's stock price moved right around the lower band, as in early June and August 2019, this may be taken as an indication that the stock is "oversold."

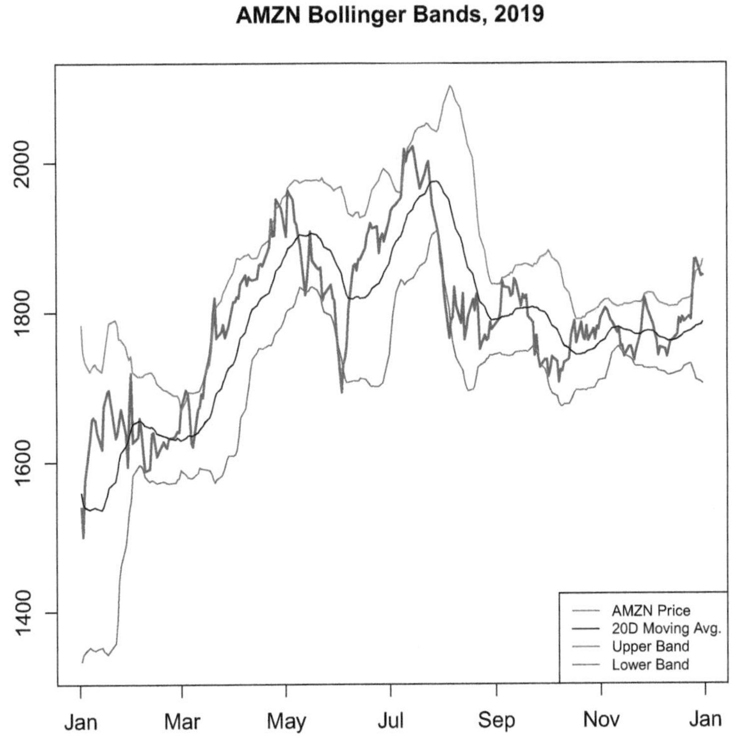

**AMZN Bollinger Bands, 2019**

**Fig. 12.2**  AMZN stock price and Bollinger Bands, 2019. Data source: Price data reproduced with permission of CSI ©2020. www.csidata.com

### 12.2.3   Momentum: Relative Strength Index

A common technical analysis momentum indicator is the Relative Strength Index (RSI). The typical calculation uses a 14-day period. The RSI is calculated as

$$RSI = 100 - \frac{100}{1 + RS},$$ (12.4)

where $RS$ is equal to the up average divided by the down average with the averages calculated using the Wilder Exponential Moving Average described below.

The RSI is used in conjunction with an overbought line and an oversold line. The overbought line is typically set at a level of 70 and the oversold line is typically set at a level of 30. A buy signal is created when the RSI rises from below the oversold line and crosses the oversold line. Conversely, a sell signal is created when the RSI falls from above the overbought line and crosses the overbought line. Below we demonstrate how to calculate the RSI for Amazon.com for the 2018–2019 period.

**Step 1: Obtain Difference in AMZN Close Price**  We save AMZN close price into a new dataset labeled *rsi*. We then use diff() to calculate the difference in the close price today from the close price the prior trading day. We call this variable delta.

```
1    > rsi <- data.amzn$AMZN.Close
2    > head.tail(rsi)
3                     AMZN.Close
4    2014-12-31          310.35
5    2015-01-02          308.52
6    2015-01-05          302.19
7                     AMZN.Close
8    2019-12-27         1869.80
9    2019-12-30         1846.89
10   2019-12-31         1847.84
11   >
12   > rsi$delta <- diff(rsi$AMZN.Close)
13   > head.tail(rsi)
14                     AMZN.Close       delta
15   2014-12-31          310.35          NA
16   2015-01-02          308.52   -1.830017
17   2015-01-05          302.19   -6.329987
18                     AMZN.Close       delta
19   2019-12-27         1869.80    1.030029
20   2019-12-30         1846.89  -22.910034
21   2019-12-31         1847.84    0.949951
```

**Step 2: Create Dummy Variables to Indicate Whether Price Went Up or Price Went Down**  Using ifelse(), we create two dummy variables: one to identify when AMZN price went up from the prior day's price and another to identify when AMZN price went down from the prior day's price.

```
 1  > rsi$up <- ifelse(rsi$delta > 0, 1, 0)
 2  > rsi$down <- ifelse(rsi$delta < 0, 1, 0)
 3  > head.tail(rsi)
 4               AMZN.Close     delta up down
 5  2014-12-31      310.35        NA NA   NA
 6  2015-01-02      308.52 -1.830017  0    1
 7  2015-01-05      302.19 -6.329987  0    1
 8               AMZN.Close     delta up down
 9  2019-12-27     1869.80  1.030029  1    0
10  2019-12-30     1846.89 -22.910034  0    1
11  2019-12-31     1847.84  0.949951  1    0
```

**Step 3: Calculate Prices for Up Days and Prices for Down Days** To construct a series of prices on up days, we multiply the AMZN close price with the up dummy variable. If it is an up day, up will equal one, so up.val will equal the AMZN close price on that day. Otherwise, up.val will equal zero, because we are multiplying a zero for the up dummy variable to the AMZN close price. down.val is calculated in a similar way. In Line 3, we clean up the *rsi* dataset. We remove the first observation (December 31, 2014), which we only used to calculate the difference in price for January 2, 2015. We also remove the up and down columns.

```
 1  > rsi$up.val <- rsi$delta * rsi$up
 2  > rsi$down.val <- -rsi$delta * rsi$down
 3  > rsi <- rsi[-1, -3:-4]
 4  > head.tail(rsi)
 5               AMZN.Close     delta up.val down.val
 6  2015-01-02      308.52 -1.830017      0 1.830017
 7  2015-01-05      302.19 -6.329987      0 6.329987
 8  2015-01-06      295.29 -6.899993      0 6.899993
 9               AMZN.Close     delta  up.val down.val
10  2019-12-27     1869.80  1.030029 1.030029  0.00000
11  2019-12-30     1846.89 -22.910034 0.000000 22.91003
12  2019-12-31     1847.84  0.949951 0.949951  0.00000
```

**Step 4: Calculate Initial Up and Down 14-Day Averages** In this step, we use rollapply() on the average function FUN = mean. We use this on both up.val and down.val.

```
 1  > rsi$up.avg1 <- rollapply(rsi$up.val,
 2  +     width = 14, FUN = mean, fill = NA, na.rm = TRUE)
 3  > rsi$down.avg1 <- rollapply(rsi$down.val,
 4  +     width = 14, FUN = mean, fill = NA, na.rm = TRUE)
 5  > head.tail(rsi)
 6               AMZN.Close     delta up.val down.val up.avg1 down.avg1
 7  2015-01-02      308.52 -1.830017      0 1.830017      NA       NA
 8  2015-01-05      302.19 -6.329987      0 6.329987      NA       NA
 9  2015-01-06      295.29 -6.899993      0 6.899993      NA       NA
10               AMZN.Close     delta  up.val down.val up.avg1 down.avg1
11  2019-12-27     1869.80  1.030029 1.030029  0.00000 10.48501 2.042149
12  2019-12-30     1846.89 -22.910034 0.000000 22.91003 10.48501 3.529297
13  2019-12-31     1847.84  0.949951 0.949951  0.00000 10.55287 2.793579
```

**Step 5: Calculate the Wilder Exponential Moving Average to Calculate Final Up and Down 14-Day Averages** To make the calculation easier, we extract the up.val and down.val columns in the *rsi* dataset into separate independent vectors. Then, we calculate the Wilder Exponential Moving Average for the up and down values. This average calculation assumes that the initial average the day before would have a weight of 13/14 days and the current average will have a weight of 1/14 days. We apply this same logic for the up average and down average. The calculated values under the Wilder Exponential Moving Average are reported under the *up.avg* and *down.avg* vectors, respectively.

```
 1  > up.val <- as.numeric(rsi$up.val)
 2  > down.val <- as.numeric(rsi$down.val)
 3  >
 4  > up.avg <- rsi$up.avg1
 5  > for (i in 15:nrow(up.avg)){
 6  +    up.avg[i] <- ((up.avg[i - 1] * 13 + up.val[i]) / 14)
 7  + }
 8  > up.avg[12:18, ]
 9               up.avg1
10  2015-01-20      NA
11  2015-01-21      NA
12  2015-01-22 2.369282
13  2015-01-23 2.347905
14  2015-01-26 2.180198
15  2015-01-27 2.024470
16  2015-01-28 1.879865
17  > tail(up.avg, 3)
18                up.avg1
19  2019-12-27 10.364727
20  2019-12-30  9.624389
21  2019-12-31  9.004786
22  >
23  > down.avg <- rsi$down.avg1
24  > for (i in 15:nrow(rsi)){
25  +    down.avg[i] <- ((down.avg[i-1] * 13 + down.val[i]) / 14)
26  + }
27  > down.avg[12:18, ]
28               down.avg1
29  2015-01-20      NA
30  2015-01-21      NA
31  2015-01-22  2.371425
32  2015-01-23  2.202037
33  2015-01-26  2.239750
34  2015-01-27  2.287625
35  2015-01-28  2.327080
36  > tail(up.avg, 3)
37                up.avg1
38  2019-12-27 10.364727
39  2019-12-30  9.624389
40  2019-12-31  9.004786
```

**Step 6: Calculate the RSI**   Before calculating the RSI, we first create a new dataset labeled *final*. We remove all the NAs when we create the *final* dataset. We do this by using `na.omit()` when we combine the *up.avg* and *down.avg* vectors. We then calculate the values in the rs column as up.avg divided by the down.avg. Then, we use the formula above to calculate rsi.

```
1  > final <- na.omit(cbind(up.avg, down.avg))
2  > names(final) <- c("up.avg", "down.avg")
3  > head.tail(final)
4               up.avg down.avg
5  2015-01-22 2.369282 2.371425
6  2015-01-23 2.347905 2.202037
7  2015-01-26 2.180198 2.239750
8               up.avg down.avg
9  2019-12-27 10.364727 3.937459
10 2019-12-30  9.624389 5.292643
11 2019-12-31  9.004786 4.914597
12 >
13 > final$rs <- final$up.avg / final$down.avg
14 > final$rsi <- 100 - (100 / (1 + final$rs))
15 > head.tail(final)
16              up.avg down.avg        rs      rsi
17 2015-01-22 2.369282 2.371425 0.9990964 49.97740
18 2015-01-23 2.347905 2.202037 1.0662423 51.60296
19 2015-01-26 2.180198 2.239750 0.9734113 49.32633
20              up.avg down.avg        rs      rsi
21 2019-12-27 10.364727 3.937459 2.632339 72.46953
22 2019-12-30  9.624389 5.292643 1.818447 64.51946
23 2019-12-31  9.004786 4.914597 1.832253 64.69242
```

**Step 7: Subset to 2018 and 2019 Data**   We now subset the data to only show the data for 2018 and 2019. We also created the data so we can chart two horizontal lines when use the xts plot. These horizontal lines are for an RSI of 30 and 70.

```
1  > final <- final["2018-01-01/2019-12-31"]
2  > final$hl1 <- rep(30, nrow(final))
3  > final$hl2 <- rep(70, nrow(final))
4  > head.tail(final)
5               up.avg down.avg        rs      rsi hl1 hl2
6  2018-01-02 5.853617 3.833006 1.527161 60.42991  30  70
7  2018-01-03 6.520497 3.559220 1.832002 64.68929  30  70
8  2018-01-04 6.439749 3.304990 1.948493 66.08437  30  70
9               up.avg down.avg        rs      rsi hl1 hl2
10 2019-12-27 10.364727 3.937459 2.632339 72.46953  30  70
11 2019-12-30  9.624389 5.292643 1.818447 64.51946  30  70
12 2019-12-31  9.004786 4.914597 1.832253 64.69242  30  70
```

**Step 8: Plot the RSI**   We create a chart of *rsi* using `plot()`. We then add horizontal lines to the chart for the RSI of 30 and 70 in Lines 8 and 9, respectively.

```
1  > plot(y = final$rsi,
2  +      x = index(final),
3  +      xlab = "",
```

```
4  +     ylab = "Relative Strength Index",
5  +     type = "l",
6  +     col = "blue",
7  +     main = "AMZN Relative Strength Index, 2018 - 2019")
8  > lines(y = final$hl1, x = index(final), col = "red")
9  > lines(y = final$hl2, x = index(final), col = "darkgreen")
```

Figure 12.3 shows the output of the above code. If the RSI was above 70 and crosses to below 70, some traders take that as an indication to purchase the stock. This appears to have occurred several times in 2018 and mid-2019. On the other hand, if the RSI was below 30 and crosses to above 30, some trades take that as an indication to sell the stock. This appears to have occurred several times during the October to December 2018 and June to August 2019 periods.

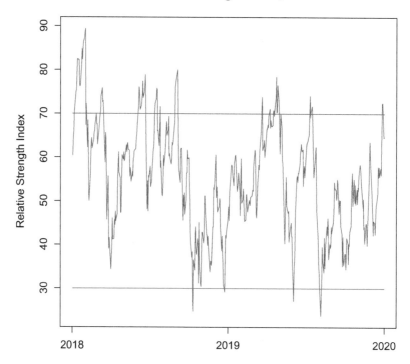

**Fig. 12.3** AMZN stock price and Relative Strength Index, 2018–2019. Data source: Price data reproduced with permission of CSI ©2020. www.csidata.com

## 12.3 Building a Simple Trading Strategy

Technical indicators like the ones we described earlier can be used to build trading strategies. In this section, we will show how to build a simple moving average trading strategy. This strategy assumes that (1) we buy the stock the next trading day if the price closes above the $n$-day high price moving average and (2) we sell the next trading day if the price closes below the $n$-day low price moving average. Note that, to be realistic, our action (i.e., buy or sell) happens *after* we observe the signal. To implement this trading strategy, we need two moving average calculations: a moving average of the high price and a moving average of the low price. For our purposes, we are calculating a 200-day moving average ($n = 200$). We first split the data to a 2-year period (2016–2017) during which we see if the strategy works and then a non-overlapping 2-year period (2018–2019) during which we implement our strategy.

**Step 1: Obtain Returns Data** We implement the above SMA strategy on AMZN data. Note that when we import the data below, instead of using the AMZN ticker as the prefix, we instead use Stock as the prefix. This is to make the rest of the code more generic and this would allow us to change only the filename of the data for the security we are importing if we wanted to see how this strategy would work for other securities. For this strategy, we need high price (Column 2), low price (Column 3), close price (Column 4), and adjusted close price (Column 6).

```
 1  > stock <- load.data("AMZN Yahoo.csv", "Stock")[, c(2:4, 6)]
 2  > rets <- Delt(stock$Stock.Adjusted)
 3  > names(rets) <- "rets"
 4  > head.tail(rets)
 5                        rets
 6  2014-12-31             NA
 7  2015-01-02 -0.005896623
 8  2015-01-05 -0.020517267
 9                        rets
10  2019-12-27  0.0005511802
11  2019-12-30 -0.0122526652
12  2019-12-31  0.0005143517
```

**Step 2: Calculate the 200-Day Moving Averages of the High and Low Prices** We create a variable $X$ to denote the number of days we would like the moving average. In our example, we set $X$ to 200. We then apply rollmean() on Stock.High and Stock.Low with k = X.

```
 1  > X <- 200
 2  > buy <- rollmeanr(stock$Stock.High, k = X)
 3  > head.tail(buy)
 4                Stock.High
 5  2015-10-15    436.7552
 6  2015-10-16    438.0450
 7  2015-10-19    439.3663
 8                Stock.High
 9  2019-12-27    1838.550
```

```
10    2019–12–30    1839.376
11    2019–12–31    1839.892
12    > sell <- rollmeanr(stock$Stock.Low, k = X)
13    > head.tail(sell)
14                  Stock.Low
15    2015–10–15    427.3774
16    2015–10–16    428.6289
17    2015–10–19    429.9310
18                  Stock.Low
19    2019–12–27    1810.569
20    2019–12–30    1811.307
21    2019–12–31    1811.905
```

**Step 3: Calculate Buy and Sell Signals** We calculate buy and sell signals using
`ifelse()`. We tell R to look at the value in the  Stock.Close of the trading day
before. If that close price is higher than the SMA high price the day before, then
a **buy signal** is generated for the day. If that close price is lower than the SMA
low price the day before, then a **sell signal** is generated for the day. The buy signal
will be a 1 and a sell signal will be a −1 for reasons that will become clear in the
subsequent steps.

```
1     > buy$buy.sig <- ifelse(Lag(stock$Stock.Close, k = 1) >
2     +     Lag(buy, k = 1), 1, 0)
3     > buy <- buy[-1, ]
4     > head(subset(buy, buy.sig == 1), 3)
5                   Stock.High buy.sig
6     2015–10–16    438.0450       1
7     2015–10–19    439.3663       1
8     2015–10–20    440.6894       1
9     >
10    > sell$sell.sig <- ifelse(Lag(stock$Stock.Close, k = 1) <
11    +     Lag(sell, k = 1), -1, 0)
12    > sell <- sell[-1, ]
13    > head(subset(sell, sell.sig == -1), 3)
14                  Stock.Low sell.sig
15    2016–02–08    531.2316      -1
16    2016–02–09    531.4066      -1
17    2016–02–10    531.6495      -1
```

**Step 4: Calculate Indicator to Determine Portfolio Returns** Based on when we
buy and sell the stock using the signals created above, we would want to create
the associated portfolio return based on that strategy. To do so, we create a column
called sign that adds the value in the buy.sig and  sell.sig columns. Since a buy
or sell signal cannot happen on the same day, summing the values in both columns
will generate either a 1 when there is a buy signal, −1 if there is a sell signal, and
zeroes if there is no buy or sell signal. We thus need to replace those zeroes with the
last signal (e.g., if the last signal was a buy, then we want to replace the signal with
continuing to hold the stock.). To do that, we create a column labeled cum, which
is shorthand for "cumulative," and use a for-loop that accomplishes what we want
to do.

```
1   > signal <- cbind(buy$buy.sig, sell$sell.sig)
2   > head.tail(signal)
3               buy.sig sell.sig
4   2015-10-16       1        0
5   2015-10-19       1        0
6   2015-10-20       1        0
7               buy.sig sell.sig
8   2019-12-27       1        0
9   2019-12-30       1        0
10  2019-12-31       1        0
11  >
12  > signal$sign <- signal$buy.sig + signal$sell.sig
13  > head.tail(signal)
14              buy.sig sell.sig sign
15  2015-10-16       1        0    1
16  2015-10-19       1        0    1
17  2015-10-20       1        0    1
18              buy.sig sell.sig sign
19  2019-12-27       1        0    1
20  2019-12-30       1        0    1
21  2019-12-31       1        0    1
22  >
23  > signal$cum <- rep(1, nrow(signal))
24  > for (t in 2:nrow(signal)){
25  +     signal$cum[t] <- ifelse(signal$sign[t] == 0,
26  +         signal$cum[t - 1], signal$sign[t])
27  + }
28  > head.tail(signal)
29              buy.sig sell.sig sign cum
30  2015-10-16       1        0    1   1
31  2015-10-19       1        0    1   1
32  2015-10-20       1        0    1   1
33              buy.sig sell.sig sign cum
34  2019-12-27       1        0    1   1
35  2019-12-30       1        0    1   1
36  2019-12-31       1        0    1   1
```

Next, we determine when we will be invested in the security. If cum equals 1, then we are invested in the stock and our portfolio return changes with the changes in the price of the stock. If cum equals 0, then we are out of the stock. This means that we will not be affected if the stock goes up or down when we are not invested in the stock.

```
1   > signal$invested <- ifelse(signal$cum == -1, 0, 1)
2   > head.tail(signal)
3               buy.sig sell.sig sign cum invested
4   2015-10-16       1        0    1   1        1
5   2015-10-19       1        0    1   1        1
6   2015-10-20       1        0    1   1        1
7               buy.sig sell.sig sign cum invested
8   2019-12-27       1        0    1   1        1
9   2019-12-30       1        0    1   1        1
10  2019-12-31       1        0    1   1        1
```

**Step 5: Calculate the In-Sample Portfolio Return**  We use the 2-year period from 2016 to 2017 as the in-sample period. We then bring in the data in the *signal* dataset and the *rets* dataset during that period in a new dataset *in.sample*. We then overwrite the first value under the rets column for December 31, 2015 as 0% return.

```
1  > in.sample <- signal["2015-12-31/2017-12-31"]
2  > in.sample$rets <- rets["2015-12-31/2017-12-31"]
3  > head.tail(in.sample)
4            buy.sig sell.sig sign cum invested        rets
5  2015-12-31      1        0    1   1         1 -0.019127218
6  2016-01-04      1        0    1   1         1 -0.057553780
7  2016-01-05      1        0    1   1         1 -0.005023646
8            buy.sig sell.sig sign cum invested        rets
9  2017-12-27      1        0    1   1         1  0.004673850
10 2017-12-28      1        0    1   1         1  0.003247988
11 2017-12-29      1        0    1   1         1 -0.014020745
12 >
13 > in.sample$rets[1] <- 0
14 > head.tail(in.sample)
15           buy.sig sell.sig sign cum invested        rets
16 2015-12-31      1        0    1   1         1  0.000000000
17 2016-01-04      1        0    1   1         1 -0.057553780
18 2016-01-05      1        0    1   1         1 -0.005023646
19           buy.sig sell.sig sign cum invested        rets
20 2017-12-27      1        0    1   1         1  0.004673850
21 2017-12-28      1        0    1   1         1  0.003247988
22 2017-12-29      1        0    1   1         1 -0.014020745
```

Next, we calculate the portfolio return port.ret by multiplying the return with the cumulative signal invested. We are either long the stock when invested equals 1 or not invested in the stock when invested equals 0.

```
1  > in.sample$port.ret <- in.sample$rets * in.sample$invested
2  > head.tail(in.sample)
3            buy.sig sell.sig sign cum invested        rets      port.ret
4  2015-12-31      1        0    1   1         1  0.000000000  0.000000000
5  2016-01-04      1        0    1   1         1 -0.057553780 -0.057553780
6  2016-01-05      1        0    1   1         1 -0.005023646 -0.005023646
7            buy.sig sell.sig sign cum invested        rets      port.ret
8  2017-12-27      1        0    1   1         1  0.004673850  0.004673850
9  2017-12-28      1        0    1   1         1  0.003247988  0.003247988
10 2017-12-29      1        0    1   1         1 -0.014020745 -0.014020745
```

We then cumulate the gross portfolio return using cumprod(). By adding one to port.ret, we are converting the net portfolio return to a gross portfolio return. As we can see, this strategy generates a portfolio return of 55.3% over the 2-year period.

```
1  > in.sample$port.cret <- cumprod(1 + in.sample$port.ret)
2  > head.tail(in.sample[, -c(1:4)])
3            invested        rets      port.ret port.cret
4  2015-12-31       1  0.000000000  0.000000000 1.0000000
5  2016-01-04       1 -0.057553780 -0.057553780 0.9424462
6  2016-01-05       1 -0.005023646 -0.005023646 0.9377117
7            invested        rets      port.ret port.cret
```

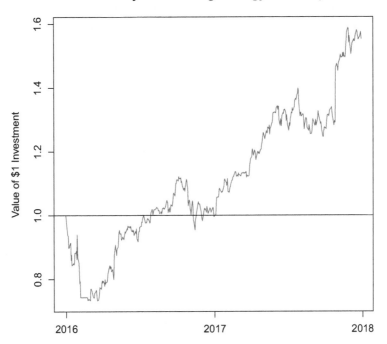

**Fig. 12.4**  200-Day SMA trading strategy example (in sample)

```
 8   2017–12–27      1   0.004673850   0.004673850   1.569832
 9   2017–12–28      1   0.003247988   0.003247988   1.574930
10   2017–12–29      1  −0.014020745  −0.014020745   1.552849
```

There are too many columns now in the *in.sample* dataset. As such, when we output
the first and last three observations above, we suppressed Columns 1 through 4.

**Step 6: Plot the Value of the Portfolio**  The code below generates a plot of the
portfolio value based on a 200-day SMA trading strategy during our in-sample
period from 2016 to 2017. The resulting chart is shown in Fig. 12.4. The chart shows
that a $1 investment on December 31, 2015 in this trading strategy generates $1.55
by December 31, 2017. That is, a 55% return over a 2-year period.

```
1   > plot(y = in.sample$port.cret,
2   +     x = index(in.sample),
3   +     xlab = "",
4   +     ylab = "Value of $1 Investment",
5   +     type = "l",
6   +     col = "blue",
7   +     main = "200–Day SMA Trading Strategy – In Sample")
8   > abline(h = 1)
```

**Step 7: Calculate Sharpe Ratio** We calculate the Sharpe ratio of this portfolio strategy. The Sharpe ratio is equal to the annualized average excess return divided by the annualized standard deviation. The excess return is the portfolio return over the risk-free rate, which we assume equals 1.55%. We calculate the average annualized return in Line 1, annualized standard deviation in Line 4, and the Sharpe ratio in Line 8.

```
1  > (avg.ret <- mean(in.sample$port.ret) * 252)
2  [1] 0.2512271
3  >
4  > (sd.ret <- sd(in.sample$port.ret) * sqrt(252))
5  [1] 0.2510118
6  >
7  > rf <- 0.0155
8  > (Sharpe <- (avg.ret - rf) / sd.ret)
9  [1] 0.9391078
```

**Step 8: Calculate Maximum Drawdown** Looking at Fig. 12.4, we can see that the biggest drawdown likely is in the first 6 months of 2016. Our initial investment is $1, so we want to look for other instances during which the portfolio value is greater than $1. We identify this using the values in the **port.cret** column. Then, we find the minimum value in **port.cret** over the period January 1, 2016 through July 11, 2016 in Line 6 to 7. The output shows the trough is on March 18, 2016 with a gross portfolio value of $0.73. This implies that the maximum drawdown is $-26.69\%$.

```
1  > head(subset(in.sample[, -c(1:4)], port.cret >= 1), 3)
2               invested        rets     port.ret port.cret
3  2015-12-31          1 0.000000000 0.000000000  1.000000
4  2016-07-11          1 0.010686409 0.010686409  1.000886
5  2016-07-29          1 0.008238016 0.008238016  1.007565
6  > min.cret <- subset(in.sample["2016-01-01/2016-07-11"],
7  +     port.cret == min(port.cret))
8  > min.cret[, -c(1:4)]
9               invested        rets      port.ret port.cret
10 2016-03-18          1 -0.01315599 -0.01315599 0.7330643
11 > (max.dd <- as.numeric(min.cret$port.cret) - 1)
12 [1] -0.2669357
```

We can verify that our calculation is correct using `table.Drawdowns()` from the `PerformanceAnalytics` package. As the output shows, the maximum drawdown is indeed during the period we were looking at above. Note that the table shows the first day is January 4, 2016, which is the first trading day of 2016. The return over that period is $-26.69\%$, which is the same as what we have calculated above.

```
1  > library(PerformanceAnalytics)
2  > table.Drawdowns(in.sample$port.ret)
3          From     Trough         To   Depth Length To Trough Recovery
4  1 2016-01-04 2016-03-18 2016-07-11 -0.2669    131        53       78
5  2 2016-10-06 2016-11-14 2017-02-17 -0.1484     93        28       65
6  3 2017-07-27 2017-09-26 2017-10-27 -0.1085     66        43       23
7  4 2017-06-06 2017-07-03 2017-07-18 -0.0570     30        20       10
8  5 2017-11-28 2017-12-04       <NA> -0.0517     24         5       NA
```

**Step 9: Calculate Out-of-Sample Portfolio Returns**  We calculate the out-of-sample portfolio returns using the same methodology as in Step 5. Note that we subset the out-of-sample data with the first date as December 29, 2017. The reason is that December 29, 2017 is the last trading day before 2018. This allows us to easily have an observation that we can use to start our calculations. As the cumulative portfolio return shows at the very bottom, this portfolio strategy generates a 2-year return of 13.1%.

```
1  > out.sample <- signal["2017-12-29/2019-12-31"]
2  > out.sample$rets <- rets["2017-12-29/2019-12-31"]
3  > head.tail(out.sample)
4             buy.sig sell.sig sign cum invested        rets
5  2017-12-29      1        0    1   1        1 -0.01402074
6  2018-01-02      1        0    1   1        1  0.01670846
7  2018-01-03      1        0    1   1        1  0.01277528
8             buy.sig sell.sig sign cum invested        rets
9  2019-12-27      1        0    1   1        1  0.0005511802
10 2019-12-30      1        0    1   1        1 -0.0122526652
11 2019-12-31      1        0    1   1        1  0.0005143517
12 >
13 > out.sample$rets[1] <- 0
14 > head.tail(out.sample)
15            buy.sig sell.sig sign cum invested        rets
16 2017-12-29      1        0    1   1        1 0.00000000
17 2018-01-02      1        0    1   1        1 0.01670846
18 2018-01-03      1        0    1   1        1 0.01277528
19            buy.sig sell.sig sign cum invested        rets
20 2019-12-27      1        0    1   1        1  0.0005511802
21 2019-12-30      1        0    1   1        1 -0.0122526652
22 2019-12-31      1        0    1   1        1  0.0005143517
23 >
24 > out.sample$port.ret <- out.sample$rets * out.sample$invest
25 > head.tail(out.sample)
26            buy.sig sell.sig sign cum invested        rets    port.ret
27 2017-12-29      1        0    1   1        1 0.00000000 0.00000000
28 2018-01-02      1        0    1   1        1 0.01670846 0.01670846
29 2018-01-03      1        0    1   1        1 0.01277528 0.01277528
30            buy.sig sell.sig sign cum invested        rets        port.ret
31 2019-12-27      1        0    1   1        1  0.0005511802  0.0005511802
32 2019-12-30      1        0    1   1        1 -0.0122526652 -0.0122526652
33 2019-12-31      1        0    1   1        1  0.0005143517  0.0005143517
34 >
35 > out.sample$port.cret <- cumprod(1 + out.sample$port.ret)
36 > head.tail(out.sample[, -c(1:4)])
37            invested       rets port.ret port.cret
38 2017-12-29        1 0.00000000 0.00000000  1.000000
39 2018-01-02        1 0.01670846 0.01670846  1.016708
40 2018-01-03        1 0.01277528 0.01277528  1.029697
41            invested       rets     port.ret port.cret
```

```
42   2019–12–27        1   0.0005511802   0.0005511802   1.144226
43   2019–12–30        1  −0.0122526652  −0.0122526652   1.130207
44   2019–12–31        1   0.0005143517   0.0005143517   1.130788
```

**Step 10: Plot Value of Portfolio**  We now plot the value of the portfolio during the out-of-sample period. Figure 12.5 shows the value of the portfolio strategy out-of-sample. We can see, there are a number of flat spots during the out-of-sample period, which means we are out of the stock several times during the investment period.

```
1   > plot(y = out.sample$port.cret,
2   +      x = index(out.sample),
3   +      xlab = "",
4   +      ylab = "Value of $1 Investment",
5   +      type = "l",
6   +      col = "blue",
7   +      main = "200–Day SMA Trading Strategy – Out–of–Sample")
8   > abline(h = 1)
```

**Step 11: Calculate Sharpe Ratio**  We calculate a Sharpe ratio of 0.32 under this portfolio strategy in the out-of-sample test. This is much lower than the in-sample Sharpe ratio of close to 1.0.

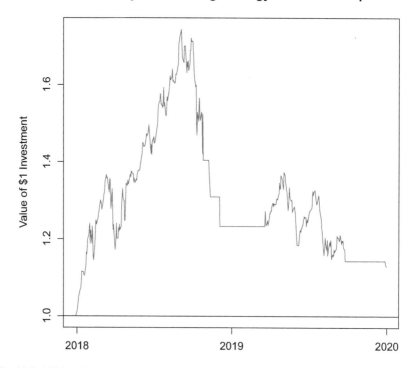

**Fig. 12.5**  200-Day SMA trading strategy example (out-of-sample)

```
1  > (avg.ret <- mean(out.sample$port.ret) * 252)
2  [1] 0.08811198
3  >
4  > (sd.ret <- sd(out.sample$port.ret) * sqrt(252))
5  [1] 0.2301935
6  >
7  > rf <- 0.0155
8  > (Sharpe <- (avg.ret - rf) / sd.ret)
9  [1] 0.3154389
```

**Step 12: Calculate Maximum Drawdown** We calculate maximum drawdown by figuring out the peak and trough. Line 1 shows the maximum portfolio value. As we can see from Fig. 12.5 shows that the portfolio would not recover back to the peak. As such, the trough is the last value of the portfolio. We then calculate the maximum drawdown as the percentage change from the peak to trough, which is a decline of 35.16%.

```
1  > (out.peak <- subset(out.sample$port.cret, port.cret == max(port.cret)))
2             port.cret
3  2018-09-04  1.743961
4  > (out.trough <- out.sample$port.cret[nrow(out.sample)])
5             port.cret
6  2019-12-31  1.130788
7  > (max.dd <- (as.numeric(out.trough) / as.numeric(out.peak)) - 1)
8  [1] -0.3515979
```

We use the `table.Drawdowns()` from the `PerformanceAnalytics` package to see what it calculates as the maximum drawdown. The output below shows that this function finds a maximum drawdown of 35.19%, which is slightly off from our calculation above.

```
1  > library(PerformanceAnalytics)
2  > table.Drawdowns(out.sample$port.ret)
3            From      Trough        To   Depth Length To Trough Recovery
4  1 2018-09-05 2019-12-30      <NA> -0.3519   334      332       NA
5  2 2018-03-13 2018-04-02 2018-05-07 -0.1416    39       14       25
6  3 2018-02-01 2018-02-09 2018-02-14 -0.0767    10        7        3
7  4 2018-06-21 2018-06-27 2018-07-11 -0.0512    14        5        9
8  5 2018-07-26 2018-07-31 2018-08-08 -0.0462    10        4        6
```

The difference between the two calculations above is that `table.Drawdowns()` somehow does not include December 31, 2019 and stops at December 30, 2019. Below, we show what our calculation of maximum drawdown would look like if we assumed the trough was on December 30, 2019. As the output below shows, we replicate the result of `table.Drawdowns()` using this calculation. Thus, we have to be careful when using `table.Drawdowns()` because it appears to be missing 1 day in the calculation.

```
1  > (out.trough2 <- out.sample$port.cret[nrow(out.sample) - 1])
2             port.cret
3  2019-12-30  1.130207
4  > (max.dd2 <- (as.numeric(out.trough2) / as.numeric(out.peak)) - 1)
5  [1] -0.3519312
```

The above is a simple implementation of a trading strategy. In practice, we would likely combine multiple trading strategies together. We could consider a number of additional factors. For example, we may want to consider commissions for every trade, which would lower the realized return from the investment. However, recently, there have been a number of brokers that offer commission-free trades. Examples of these include Schwab and Alpaca (algorithmic trading). There are other issues to consider as well, such as buying and selling at the bid-ask prices. These considerations depend on the strategies we would like to implement.

## 12.4  Machine Learning Techniques

**Machine learning** (ML) is premised on the idea that machines can learn from data and make predictions (decisions) with little human intervention. There are many things that we can apply ML to. One argument is that a machine can likely learn anything a human can do if it takes a human to do it in 1 s. The idea is that something a human does in 1 s is something that the person learned and implemented. For example, pattern recognition in investments is an attractive candidate for ML because we can teach a machine to identify and learn different patterns. Consider the following hypothetical. Suppose you follow four stocks. A machine may be able to identify that stock A moves up 70% of the time after both stocks B, C, and D move up by at least 3% the prior trading day. Then it may be possible to create a model that recognizes such patterns and triggers a decision to buy stock A whenever we observe stocks B, C, and D all move up by at least 3% each. Of course the patterns in the real world are rarely this straightforward or the likelihoods are usually much lower.

However, finding similarities based on fundamentals of different investment options is not pattern recognition. Thus, it is harder for machines to completely replace fundamental investing. Despite this, part of the fundamental investing process may be replaced by machines. On the other hand, technical analysis, which primarily relies on exploiting patterns in past prices and data, can potentially be replaced by machines because it is somewhat based on pattern recognition. ML can also be helpful in recognizing patterns not easily detectable by humans. One example of where using ML can yield greater efficiencies is that machines can be trained to learn text of company SEC filings to see if they contain certain word patterns. Instead of having to manually look at all SEC filings, a machine can be taught to identify the key words and any pattern of those words.

We already have some tools that recognize patterns. For example, linear regressions are good at recognizing patterns in data, but linear regression has trouble when patterns are complex (e.g., non-linearities). In many finance applications, there may not be enough data to use very complex methods. For example, when we look at earnings data for a publicly traded company, we would typically only observe four data points each year. We can increase the number of observations by going further back in time, but we quickly encounter the issue of relevance of the information that far back to predictions of the future. For example, the company may be substantially different today than they were 5 years ago, so at most we could only have 20 earnings disclosures over the date range if the company announces earnings every quarter. The problem is worse if we are dealing with private companies, as it is uncertain if data would be available. The problem with not having enough data is that the machine ends up learning the noise and not the signal. The key to ML is reasonable amounts of high quality data with sufficient variation.

The biggest problem in applying ML techniques is that of **overfitting**. Data typically consists of a noise component and a predictive component (i.e., signal). Overfitting is when a model tries to learn something from the noise component, which by definition is random (i.e., there is no signal to learn). From an investment perspective, overfitting often arises because of the bias we may have in finding a very good fit in our training dataset (in sample), but we end up generating terrible predictions in our test data (out-of-sample) or when the model is applied to the real world.

In theory, factors used in the models are selected to boost the signal-to-noise ratio. For this, we would require knowledge and expertise by the modeler as to the theory of why certain factors should be included. For example, if we are trying to predict future stock returns, we would only want to include factors that are related to future stock returns based on, say, empirical studies or theory. This means that if we apply ML techniques that tell us that the change in temperature on the moon explains the variation in stock returns, we will likely not include that in our model. Popular choices for factors are either fundamental factors (e.g., determinants of the future prospects of the stock) and technical indicators (i.e., factors related to future supply and future demand).

There are generally two types of ML algorithms: supervised and unsupervised. Supervised learners are used to construct predictive models, and unsupervised learners are used to build descriptive models. Below, we will discuss models that attempt to predict the future stock price and, thus, we will be discussing examples of **predictive models**. We often think of predictive models as forward-looking predictions, but such models could also be used for backward-looking predictions. For example, we may want to use a predictive model to know what a company's stock price would have been 1 year earlier given the stock's current price and certain pieces of information that could have contributed to the evolution in the company's price from 1 year ago through today. In general, predictive models are given instructions on what they should learn and how they should learn that information. Thus, given a set of data, the algorithm attempts to optimize a model to find the combination of features that results in the desired output.

Below we discuss examples of two types of predictive models. The first type is **classification**, which assigns the predicted value to one or more categories. In our example using $k$-nearest neighbor algorithm, we predict whether a company's stock price would increase or decrease. The second type is **numeric prediction**, which as the name indicates predicts a numerical value. In our example below, we fit a linear regression model to our stock price data and the model generates numerical predictions. We conclude the ML section with an example that uses both numeric prediction and classification. In our example, we use artificial neural networks (ANN) to make predictions about the future price of a company's stock. Although fancier, neural network models are difficult to explain without resorting to some heavy mathematical machinery and, consequently, the interpretation of the process becomes difficult. In other words, it may seem to many that neural networks are black boxes. Thus, if one is required to explain the details of the methodology that led to a particular outcome, neural network models may not be the best. For example, when a lender turns down a person that seeks credit, by law the lender is required to give an explanation of why the applicant was turned down. The difficulty to interpret ML algorithms and the difficulty to identify why and how the prediction led to the denial of credit makes the applicability of ML to such markets challenging. Fortunately, some of the current research have been focused on mitigating this interpretability issue.

### 12.4.1   General Steps to Apply ML

In general, developing an ML algorithm requires five steps.

1. **Collect Data.** We need to collect the data needed for the ML algorithm. Then, we would need enough data so we can separate the data into a training set for Step 3 and a test set for Step 4.
2. **Analyze Data.** As we have discussed throughout this book, real world data is typically not setup for a particular purpose and may contain errors and other issues. Thus, a lot of time could be spent on getting the data into its proper shape. During this step, we have to understand the data, including its nuances. This becomes harder for a generalist at programming because these nuances and the meaning of certain numbers may not be apparent simply by looking at the data. Therefore, a mix of programming skills and actual experience working with the data is best, as this will give us a sense of what we can "learn" from the data we have. It may be that the data we have is sufficient or we may realize that we need to supplement the data.
3. **Train the Model.** We want to allocate a sufficient amount of the data to the training set. The end result of this step is in the ML algorithm generating a representation of the data in the form of a model. There is no single approach that is best for every scenario. Thus, the process of choosing an ML algorithm relies on the characteristics of the data we are using.
4. **Test the Model.** Using the model developed in Step 3, we test how well the model learned from the data by seeing how well the model performs on a separate dataset, which we call the test set. To measure performance below, we use the root mean square errors (RMSE) as our metric. However, RMSE is not the only way to measure performance and we could also select other metrics.
5. **Improve the Model.** If the model performance is unsatisfactory, we can try to make modifications to improve the performance of the model. This may mean

tweaking the current model or using a completely different model. This may also require using additional data or alternative sources of data. However, we have to be careful that we do not end up overfitting.

## 12.4.2 k-Nearest Neighbor Algorithm

The k-**nearest neighbor** (kNN) algorithm is a **non-parametric** method that can be used to predict whether, say, a stock will increase or decrease in price. Think of the saying "birds of the same feather flock together." The nearest neighbor algorithm is like identifying who our closest $k$ friends are and determines whether they either belong to, say, either a type 1 or type 2 personality. For example, if $k = 5$, and our closest five friends are four type 1s and one type 2, then the kNN algorithm would classify us as having an 80% chance we are a type 1 person.

As one may have gathered from the above, $k$ is the number of neighbors that is selected. For example, if $k = 1$, then the algorithm finds the "closest" neighbor. Closest is typically defined as the points with the smallest Euclidean distance. Basically, the Euclidean distance is what you get if you use a ruler to connect two points. In our example below, if the closest neighbor is a gain, then we classify the predicted outcome as a gain. On the other hand, if the closest neighbor is a loss, we classify the predicted outcome as a loss.

Now, assume $k = 5$. In that case, the algorithm finds the five closest neighbors regardless of type. For example, the five closest neighbors could be 4 gains and 1 loss, in which case we say that there is an 80% chance of a gain and classify the predicted outcome as a gain. Alternatively, the five closest neighbors could be 2 gains and 3 losses, in which case we say that there is a 60% chance of a loss and classify the predicted outcome as a loss.

As the above shows, by choosing an odd numbered $k$, we avoid any ties. If we choose $k = 4$, for example, we may end up with two gains and two losses as the nearest four neighbors. The tie then has to be broken using some likely subjective criteria or by random. Thus, to avoid the problem of ties altogether, we use an odd numbered $k$.

Also, there is trade-off between choosing a smaller $k$ and a larger $k$. We can think of the two extremes. If $k$ equals all observations, then the majority of the population is chosen regardless of who the nearest neighbor is. Thus, if we have a situation with a total of 30 observations, where the 10 nearest neighbors are gains and the next 20 neighbors are losses, the model would predict that it will be loss. In the other extreme, if we select $k = 1$, then we are solely relying on the closest neighbor, but being the closest neighbor could be due to random chance, an outlier, or classified by error. Therefore, the best value for $k$ would be an amount that balances these two extremes. Although there is no hard and fast rule on what $k$ to select, in practice it is common to set $k$ somewhere between 3 and 10.

The nearest neighbor algorithm is better suited when the relationships among factors (often called "features") and the independent variable (often called "target classes") are numerous or complicated, but we know that stocks of the same type tend to be homogeneous. We use a simple example below to highlight the mechanics of using the kNN algorithm to predict whether AMZN will gain or lose. In our example, we use SPY as a factor. We can include a number of other factors as well, like more stocks. However, for simplicity, we only use one factor.

**Step 1: Load Price Data**  Using `load.data()` we import AMZN and SPY prices and combine the closing prices of the two price series.

```
1  > amzn <- load.data("AMZN Yahoo.csv", "AMZN")
2  > spy <- load.data("SPY Yahoo.csv", "SPY")
3  >
4  > prices <- cbind(amzn$AMZN.Close, spy$SPY.Close)
5  > names(prices) <- c("AMZN", "SPY")
6  > head.tail(prices)
7                AMZN    SPY
8   2014-12-31 310.35 205.54
9   2015-01-02 308.52 205.43
10  2015-01-05 302.19 201.72
11                AMZN    SPY
12  2019-12-20 1786.50 320.73
13  2019-12-23 1793.00 321.22
14  2019-12-24 1789.21 321.23
```

**Step 2: Create Required Variables**  In this step, we will create two columns in the *prices* dataset: date and Lag.AMZN. We also create a column Lag.AMZN for the one trading day lagged price. We use this variable to determine whether AMZN price increased or decreased relative to the prior trading day's price.

```
1  > date <- data.frame(index(prices[-1,]))
2  > names(date) <- "date"
3  > prices$Lag.AMZN <- Lag(prices$AMZN, k = 1)
4  > prices <- data.frame(prices[-1, ])
5  > prices <- cbind(date, prices)
6  > rownames(prices) <- seq(1, nrow(prices), 1)
7  > head.tail(prices)
8            date    AMZN    SPY Lag.AMZN
9   1 2015-01-02 308.52 205.43   310.35
10  2 2015-01-05 302.19 201.72   308.52
11  3 2015-01-06 295.29 199.82   302.19
12            date    AMZN    SPY Lag.AMZN
13  1252 2019-12-20 1786.50 320.73  1792.28
14  1253 2019-12-23 1793.00 321.22  1786.50
15  1254 2019-12-24 1789.21 321.23  1793.00
```

**Step 3: Identify Break in Data**  To simplify our coding going forward, we can identify the observation number that separates out the training data, which will be data prior to 2019, and the test data, which will be data in 2019. So we use `subset()` to identify dates around December 31, 2018. As we can see, the training dataset will contain observations 1 through 1006 and the test data will contain observations 1007–1254. We obtained the last observation number using `nrow()`.

```
1   > subset(prices, date >= "2018-12-28" & date <= "2019-01-04")
2              date     AMZN     SPY Lag.AMZN
3   1005 2018-12-28 1478.02 247.75  1461.64
4   1006 2018-12-31 1501.97 249.92  1478.02
5   1007 2019-01-02 1539.13 250.18  1501.97
6   1008 2019-01-03 1500.28 244.21  1539.13
7   1009 2019-01-04 1575.39 252.39  1500.28
8   > nrow(prices)
9   [1] 1254
```

**Step 4: Create Logical Vector for Gains and Losses** By comparing the Amazon price on the day to Amazon's price yesterday, we can determine if the stock gained in value or not. We can thus create a logical vector of TRUE or FALSE that contains those values for each observation in *prices* (i.e., includes both data we will use in training and in testing). We use as.logical() to do this.

```
1   > gain <- ifelse(prices$AMZN > prices$Lag.AMZN, "TRUE", "FALSE")
2   > gain <- as.logical(gain)
3   > class(gain)
4   [1] "logical"
```

**Step 5: Normalize Prices** In applying kNN, it is typical to standardize the range of values. This is because the distance formula is dependent on how the factors are measured. We want to avoid the factors with larger values relative to others because that would bias the distance calculation. In our example, the values of Amazon and SPY are quite different in scale. It may be better to normalize the values. We do a simple **min-max method** of normalizing the values of each series between 0 and 1. We do the min-max method in a function called normalize(). As shown in Lines 5 through 8, this function normalizes the values between 0 and 1. Each element in the vector in Line 7 is 5x the value of the corresponding element in the vector in Line 5. The normalization function essentially puts them in the same scale. We can see this when we compare Line 6 with Line 8. Then, in Line 10, we apply this normalization to the AMZN and SPY prices.

```
1   > normalize <- function(x) {
2   +    return ((x - min(x)) / (max(x) - min(x)))
3   + }
4   >
5   > normalize(c(1, 2, 3, 4, 5))
6   [1] 0.00 0.25 0.50 0.75 1.00
7   > normalize(c(5, 10, 15, 20, 25))
8   [1] 0.00 0.25 0.50 0.75 1.00
9   >
10  > prices <- as.data.frame(lapply(prices[2:3], normalize))
11  > head.tail(prices)
12            AMZN        SPY
13  1 0.012307697 0.1631133
14  2 0.008695845 0.1363012
15  3 0.004758751 0.1225700
16            AMZN        SPY
17  1252 0.8556340 0.9963865
```

```
18   1253 0.8593429 0.9999277
19   1254 0.8571803 1.0000000
20   > min(prices)
21   [1] 0
22   > max(prices)
23   [1] 1
```

**Step 6: Create Datasets for Use in kNN**  This is where identifying the observation number will be helpful. knn() requires inputs for the training dataset, test dataset, and a factor of true classifications of the training dataset.

```
1   > train.data <- prices[1:1006, ]
2   > test.data <- prices[1007:1254, ]
3   > cl.data <- gain[1:1006]
```

**Step 7: Create Prediction Using knn()**  We first load the class package, which contains knn(). The three datasets we created in the prior step slots into the arguments needed by knn(). We start with using a $k = 1$. R does the heavy lifting here.

```
1   > library(class)
2   > predict <- knn(train = train.data,
3   +     test = test.data,
4   +     cl = cl.data,
5   +     k = 1)
```

**Step 8: Look at Output**  The output can be generated using table(). The first argument is the dataset that contains the prediction, which we have labeled *predict*. The second argument is the classification of gain or loss but this time we are using the observations in the test dataset. According to the output, we have 71 false negatives and 49 false positives. Put differently, 120 values were predicted incorrectly out of a total of 248. That is 48.4% which was incorrectly classified, while 51.6% was correctly classified. We can also get the percentage of correct predictions using the code in Line 7.

```
1   > table(predict,
2   +     gain[1007:1254])
3
4   predict FALSE TRUE
5     FALSE   66   71
6     TRUE    49   62
7   > mean(predict == gain[1007:1254])
8   [1] 0.516129
```

**Step 9: Test Other Values of $k$**  In the above, we used $k = 1$. However, we can use other values of $k$. To avoid ties, it is recommended to only use odd values of k, so we first create a sequence of odd numbers between 1 and 40. We then create a second variable *pct*. We first create temporary values of zero for *pct*, but we will fill it in with the percentage of accurate predictions later. We then combined the two in the dataset labeled *accuracy*.

```
 1   > k <- seq(1, by = 2, 40)
 2   > pct <- rep(0, length(k))
 3   > accuracy <- data.frame(k, pct)
 4   > accuracy
 5       k pct
 6   1   1   0
 7   2   3   0
 8   3   5   0
 9   4   7   0
10   5   9   0
11   6  11   0
12   7  13   0
13   8  15   0
14   9  17   0
15  10  19   0
16  11  21   0
17  12  23   0
18  13  25   0
19  14  27   0
20  15  29   0
21  16  31   0
22  17  33   0
23  18  35   0
24  19  37   0
25  20  39   0
```

We then run a for-loop using the 20 values of $k$. We also use the set.seed(1) in order to replicate the results. We find that the most accurate result occurs when $k = 33$ in which case the accuracy percentage is 56.0%. We can also visually see this in Fig. 12.6.

```
 1   > set.seed(1)
 2   > for (i in 1:20){
 3   +    predict <- knn(train.data, test.data, cl.data, k = accuracy[i, 1])
 4   +    accuracy[i, 2] <- mean(predict == gain[1007:1254])
 5   + }
 6   > plot(x = accuracy$k,
 7   +      y = accuracy$pct,
 8   +      xlab = "k",
 9   +      ylab = "Accuracy",
10   +      type = "l",
11   +      col = "blue",
12   +      main = "Accuracy Based on Different Values of k")
13   >
14   > subset(accuracy,
15   +      accuracy$pct == max(accuracy$pct))
16      k       pct
17   17 33 0.5604839
```

We can put the above results in context. A fair coin will give you a 50-50 chance. Using $k = 1$, our algorithm was accurate 51.6% of the time. Even at the maximum accuracy percentage of 56.0% when $k = 33$, we are only a few percentage points better than when we used $k = 1$ and have only increased our accuracy by 6% over

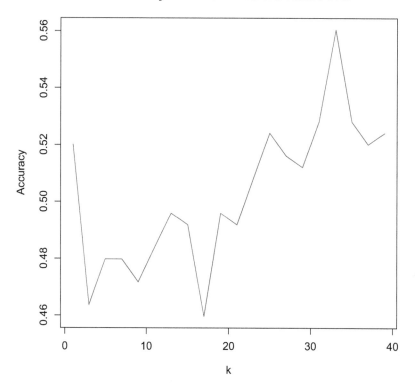

**Fig. 12.6**  Accuracy of kNN Algorithm Based on Different Values of k

flipping a coin. These small percentage increases though may still result in non-trivial gains depending on how we implement our trading strategy and how much money and leverage we incorporate.

The model above is a simple model to show how kNN could be implemented. More complex models involving many more factors can be applied to improve the accuracy percentages, but we have to be careful that we do not overfit.

### 12.4.3   *Regression and k-Fold Cross Validation*

We may not immediately think of **regressions** as a machine learning technique, but they are actually a form of supervised ML algorithm. We can use regression models to learn from various factors and then use those to predict future values. Currently, one of the most popular areas of research is in determining the relevant factors for investment purposes. In Chap. 5, we discussed what has been historically the time-tested factor models: CAPM and Fama-French models. However, research on factors have expanded to potentially include a wide variety of factors. For example, aside from market, size, and value, other factors that have been found to potentially be relevant include momentum, quality, economic growth, and minimum volatility.

Below, we show how to choose the "best" model based on the training dataset by combining regressions with $k$-fold **cross validation**. For our regressions, we use as the measure of accuracy the root mean square error (RMSE). In cross validation, we first divide the training data into $k$ folds (i.e., groups) of approximately equal size. The model is then fit on $k - 1$ folds and then the remaining fold (i.e., where the model was not fit) is used to compute model performance. This procedure is repeated $k$ times (i.e., cross validation is a resampling method). In each **resampling**, a different fold is held out as the validation set. We then end up with $k$ estimates of the RMSE and the $k$-fold cross validation RMSE is computed by averaging the $k$ RMSEs. This provides us with an approximation of the error we might expect to see in the future (e.g., test data). For example, if we use 10-fold cross validation, the data will be divided into 90% training set and 10% test set for each run. Then, for the second run, we will divide the data again 90% training set and 10% test set but with different data in each of the two sets. We repeat this 10 times. Each of the 10 runs will produce 10 RMSE estimates and the average of the 10 RMSEs will be the 10-fold cross validation RMSE. Although this approach sounds like it may be challenging to implement, we can use `train()` in the `caret` package to simplify our task substantially.

Below, for our example, we want to select between three models to predict the return of AMZN. Model 1 only uses the market. Model 2 adds to Model 1 the Fama-French SMB and HML factors. Finally, Model 3 adds to Model 2 the Fama-French RMW and CMA factors. We want to know which model does "best" in the training dataset, and then use that model to predict AMZN price using the test dataset. We can then test the prediction accuracy of that model using the RMSE during the period of the test dataset. Let us see how to implement this.

**Step 1: Import Returns Data**   We first load the AMZN data using `load.data()` and then calculate log returns off the AMZN adjusted close prices. We then import the FF 5 Factor Model data. Because Fama-French returns data uses **arithmetic returns**, we convert them to log returns in Line 20. We then check the number of observations in *rets.amzn* and *ff.xts* are the same using `dim()`. After confirming that they both have 1258 observations, we use `cbind()` to combine the two datasets. I also calculate the AMZN return in excess of the risk-free rate  exret.

```
 1   > amzn <- load.data("AMZN Yahoo.csv", "AMZN")
 2   > rets.amzn <- diff(log(amzn$AMZN.Adjusted))
 3   > rets.amzn <- rets.amzn["2015-01-01/2019-12-31"]
 4   >
 5   > ff <- read.csv("F-F_Research_Data_5_Factors_2x3_daily.csv",
 6   +      skip = 3, header = TRUE)
 7   > library(lubridate)
 8   > ff$date <- ymd(ff$X)
 9   > ff.xts <- xts(ff[, -c(1, 8)], order = ff$date)
10   > head.tail(ff.xts)
11                 Mkt.RF    SMB    HML    RMW    CMA    RF
```

```
12   1963–07–01  −0.67  0.00 −0.32 −0.01  0.15 0.012
13   1963–07–02   0.79 −0.27  0.27 −0.07 −0.19 0.012
14   1963–07–03   0.63 −0.17 −0.09  0.17 −0.33 0.012
15               Mkt.RF    SMB    HML   RMW CMA    RF
16   2019–12–27 −0.09 −0.54 −0.07  0.24 0.16 0.007
17   2019–12–30 −0.57  0.27  0.58  0.15 0.45 0.007
18   2019–12–31  0.28  0.02  0.14 −0.11 0.22 0.007
19   >
20   > ff.xts <- log(1 + ff.xts / 100)
21   > ff.xts <- ff.xts["2015–01–01/2019–12–31"]
22   > options(digits = 3)
23   > head.tail(ff.xts)
24                 Mkt.RF       SMB      HML       RMW       CMA RF
25   2015–01–02 −1.10e–13 −5.62e–13  8.99e–14 −2.50e–13  9.99e–14  0
26   2015–01–05 −1.86e–12  2.60e–13 −6.32e–13  1.60e–13 −1.20e–13  0
27   2015–01–06 −1.05e–12 −7.83e–13 −2.60e–13  5.29e–13  4.00e–14  0
28                 Mkt.RF       SMB      HML       RMW       CMA       RF
29   2019–12–27 −9.00e–14 −5.41e–13 −7.01e–14  2.4e–13 1.60e–13 7.11e–15
30   2019–12–30 −5.72e–13  2.70e–13  5.78e–13  1.5e–13 4.49e–13 7.11e–15
31   2019–12–31  2.80e–13  2.00e–14  1.40e–13 −1.1e–13 2.20e–13 7.11e–15
32   >
33   > dim(rets.amzn)
34   [1] 1258    1
35   > dim(ff.xts)
36   [1] 1258    6
37   >
38   > rets <- cbind(rets.amzn, ff.xts)
39   > names(rets)[1] <- "AMZN"
40   > rets$exret <- rets.amzn$AMZN.Adjusted − rets$RF
41   > head.tail(rets[, c(1, 8)])
42                AMZN      exret
43   2015–01–02 −0.00591 −0.00591
44   2015–01–05 −0.02073 −0.02073
45   2015–01–06 −0.02310 −0.02310
46                AMZN       exret
47   2019–12–27  0.000551  0.000551
48   2019–12–30 −0.012328 −0.012328
49   2019–12–31  0.000514  0.000514
50   > options(digits = 7)
```

**Step 2: Split the Data into Training and Test Datasets** For the training dataset, we use data in 2015–2018. For the test dataset, we use data in 2019.

```
1    > train.data <- rets["2015–01–01/2018–12–31"]
2    > head.tail(train.data[, c(1, 8)])
3                 AMZN     exret
4    2015–01–02 −0.00591 −0.00591
5    2015–01–05 −0.02073 −0.02073
6    2015–01–06 −0.02310 −0.02310
7                 AMZN     exret
8    2018–12–27 −0.00632 −0.00632
9    2018–12–28  0.01114  0.01114
10   2018–12–31  0.01607  0.01607
11   >
```

```
12  > test.data <- rets["2019-01-01/2019-12-31"]
13  > head.tail(test.data[, c(1, 8)])
14                AMZN    exret
15  2019-01-02  0.0244   0.0244
16  2019-01-03 -0.0256  -0.0256
17  2019-01-04  0.0489   0.0489
18                AMZN    exret
19  2019-12-27  0.000551  0.000551
20  2019-12-30 -0.012328 -0.012328
21  2019-12-31  0.000514  0.000514
```

**Step 3: Use 10-Fold Cross Validation** In the `caret` package, we can use `train()` to easily apply cross validation. In the first model *cv1*, we only use the market return. The average RMSE is 1.51%. The second model *cv2* adds SMB and HML and the average RMSE is 1.38%. In the last model *cv3*, we add RMW and CMA and the average RMSE is 1.31%. Thus, the third model has the lowest RMSE and what we consider the "best" model.

```
 1  > library(caret)
 2  > set.seed(1)
 3  > (cv1 <- train(
 4  +    form = exret ~ Mkt.RF,
 5  +    data = train.data,
 6  +    method = "lm",
 7  +    trControl = trainControl(method = "cv", number = 10)
 8  + ))
 9  Linear Regression
10
11  1006 samples
12     1 predictor
13
14  No pre-processing
15  Resampling: Cross-Validated (10 fold)
16  Summary of sample sizes: 905, 906, 906, 904, 905, 906, ...
17  Resampling results:
18
19    RMSE    Rsquared  MAE
20    0.0151  0.379     0.00961
21
22  Tuning parameter 'intercept' was held constant at a value of TRUE
23  >
24  > set.seed(1)
25  > (cv2 <- train(
26  +    form = exret ~ Mkt.RF + SMB + HML,
27  +    data = train.data,
28  +    method = "lm",
29  +    trControl = trainControl(method = "cv", number = 10)
30  + ))
31  Linear Regression
32
33  1006 samples
34     3 predictor
35
```

```
36   No pre-processing
37   Resampling: Cross-Validated (10 fold)
38   Summary of sample sizes: 905, 906, 906, 904, 905, 906, ...
39   Resampling results:
40
41     RMSE    Rsquared  MAE
42     0.0138  0.478     0.00874
43
44   Tuning parameter 'intercept' was held constant at a value of TRUE
45   >
46   > set.seed(1)
47   > (cv3 <- train(
48   +    form = exret ~ Mkt.RF + SMB + HML + RMW + CMA,
49   +    data = train.data,
50   +    method = "lm",
51   +    trControl = trainControl(method = "cv", number = 10)
52   + ))
53   Linear Regression
54
55   1006 samples
56      5 predictor
57
58   No pre-processing
59   Resampling: Cross-Validated (10 fold)
60   Summary of sample sizes: 905, 906, 906, 904, 905, 906, ...
61   Resampling results:
62
63     RMSE    Rsquared  MAE
64     0.0131  0.526     0.00848
65
66   Tuning parameter 'intercept' was held constant at a value of TRUE
67   >
68   > summary(resamples(list(
69   +    model1 = cv1,
70   +    model2 = cv2,
71   +    model3 = cv3
72   + )))
73
74   Call:
75   summary.resamples(object = resamples(list(model1 = cv1, model2 = cv2 ...
76
77   Models: model1, model2, model3
78   Number of resamples: 10
79
80   MAE
81              Min. 1st Qu. Median   Mean 3rd Qu.   Max. NA's
82   model1 0.00814 0.00911 0.00931 0.00961 0.00995 0.0115   0
83   model2 0.00719 0.00803 0.00866 0.00874 0.00943 0.0106   0
84   model3 0.00727 0.00782 0.00846 0.00848 0.00915 0.0099   0
85
86   RMSE
87              Min. 1st Qu. Median   Mean 3rd Qu.   Max. NA's
88   model1 0.01121  0.0137 0.0147 0.0151  0.0163 0.0212     0
89   model2 0.00933  0.0124 0.0138 0.0138  0.0148 0.0203     0
```

```
90    model3 0.00940  0.0118 0.0127 0.0131   0.0143 0.0191    0
91
92    Rsquared
93             Min. 1st Qu. Median  Mean 3rd Qu.  Max. NA's
94    model1 0.0915   0.333  0.386 0.379   0.409 0.623    0
95    model2 0.1510   0.399  0.471 0.478   0.528 0.729    0
96    model3 0.2346   0.476  0.521 0.526   0.567 0.741    0
```

**Step 4: Predict Returns Using "Best" Model**  We run the "best" model on the data and store the result in *train.reg*. We then use `predict()` to use the parameters of the regression on the *test.data* dataset.

```
1   > train.reg <- lm(exret ~ Mkt.RF + SMB + HML + RMW + CMA,
2   +      data = train.data)
3   > summary(train.reg)
4
5   Call:
6   lm(formula = exret ~ Mkt.RF + SMB + HML + RMW + CMA, data = train.data)
7
8   Residuals:
9        Min       1Q   Median       3Q      Max
10   −0.09662 −0.00687 −0.00093  0.00537  0.13053
11
12   Coefficients:
13                 Estimate Std. Error t value Pr(>|t|)
14   (Intercept)  1.04e−03   4.21e−04    2.46  0.01391 *
15   Mkt.RF       1.10e+10   5.17e+08   21.32  < 2e−16 ***
16   SMB         −4.46e+09   8.61e+08   −5.18 2.7e−07 ***
17   HML         −3.57e+09   1.01e+09   −3.52  0.00045 ***
18   RMW         −2.35e+09   1.30e+09   −1.81  0.07094 .
19   CMA         −1.70e+10   1.63e+09  −10.46  < 2e−16 ***
20   ——
21   Signif. codes:  0 '***' 0.001 '**' 0.01 '*' 0.05 '.' 0.1 ' ' 1
22
23   Residual standard error: 0.0133 on 1000 degrees of freedom
24   Multiple R-squared:  0.514, Adjusted R-squared:  0.512
25   F-statistic:  212 on 5 and 1000 DF,  p-value: <2e−16
26
27   >
28   > predicted <- predict(train.reg, test.data)
29   > pred.data <- data.frame(predicted)
30   > head.tail(pred.data)
31              predicted
32   2019−01−02  −0.00786
33   2019−01−03  −0.04780
34   2019−01−04   0.05050
35              predicted
36   2019−12−27 −5.70e−04
37   2019−12−30 −1.65e−02
38   2019−12−31  5.13e−05
```

**Step 5: Calculate Prediction Error**  As our measure of the prediction error, we will use the root mean square error (RMSE). To do this, we first combine the actual

AMZN return with the predicted AMZN return. Since our prediction is in excess
returns, we have to add back the risk-free rate. We then calculate the resid, which
is the difference between the actual and predicted AMZN returns. We then calculate
the RMSE by (1) squaring the residual, (2) calculating the mean of the squared
residuals, and (3) taking the square root of that average squared error. As we can
see, the RMSE is 0.970%.

```
 1   > predicted <- data.frame(test.data[, c(1, 7)], pred.data)
 2   > head.tail(predicted)
 3                  AMZN        RF predicted
 4   2019-01-02  0.0244 9.99e-15  -0.00786
 5   2019-01-03 -0.0256 9.99e-15  -0.04780
 6   2019-01-04  0.0489 9.99e-15   0.05050
 7                  AMZN        RF predicted
 8   2019-12-27  0.000551 7.11e-15 -5.70e-04
 9   2019-12-30 -0.012328 7.11e-15 -1.65e-02
10   2019-12-31  0.000514 7.11e-15  5.13e-05
11   >
12   > predicted$rhat <- predicted$predicted + predicted$RF
13   > head.tail(predicted)
14                  AMZN        RF predicted       rhat
15   2019-01-02  0.0244 9.99e-15  -0.00786 -0.00786
16   2019-01-03 -0.0256 9.99e-15  -0.04780 -0.04780
17   2019-01-04  0.0489 9.99e-15   0.05050  0.05050
18                  AMZN        RF predicted       rhat
19   2019-12-27  0.000551 7.11e-15 -5.70e-04 -5.70e-04
20   2019-12-30 -0.012328 7.11e-15 -1.65e-02 -1.65e-02
21   2019-12-31  0.000514 7.11e-15  5.13e-05  5.13e-05
22   >
23   > predicted$resid <- predicted$AMZN - predicted$rhat
24   > head.tail(predicted)
25                  AMZN        RF predicted       rhat    resid
26   2019-01-02  0.0244 9.99e-15  -0.00786 -0.00786  0.03230
27   2019-01-03 -0.0256 9.99e-15  -0.04780 -0.04780  0.02223
28   2019-01-04  0.0489 9.99e-15   0.05050  0.05050 -0.00165
29                  AMZN        RF predicted       rhat    resid
30   2019-12-27  0.000551 7.11e-15 -5.70e-04 -5.70e-04 0.001121
31   2019-12-30 -0.012328 7.11e-15 -1.65e-02 -1.65e-02 0.004188
32   2019-12-31  0.000514 7.11e-15  5.13e-05  5.13e-05 0.000463
33   > options(digits = 7)
34   >
35   > (rmse <- sqrt(mean(predicted$resid^2)))
36   [1] 0.009696332
```

### 12.4.4   Artificial Neural Networks

Artificial neural networks are often described as **black boxes**. The way the output is
generated may not be easily explainable given the inputs, models, training dataset,
and how the machine learns. With that caveat, we will attempt to give a high-level

discussion of how neural networks are implemented with the hopes that we can at least gain a little bit of insight into what is inside the black box.

We can think of a neural network acting like a human brain. The human brain has approximately 100 billion neurons. According to the University of Queensland's Brain Institute, neurons are "cells responsible for receiving sensory input from the external world, for sending motor commands to our muscles, and for transforming and relaying the electrical signals at every step in between" and "their interactions define who we are as people." In other words, neurons are an interconnected network of cells that converts some input signal to some output signal. Neural networks simulate this function of neurons.

In general, to "learn," we would require plenty of data and many parameters. However, for our illustration and to help gain intuition (as well as due to data permission limitations), we only use a relatively small amounts of data with only few parameters. This is sufficient to show how to implement the models. Once we have the input data, we would need to worry about the (1) layers and nodes and (2) activation. This is known as the **network topology** or **network architecture**. The **layers** and **nodes** are the building blocks of neural networks. The more layers and nodes we add, the more opportunities to learn new features. There is no preset rule as to how many layers we would need. A rule of thumb often used for tabular data is to have between two and five **hidden layers** and, when in doubt, we should opt for more layers. Then, the number of **nodes** is usually selected to be less than the number of layers. The one practical consideration is that the more layers and nodes we have, the more complex the model becomes and the more computing power may be needed. This is a bigger issue that does not concern our example but is very relevant when working with large amounts of data in practice. After passing through the layers and nodes, the neural network goes through an **output layer**. For regression problems, like the one we will show below, the output layer will contain one node. For other problems (i.e., classification problems), there could be one or more nodes depending on the problem. If the problem has a binary output (e.g., profit or loss, true or false), then we will likely only need one node. If, on the other hand, we are predicting the outcome of five different potential outcomes, then we would have five nodes and the output would be the probabilities of each potential outcome.

Neural networks are also characterized by an **activation function**, which are functions that determine whether or not there is enough informative input at a node to send a signal to the next layer. For example, for regression problems, we will use a linear activation function. All nodes in one layer are connected to the nodes in the previous layer and each connection gets a weight. Then, that node adds the incoming inputs multiplied by its weight and then adds an extra bias parameter. The total of these inputs become an input to the activation function.

As we show below, the `neuralnet` package allows us to do a simple layered network approach. On the first run, the neural network will select a group of observations, randomly assign weights across all the node connections, and predict the output. The neural network assesses its own accuracy and automatically adjusts the weights across all the node connections to improve that accuracy. This process is called **backpropagation**. To perform backpropagation, we require (1) an objective function and (2) an optimizer. For regression problems, an objective function to measure performance could be the root mean square error (RMSE). On each forward pass, the neural network will measure its performance based on RMSE. The neural network will then work backwards through the layers, compute the gradient of the loss (i.e., derivative of the loss curve) with regards to the network weights, adjust the weights by a small amount in the opposite direction of the gradient, and then grab another group of observations to run through the model. The process repeats until the objective function is minimized or below a certain pre-defined threshold. This process is called **gradient descent** and is the optimizer used.

**Step 1: Scale the Data** We will use the same variables in the above regression discussion to predict AMZN returns. Thus, we can use the dataset *rets*, but what we have to do is develop a methodology to scale the data. The reason for this is that common activation functions of the network's neurons are defined over some common interval, like −1 to +1 or from 0 to 1. For our purposes, we choose to normalize the data using the **min-max method** such that they are scaled between 0 and 1, inclusive. We call this function `normalize()`, which is the same function we used earlier. We then use `lapply()` to apply this function to every column in the *full.data* dataset.

```
1  > normalize <- function(x){
2  +    return ((x - min(x)) / (max(x) - min(x)))
3  + }
4  > maxmindf <- as.data.frame(
5  +    lapply(rets, normalize))
6  + )
7  > head.tail(maxmindf)
8                    AMZN      SPYV      SPYG       CWI      IGLB
9  2015-01-02 0.3535063 0.4745962 0.4206121 0.6375187 0.7666175
10 2015-01-05 0.2841406 0.2830034 0.2802130 0.5066617 0.6516810
11 2015-01-06 0.2730577 0.3978058 0.3635419 0.6185734 0.8238683
12                    AMZN      SPYV      SPYG       CWI      IGLB
13 2019-12-27 0.3837734 0.4951686 0.4488421 0.7231575 0.6352690
14 2019-12-30 0.3234771 0.4416833 0.3734090 0.6456770 0.6250283
15 2019-12-31 0.3836011 0.5286278 0.4670977 0.7408653 0.4814883
```

**Step 2: Create the Training and Test Datasets** The above is now a data frame object, so we have to create a date variable that R will read-in as a date. To do so, we first extract the dates from an xts object that has the date series we want. In our illustration, that xts object is the *rets* dataset. We then combine the *date* vector and the *maxmindf* dataset from Step 1. We can then use date to subset the data into the training dataset *train.set* and test dataset *test.set*.

```
1  > date <- as.Date(index(rets))
2  > head.tail(date)
3  [1] "2015-01-02" "2015-01-05" "2015-01-06"
4  [1] "2019-12-27" "2019-12-30" "2019-12-31"
5  >
6  > rownames(maxmindf) <- seq(1, nrow(maxmindf), 1)
```

```
7   > maxmindf <- data.frame(date, maxmindf)
8   > head.tail(maxmindf)
9          date      AMZN      SPYV      SPYG      CWI      IGLB
10  1 2015-01-02 0.3535063 0.4745962 0.4206121 0.6375187 0.7666175
11  2 2015-01-05 0.2841406 0.2830034 0.2802130 0.5066617 0.6516810
12  3 2015-01-06 0.2730577 0.3978058 0.3635419 0.6185734 0.8238683
13          date      AMZN      SPYV      SPYG      CWI      IGLB
14  1256 2019-12-27 0.3837734 0.4951686 0.4488421 0.7231575 0.6352690
15  1257 2019-12-30 0.3234771 0.4416833 0.3734090 0.6456770 0.6250283
16  1258 2019-12-31 0.3836011 0.5286278 0.4670977 0.7408653 0.4814883
17  >
18  > options(digits = 3)
19  > train.set <- subset(maxmindf, date <= "2018-12-31")
20  > head.tail(train.set)
21          date AMZN Mkt.RF   SMB   HML   RMW   CMA RF exret
22  1 2015-01-02 0.354  0.442 0.261 0.410 0.427 0.437  0 0.354
23  2 2015-01-05 0.284  0.249 0.460 0.266 0.553 0.370  0 0.284
24  3 2015-01-06 0.273  0.339 0.207 0.340 0.666 0.419  0 0.273
25          date AMZN Mkt.RF   SMB   HML   RMW   CMA RF exret
26  1004 2018-12-27 0.352  0.540 0.224 0.374 0.467 0.437  1 0.351
27  1005 2018-12-28 0.433  0.451 0.574 0.440 0.397 0.385  1 0.433
28  1006 2018-12-31 0.456  0.554 0.366 0.298 0.480 0.370  1 0.456
29  >
30  > test.set <- subset(maxmindf, date >= "2019-01-01")
31  > head.tail(test.set)
32          date AMZN Mkt.RF   SMB   HML   RMW   CMA RF exret
33  1007 2019-01-02 0.496   0.48 0.574 0.621 0.464 0.486  1 0.495
34  1008 2019-01-03 0.262   0.18 0.523 0.637 0.427 0.683  1 0.261
35  1009 2019-01-04 0.610   0.84 0.492 0.244 0.476 0.228  1 0.610
36          date AMZN Mkt.RF   SMB   HML   RMW   CMA RF exret
37  1256 2019-12-27 0.384  0.445 0.265 0.378 0.578 0.455 0.7 0.384
38  1257 2019-12-30 0.323  0.391 0.463 0.508 0.550 0.544 0.7 0.323
39  1258 2019-12-31 0.384  0.485 0.402 0.420 0.470 0.474 0.7 0.384
```

**Step 3: Fama-French Model Based Prediction**  We start with performing predictions using the Fama-French 5 Factor Model. We load the `neuralnet` package. This will allow us to use `neuralnet()`. The argument `hidden` is a vector of integers that specifies the number of neurons in each hidden layer. As we can see in Line 5 below, we are using 5 layers and 3 neurons. Graphically, we see this in Fig. 12.7. Then, we set the `threshold` argument to 0.01, which specifies the threshold for the stopping criteria. The value of the `result.matrix` below tells us various information regarding the threshold. For example, how many steps were needed to reach the threshold.

```
1   > library(neuralnet)
2   > set.seed(1)
3   > nn.ff <- neuralnet(exret ~ Mkt.RF + SMB + HML + RMW + CMA,
4   +     data = train.set,
5   +     hidden = c(5, 3),
6   +     linear.output = TRUE,
7   +     threshold = 0.01)
8   > nn.ff$result.matrix
```

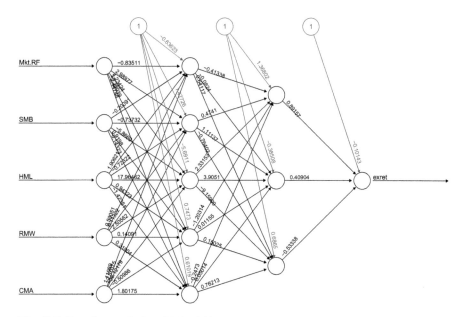

**Fig. 12.7** Neural network plot with five hidden layers and three nodes

| 9 | | [,1] |
|---|---|---|
| 10 | error | 1.79e+00 |
| 11 | reached.threshold | 8.63e−03 |
| 12 | steps | 1.06e+03 |
| 13 | Intercept.to.1layhid1 | −6.36e−01 |
| 14 | Mkt.RF.to.1layhid1 | −8.35e−01 |
| 15 | SMB.to.1layhid1 | −2.01e−01 |
| 16 | HML.to.1layhid1 | 1.91e+00 |
| 17 | RMW.to.1layhid1 | 5.91e−01 |
| 18 | CMA.to.1layhid1 | 1.11e+00 |
| 19 | Intercept.to.1layhid2 | 1.43e+00 |
| 20 | Mkt.RF.to.1layhid2 | 2.99e+00 |
| 21 | SMB.to.1layhid2 | −7.37e−01 |
| 22 | HML.to.1layhid2 | −7.23e−01 |
| 23 | RMW.to.1layhid2 | 1.13e−01 |
| 24 | CMA.to.1layhid2 | −2.57e+00 |
| 25 | Intercept.to.1layhid3 | −5.69e+00 |
| 26 | Mkt.RF.to.1layhid3 | 4.23e+00 |
| 27 | SMB.to.1layhid3 | −5.87e+00 |
| 28 | HML.to.1layhid3 | 1.80e+01 |
| 29 | RMW.to.1layhid3 | 2.51e+00 |
| 30 | CMA.to.1layhid3 | −3.61e+01 |
| 31 | Intercept.to.1layhid4 | 7.47e−01 |
| 32 | Mkt.RF.to.1layhid4 | −3.38e+00 |
| 33 | SMB.to.1layhid4 | 9.20e−01 |
| 34 | HML.to.1layhid4 | 9.43e−01 |
| 35 | RMW.to.1layhid4 | 1.41e−01 |
| 36 | CMA.to.1layhid4 | −5.10e−01 |

```
37   Intercept.to.1layhid5   6.10e−01
38   Mkt.RF.to.1layhid5     −2.45e+00
39   SMB.to.1layhid5        −1.43e−01
40   HML.to.1layhid5        −1.42e+00
41   RMW.to.1layhid5         3.19e−01
42   CMA.to.1layhid5         1.80e+00
43   Intercept.to.2layhid1   1.37e+00
44   1layhid1.to.2layhid1   −4.13e−01
45   1layhid2.to.2layhid1    4.74e−01
46   1layhid3.to.2layhid1    2.33e+00
47   1layhid4.to.2layhid1   −1.20e+00
48   1layhid5.to.2layhid1   −2.47e−01
49   Intercept.to.2layhid2  −3.85e−01
50   1layhid1.to.2layhid2   −8.24e−02
51   1layhid2.to.2layhid2    1.11e+00
52   1layhid3.to.2layhid2    3.91e+00
53   1layhid4.to.2layhid2    1.16e−02
54   1layhid5.to.2layhid2   −5.01e−02
55   Intercept.to.2layhid3   6.87e−01
56   1layhid1.to.2layhid3    6.41e−01
57   1layhid2.to.2layhid3   −7.64e−01
58   1layhid3.to.2layhid3   −9.16e+00
59   1layhid4.to.2layhid3    1.50e−01
60   1layhid5.to.2layhid3    7.62e−01
61   Intercept.to.exret     −1.01e−01
62   2layhid1.to.exret       8.92e−01
63   2layhid2.to.exret       4.09e−01
64   2layhid3.to.exret      −5.33e−01
65   > plot(nn.ff)
```

**Step 4: Apply Results to Test Dataset**  We apply the results to the test dataset using compute() with *nn.ff* and *test.set* as the two arguments. The neural network predictions are in net.result.

```
1    > nn.result <- compute(nn.ff, test.set)
2    > str(nn.result)
3    List of 2
4     $ neurons   :List of 3
5      ..$ : num [1:252, 1:6] 1 1 1 1 1 1 1 1 1 1 ...
6      .. ..− attr(*, "dimnames")=List of 2
7      .. .. ..$ : chr [1:252] "1007" "1008" "1009" "1010" ...
8      .. .. ..$ : chr [1:6] "" "Mkt.RF" "SMB" "HML" ...
9      ..$ : num [1:252, 1:6] 1 1 1 1 1 1 1 1 1 1 ...
10     .. ..− attr(*, "dimnames")=List of 2
11     .. .. ..$ : chr [1:252] "1007" "1008" "1009" "1010" ...
12     .. .. ..$ : NULL
13     ..$ : num [1:252, 1:4] 1 1 1 1 1 1 1 1 1 1 ...
14     .. ..− attr(*, "dimnames")=List of 2
15     .. .. ..$ : chr [1:252] "1007" "1008" "1009" "1010" ...
16     .. .. ..$ : NULL
17    $ net.result: num [1:252, 1] 0.342 0.146 0.557 0.464 0.441 ...
18     ..− attr(*, "dimnames")=List of 2
19     .. ..$ : chr [1:252] "1007" "1008" "1009" "1010" ...
```

```
20     .. ..$ : NULL
21   > head.tail(nn.result$net.result)
22        [,1]
23   1007 0.342
24   1008 0.146
25   1009 0.557
26        [,1]
27   1256 0.383
28   1257 0.305
29   1258 0.386
```

**Step 8: Reverse Scaling in Predicted and Actual Data**   To calculate the prediction error, we have to first remove the effects of the scaling we did earlier. We undo the effect of the **min-max method** in Step 4 by multiplying the *net.result* by the difference of the max and min of the *full.data* dataset and then adding the minimum of the *full.data* dataset. We do this for both the predicted and test dataset.

```
1   > ff.predicted <- nn.result$net.result * (max(rets$AMZN) -
2   +    min(rets$AMZN)) + min(rets$AMZN)
3   > head.tail(ff.predicted)
4            [,1]
5   1007 -0.00838
6   1008 -0.05022
7   1009  0.03761
8            [,1]
9   1256  0.000331
10  1257 -0.016268
11  1258  0.001024
12  >
13  > ff.predicted <- as.numeric(ff.predicted)
14  > head.tail(ff.predicted)
15  [1] -0.00838 -0.05022  0.03761
16  [1]  0.000331 -0.016268  0.001024
17  >
18  > ff.test <- test.set$AMZN * (max(rets$AMZN) -
19  +    min(rets$AMZN)) + min(rets$AMZN)
20  > head.tail(ff.test)
21  [1]  0.0244 -0.0256  0.0489
22  [1]  0.000551 -0.012328  0.000514
```

We can verify that we scaled the valued back correctly when we compare the AMZN values in *rets* with the values in *capm.results*. For example, the actual AMZN returns for December 27, 30, and 31 are 0.055%, −1.233%, and 0.051%, respectively. These are the exact values in the last line of output above.

**Step 9: Combine Actual and Predicted Values**   We combine the two vectors *ff.test* and *ff.predicted* into one dataset we label *ff.results*.

```
1   > ff.results <- data.frame(ff.test, ff.predicted)
2   > names(ff.results) <- c("actual", "predicted")
3   > head.tail(ff.results)
4     actual predicted
5   1  0.0244  -0.00838
```

```
 6    2 −0.0256  −0.05022
 7    3  0.0489   0.03761
 8            actual predicted
 9   250  0.000551  0.000331
10   251 −0.012328 −0.016268
11   252  0.000514  0.001024
```

**Step 10: Calculate Prediction Error** We will use the root mean square error (RMSE) as our metric to determine the accuracy of the prediction. We first calculate the squared deviation of the predicted from the actual. We then take the average of the squared deviation using mean() and then take the square root using sqrt(). As the output below shows, the FF-based prediction had an RMSE of 0.979%. Note that every run of the neural network will result in a different result. Thus, the output above may not be exactly what you will get, but it should be reasonably close.

```
 1  > sq_deviation <- (ff.results$actual − ff.results$predicted)^2
 2  > comparison <- cbind(ff.results, sq_deviation)
 3  > head.tail(comparison)
 4      actual predicted sq_deviation
 5  1   0.0244 −0.00838     0.001077
 6  2 −0.0256 −0.05022     0.000608
 7  3  0.0489   0.03761     0.000126
 8            actual predicted sq_deviation
 9  250  0.000551  0.000331     4.82e−08
10  251 −0.012328 −0.016268     1.55e−05
11  252  0.000514  0.001024     2.60e−07
12  >
13  > (rmse <- sqrt(mean(sq_deviation)))
14  [1] 0.00979
15  > options(digits = 7)
```

The RMSE from the above is 0.979%. Recall from the results of our prediction using regressions the RMSE was 0.970%, which is similar and, in fact, slightly lower (i.e., better) than the RMSE from the artificial neural network prediction. Thus, there does not appear to be any benefit of using neural networks versus a linear model in our example. Although using regressions do not guarantee outperformance all the time, what this highlights is that using a more complex model does not always yield better results. A simpler model may do just as well, if not better, in certain circumstances. Thus, we should not be enamored by the complexity of certain models but the focus has to be on the effectiveness of the models for the given task.

## 12.5 Further Reading

This chapter provides a high-level introduction to trading strategies, including machine learning. We began with a discussion of market efficiency and presented some empirical tests that have been used to test market efficiency. More tests and a more thorough explanation of these tests are found in [1, Campbell, Lo, and

MacKinlay (1997)]. Next, we presented some technical indicators. Many other technical indicators can be generated using the `chartSeries`. For more details, visit the web address http://www.quantmod.com/examples/charting/. To learn more about technical analysis in general, a good book is [9, Kirkpatrick and Dahlquist (2015)]. There are several excellent texts that are available that delve into the details of these different topics. For trading strategies, a book that follows a quantitative approach is [8, Kestner (2003)] and [6, Gray and Carlisle (2013)]. There are also books that discuss trading systems at a more general level, such as [2, Chan (2009)] and [3, Chan (2013)]. There are also several books that take us into a lot of detail of building trading systems. Examples of these are [5, Fitschen (2013)] and [7, Kaufman (2013)]. With respect to machine learning, a recent text that goes through financial ML topics in more depth is [11, Lopez de Prado (2018)]. We can complement this text with more general ML books that are written for R, such as [4, Ciaburro and Venkateswaran (2017)], [10, Lantz (2013)], and [12, Wei (2015)].

# References

1. Campbell, J., Lo, A., & MacKinlay, A. C. (1997). *The Econometrics of financial markets*. Princeton: Princeton University.
2. Chan, E. (2009). *Quantitative trading: How to build your own algorithmic trading business*. New York: Wiley.
3. Chan, E. (2013). *Algorithmic trading: winning strategies and other rationale*. New York: Wiley.
4. Ciaburro, G., & Venkateswaran, B. (2017). *Neural networks with R*. London: Packt
5. Fitschen, K. (2013). *Building reliable trading systems*. New York: Wiley.
6. Gray, W., & Carlisle, T. (2013). *Quantitative value*. New York: Wiley.
7. Kaufman, P. (2013). *Trading systems and methods* (5th ed.). New York: Wiley.
8. Kestner, L. (2003). *Quantitative trading strategies*. New York: McGraw-Hill.
9. Kirkpatrick, C., & Dahlquist, J. (2015). *Technical analysis: The complete resource for financial market technicians* (3rd ed.). New York: Wiley.
10. Lantz, B. (2013). *Machine learning with R*. London: Packt
11. Lopez de Prado, M. (2018). *Advances in financial machine learning*. New York: Wiley.
12. Wei, Y.-W. (2015). *Machine learning with R cookbook*. London: Packt.

# Appendix A
# Getting Started with R

## A.1 Installing R

To get started with R, we first have to download and install R on our local system. We can go to the Comprehensive R Archive Network (CRAN) website (http://cran. r-project.org) and download the latest version. There are versions of R for Windows, OS X, and Linux. As of the writing of this book, the latest version of R available is version 3.6.2, named "Dark and Stormy Night" and dated December 12, 2019. The code in this book has been tested to run on the 64-bit Windows version of R 3.6.2, but most, if not all, of the programs should work with other versions of R as well.

## A.2 The R Working Directory

In this book, the programs and any external CSV files we will use should be placed in the R working directory. To determine which directory on your hard drive is used by R as the **working directory**, type in getwd() in the command prompt. The command prompt is the line that begins with ">" in the R console.

```
1  > getwd()
2  [1] "C:/Users/cliff"
```

Note that the line that does not begin with the > sign is the output of what you typed. In the above, the output tells us that the working directory in my system is /Users/cliff. We can then create and save all relevant programs and files into this R working directory or any other wording directory that we choose.

Although we can change the working directory, this is not essential for this book and we can simply use whatever default working directory R outputs. To select a different working directory, we use setwd(). Suppose we want to use the directory c:/Users/cliff/R instead, we can enter the following in the command prompt:

```
1   > setwd(''C:/Users/cliff/R'')
```

## A.3   R Console Output

Every line of output in this book that begins with the command prompt (>) or a plus sign (+) is a line of code. The plus sign (+) denotes that the particular line of code and the preceding line(s) of code are being treated by R as one continuous string of code that for whatever reason was separated into multiple lines. For example, when we subset data, we can type the following:

```
1   > subset(data, index(data) > "2018–12–31")
```

or

```
1   > subset(data,
2   +   index(data) > "2018–12–31")
```

The second type of code makes it easier to read, but it uses several lines instead of the first type of code that puts all the arguments in one line. Both lines will generate the same output, which is to subset *data* to only contain values after December 31, 2018. For longer lines of code, experienced programmers may prefer to use the second form of the code because it makes it easier to read which lines of code go together and what are the different actions the code is using. In this book, we will use both forms.

## A.4   R Editor

When writing code in R, we can directly type in the code in the R console, and when we hit ENTER, the code we just typed will execute. However, it is generally better for us to program using the R Editor. The R Editor looks like an ordinary text editor (e.g., Notepad in Windows or TextEdit in OS X). We can call up the R Editor by clicking on the File > New Script or CTRL+N in R for Windows. In OS X, a new script can be created by clicking on File > New Document or Command+N.

We would then type in the R Editor the text in the output that follows the command prompt (>) or plus sign (+), excluding these symbols. In other words, for the subsetting example above, we would type the following in the R Editor:

```
1   subset(data,
2      index(data) > "2018–12–31")
```

We then highlight the portion of the code we want to execute and type CTRL+R in Windows or Command+ENTER in OS X. Doing so in the code above will execute the code and generate the output in which we break the code into two lines with a plus (+) sign preceding the second line. Note also that the second line is indented. Indenting or not indenting code does not generally affect how the program is executed. Indenting only makes it easier to identify which lines belong together (and is likely good practice for programming). In our example above, the indented second line can easily be identified as being one set of code together with the first line.

## A.5 Packages

Packages are like toolboxes that contain a lot of functions that help us simplify our tasks. The base installation of R comes with a set of built-in packages to help us do many things. However, for some of the calculations in this book, we will need to install additional packages. We will use primarily quantmod, which also loads another package that we will often use and that is xts package.

To load a package, we use library(). For example, to load the quantmod package, we enter the following:

```
1  > library(quantmod)
2  Loading required package: Defaults
3  Loading required package: xts
4  Loading required package: zoo
5
6  Attaching package: 'zoo'
7
8  The following objects are masked from 'package:base':
9
10     as.Date, as.Date.numeric
11
12  Loading required package: TTR
13  Version 0.4-0 included new data defaults. See ?getSymbols.
```

During each R session, the first time we load a package, we may see a series of messages like the above for quantmod. We only need to load packages once during each R session, so to make sure codes get executed smoothly, we should always load quantmod, which would include the xts package as well, every time we start R. We discuss some other code that we should pre-load in Appendix B. These will help us simplify our discussion and leverage what we have previously discussed.

Reloading a package that has already been previously loaded does not do any harm. However, no warning messages appear for the subsequent instances we load the package, so do not be alarmed when you do not see the flurry of warning messages like the one for quantmod above. For example, if we load quantmod again, the output will be as shown below (i.e., without any warning messages):

```
1  > library(quantmod)
```

If the package has not been installed on our local system, R will generate an error message. For example, if we try to load the package tseries and it has not been installed on our local machine, we would observe the following output:

```
1  > library(tseries)
2  Error in library(tseries) : there is no package called 'tseries'
```

We can then install the package using `install.packages()`. Note that we should put the package name inside the parentheses and *in between quotes* or else R will generate an error stating it could not find the package.

```
1  > install.packages(tseries)
2  Error in install.packages(tseries) : object 'tseries' not found
```

Aside from putting the package name in quotes, we can also include `dependencies = TRUE`, which also installs additional packages that the current package we are installing requires.

```
1   > install.packages("tseries", dependencies = TRUE)
2   Installing package into 'C:/Users/cang/Documents/R/win-library/3.6'
3   (as 'lib' is unspecified)
4   —— Please select a CRAN mirror for use in this session ——
5   trying URL 'https://cran.cnr.berkeley.edu/bin/windows/contrib/3.6/ ...
6   Content type 'application/zip' length 417283 bytes (407 KB)
7   downloaded 407 KB
8
9   package 'tseries' successfully unpacked and MD5 sums checked
10
11  The downloaded binary packages are in
12          C:\Users\cang\AppData\Local\Temp\RtmpqKcNnm\downloaded_packages
```

Unless we have previously selected a CRAN mirror site from where we download the package, R will open a window that would require us to select a CRAN mirror site. We can generally choose any site and the package will still install. Each time we open an R session and have to use a CRAN mirror, this window will always pop up. As such, we are not tied to the choice we make as to which CRAN mirror to use. From a practical perspective, the choice of CRAN mirror may not be a big deal.

In addition, we can use `help()` to pull up the help file for that package. A new window will open up in which a web page-styled help document specific to that package will load.

```
1  > help(package="tseries")
2  starting httpd help server ... done
```

You can also check whether a package is already installed using `(.packages())`. When we start R, there are already some packages that are installed. In my system, the following output is displayed:

```
1  > (.packages())
2  [1] "stats"     "graphics" "grDevices" "utils"     "datasets"  "methods"
3  "base"
```

After I load `quantmod`, the following output is displayed:

```
1  > (.packages())
2   [1] "quantmod" "TTR"      "xts"      "zoo"      "stats"
3  "graphics" "grDevices" "utils"  ·
4   [9] "datasets" "methods"  "base"
```

As we can see, loading `quantmod` also loads the `TTR` and `xts` packages.

## A.6 Basic Commands

In this book, we will use the `teletype` font to denote R-related commands. Hopefully, this will aid in separating R-related inputs from normal text. We *italicize* datasets, matrices, vectors, and variables that we created. We use serif for column headers/labels and filenames.

R is interactive, so we can directly type in expressions in the command prompt and see the results just like we would in a calculator. For example, suppose we want to add 2 and 3 together, we can use R like a calculator.

```
1  > 2 + 3
2  [1] 5
```

Alternatively, we can also multiply 2 and 3.

```
1  > 2 * 3
2  [1] 6
```

Using R as a calculator may not be the best use of R. We can store things in **objects** by using the assignment operator "←," which is comprised of a less than sign ($<$) and a minus sign (-). Suppose we let $x = 2$ and $y = 3$ and add and multiply the two objects. Note that to see what the object holds, we have to type in the object name in the command prompt. That is, the first line merely generates an empty command prompt, and typing $x$ again shows us the output of 2.

```
1   > x <- 2
2   > x
3   [1] 2
4   > y <- 3
5   > y
6   [1] 3
7   > x + y
8   [1] 5
9   > x * y
10  [1] 6
```

Alternatively, we can show what is stored in the object by wrapping the code in parentheses. For example, to show x  <-  2 has 2 stored in $x$ in one step, we enter

```
1  > (x <- 2)
2  [1] 2
```

Although we will primarily use the assignment operator above, we sometimes will also assign values using the equal sign ($=$). The choice is a matter of preference and switching between the two methods does not affect any results. A third alternative way to assign values is to use the assignment operator with the arrow pointing right ($\rightarrow$). However, this may be too confusing to use, so we will systematically make assignments only having the arrow pointing to the left or using the equal sign.

```
1  > 2 -> x
2  > x
3  [1] 2
```

Note that R is case-sensitive. For example, the object *x* above is in lowercase. An error would be generated if we typed in an uppercase *X* instead.

```
1  > X
2  Error: object 'X' not found
```

The above is clearly not an exhaustive list of issues, but merely the most basic of commands. We will go through the necessary operations, functions, and commands in the text of the book, which is used for specific applications. This will hopefully aid in recalling how to use these terms in the future.

## A.7   The R Workspace

The R workspace contains a list of the objects that are in the R memory. To view the R workspace, we use `ls()`. Since we created the objects *x* and *y* previously, those should be the variables in the R workspace and this is confirmed below:

```
1  > ls()
2  [1] "x" "y"
```

Within each chapter, we may use certain objects created in earlier sections of the chapter. For example, we may retrieve stock price data in the first section of the chapter and use the same stock price data in the fifth section. As such, having objects in the R workspace is helpful as it helps avoid repulling data. This may not be a meaningful issue at this point but could be an issue as you construct more complex models.

However, when we move to a different chapter or perform a completely independent analysis, we may want to clear up the memory of any old objects. We may reuse variable and dataset names to make things more intuitive and easy to follow. For example, we may use *data.amzn* in all the chapters to refer to the Amazon stock price data. However, we may manipulate that dataset in a different way. Therefore, to avoid inadvertently calling an object that has been manipulated and consequently result in incorrect calculations, we may want to consider clearing the workspace (i.e., wiping out anything in memory). We do this using `rm(list=ls())`. As the below shows, after running this code, the datasets *x* and *y* that are used in the R workspace no longer exist.

```
1  > rm(list=ls())
2  > ls()
3  character(0)
```

An important thing to note with using `rm(list=ls())` is that we have to be careful when we use this code as it will delete any object in the workspace. Make sure you save any temporary data or function prior to using this code. In addition, if we were to share our code with others, we should make a conscious choice whether we should include `rm(list=ls())` or not. The person running the code may inadvertently erase important objects from their workspace if we send the code with the `rm(list=ls())` command and they simply run it without removing or commenting out `rm(list=ls())`.

## A.8 Vectors

We can create **vectors** in R, which can contain numbers or text. The vector *a* contains a vector of numbers, vector *b* contains a vector of text, and vector *c* contains a vector of numbers and text.

```
1  > a <- c(1, 2, 3, 5, 6, 7)
2  > a
3  [1] 1 2 3 5 6 7
4  > (b <- c("Sun", "Mon", "Tue", "Thu", "Fri" ,"Sat"))
5  [1] "Sun" "Mon" "Tue" "Thu" "Fri" "Sat"
6  > (c <- c(1, 2, 3, "Sun", "Mon", "Tue"))
7  [1] "1"   "2"   "3"   "Sun" "Mon" "Tue"
```

Note that the fully numeric vector *a* lists out the numbers without quotes, whereas any vector that partially has text or characters lists out the values with quotes. In the above example, we created a vector using `c()`. Note that many things that need to be strung together in R make use of `c()`.

We can select elements in a vector by using square brackets and identifying the element number. Suppose we want to choose the fourth element in vector *a* and vector *b*. We thus put "4" inside square brackets after calling the vector name.

```
1  > a[4]
2  [1] 5
3  > b[4]
4  [1] "Thu"
```

## A.9 Combining Vectors

We can combine vectors by rows or columns. To combine vectors as columns, we use `cbind()`.

```
1  > (col.vectors <- cbind(a, b))
2         a   b
3  [1,] "1" "Sun"
4  [2,] "2" "Mon"
5  [3,] "3" "Tue"
6  [4,] "5" "Thu"
7  [5,] "6" "Fri"
8  [6,] "7" "Sat"
```

To combine vectors as rows, we use `rbind()`.

```
1  > (row.vectors <- rbind(a ,b))
```

```
2     [,1]   [,2]   [,3]   [,4]   [,5]   [,6]
3   a "1"    "2"    "3"    "5"    "6"    "7"
4   b "Sun"  "Mon"  "Tue"  "Thu"  "Fri"  "Sat"
```

The *col.vectors* object looks like a 6 (row) × 2 (column) matrix. We can add column vector *c* to the left side of this object by applying another cbind() and putting the *c* vector as the first argument, which now creates a 6 (row) × 3 (column) object.

```
1  >( leftcol.vectors <- cbind(c, col.vectors))
2        c     a    b
3  [1,] "1"   "1"  "Sun"
4  [2,] "2"   "2"  "Mon"
5  [3,] "3"   "3"  "Tue"
6  [4,] "Sun" "5"  "Thu"
7  [5,] "Mon" "6"  "Fri"
8  [6,] "Tue" "7"  "Sat"
```

Alternatively, we could have added the column vector *c* to the right side of the object by putting the *c* vector as the second object.

```
1  > (rightcol.vectors <- cbind(col.vectors, c))
2        a    b    c
3  [1,] "1"  "Sun" "1"
4  [2,] "2"  "Mon" "2"
5  [3,] "3"  "Tue" "3"
6  [4,] "5"  "Thu" "Sun"
7  [5,] "6"  "Fri" "Mon"
8  [6,] "7"  "Sat" "Tue"
```

The above examples demonstrate that the order in which the columns appear depends on the order that they appear inside cbind(). This adds flexibility in terms of how we want to show the data in the object. In the main text, we will go through in the context of applications other ways of combining data.

## A.10  Matrices

When we combined the vectors above, we converted its data structure into a matrix. We show the class of the object using class().

```
1  > class(rightcol.vectors)
2  [1] "matrix"
3  > dim(rightcol.vectors)
4  [1] 6 3
```

A **matrix** is a vector with dimensions. In our example, *righcol.vectors* has dimensions of 6 rows × 3 columns and is often described as a 6 × 3 matrix. We can show the dimensions of a matrix using dim(). We will deal with matrices when we deal with portfolio calculations as this greatly simplifies the implementation.

## A.11   data.frame

The **data.frame** object is a flexible data structure that we will use when performing more involved calculations. A data.frame object has rows and columns.
To convert the *rightcol.vectors* matrix above into a data.frame object, we use
data.frame().

```
 1   > df <- data.frame(rightcol.vectors)
 2   > class(df)
 3   [1] "data.frame"
 4   > df
 5     a  b   c
 6   1 1 Sun   1
 7   2 2 Mon   2
 8   3 3 Tue   3
 9   4 5 Thu Sun
10   5 6 Fri Mon
11   6 7 Sat Tue
```

Now that the object is a data.frame, we can pick out certain columns by using the
dollar sign ($) followed by the column label. For example, if we want to pick out
the second column of $df$, we type df$b.

```
 1   > df$b
 2   [1] Sun Mon Tue Thu Fri Sat
 3   Levels: Fri Mon Sat Sun Thu Tue
```

We can also accomplish the same thing by placing square brackets after $df$ and
typing 2 after the comma, as in df[, 2].

```
 1   > df[, 2]
 2   [1] Sun Mon Tue Thu Fri Sat
 3   Levels: Fri Mon Sat Sun Thu Tue
```

   In a data.frame, we can also identify rows by their row number. For example,
if we want to view only the fourth row, we use the square brackets again, but this
time, putting a number 4 before the comma, such as df[4, ]. Note that the 4 at
the beginning of Line 1 is the row number, which is the row that we want.

```
 1   > df[4, ]
 2     a  b   c
 3   4 5 Thu Sun
```

If we want to know the value that is in a specific row/column, we can do that by
entering values on both sides of the comma. Suppose we want to know the fourth
value in the second column of $df$, we would type in the command prompt df[4,
2].

```
 1   > df[4, 2]
 2   [1] Thu
 3   Levels: Fri Mon Sat Sun Thu Tue
```

These techniques will come in handy when we have to subset data and extract
elements from a data frame object throughout the text.

## A.12   Date Formats

R has specific symbols it uses for its **date formats**. The table below describes these symbols and provides examples (Table A.1).

**Table A.1** Symbols and description of R Date Formats for December 31, 2019

| Symbol | Description | Examples |
| --- | --- | --- |
| %Y | 4-digit year | 2019 |
| %y | 2-digit year | 19 |
| %m | 2-digit month | 12 |
| %b | Short month name | Dec |
| %B | Long month name | December |
| %d | day of the week (0 to 31) | 31 |
| %a | Short weekday | Tue |
| %A | Long weekday | Tuesday |

# Appendix B
# Pre-Loaded Code

In this book, we will use three functions repeatedly. Thus, it may be beneficial to include the code for these functions at the beginning of your scripts for the different chapters.

The first of these functions is the load.data() function. This function imports the CSV file for prices of stocks and ETFs we obtained from Yahoo Finance. We discuss in detail the steps of this function in 1. Although we do not need all of the quantmod package for this specific function, quantmod also loads the xts package that we need. We will also likely need quantmod for many applications, so it is more efficient to load the package from the get-go.

```
1   > library(quantmod)
2   > load.data <- function(rawdata, ticker){
3   +   data.raw <- read.csv(rawdata, header = TRUE)
4   +   Date <- as.Date(data.raw$Date, format = "%Y-%m-%d")
5   +   data.raw <- cbind(Date, data.raw[, -1])
6   +   data.raw <- data.raw[order(data.raw$Date),]
7   +   data.raw <- xts(data.raw[, 2:7],order.by = data.raw[, 1])
8   +   A <- paste(ticker, ".Open", sep = "")
9   +   B <- paste(ticker, ".High", sep = "")
10  +   C <- paste(ticker, ".Low", sep = "")
11  +   D <- paste(ticker, ".Close", sep = "")
12  +   E <- paste(ticker, ".Adjusted", sep="")
13  +   F <- paste(ticker, ".Volume", sep="")
14  +   names(data.raw) <- paste(c(A, B, C, D, E, F))
15  +   data.raw <- cbind(data.raw[, 1:4], data.raw[, 6], data.raw[, 5])
16  +   return(data.raw)
17  + }
18  }
```

The second function we will often use is the head.tail() function. This function outputs the first three and last three observations of an object. We use head() to output the first six observations of an object and tail() to output the last six observations of an object. For purposes of this book, we show intermediate

© The Author(s), under exclusive license to Springer Nature Switzerland AG 2021
C.S. Ang, *Analyzing Financial Data and Implementing Financial Models Using R*,
Springer Texts in Business and Economics,
https://doi.org/10.1007/978-3-030-64155-9

steps, and to save on space without any loss of effectiveness, we can show the first
three and last three observations of each object. To do so, we use a combination
of head() and tail() but add a "3" to the second argument of each. We add
print() at the beginning of Lines 2 and 3. Otherwise, running the function will
not output any observations.

```
1   > head.tail <- function(dataset){
2   +    print(head(dataset, 3))
3   +    print(tail(dataset, 3))
4   + }
```

The third function is load.fred(). This function imports data from the
Federal Reserve Electronic Database (FRED). This function is extensively used in
Chap. 9.

```
1    > load.fred <- function(rawdata, symbol){
2    +    temp <- read.csv(rawdata, header = TRUE)
3    +    temp <- subset(temp, temp[, 2] != ".")
4    +    library(lubridate)
5    +    date <- ymd(temp$DATE)
6    +    value <- as.numeric(as.character(temp[, 2]))
7    +    library(xts)
8    +    temp2 <- xts(value, order.by = date)
9    +    names(temp2) <- symbol
10   +    return(temp2)
11   + }
```

# Appendix C
# Constructing a Hypothetical Portfolio (Monthly Returns)

In this appendix, we construct a hypothetical portfolio assuming portfolio returns are equal to that of an equal-weighted portfolio of Amazon.com (AMZN), Alphabet (GOOG), and Apple (AAPL) from January 2015 to December 2019 with monthly rebalancing. To implement this, we need to make sure that we have data from December 2014 in order to calculate a monthly return for the month of January 2015. Note that this is merely a hypothetical portfolio with constructed returns. This is meant as a proxy for our actual portfolio returns used in this book.

**Step 1: Calculate Monthly Returns** To construct the portfolio returns, we first calculate the monthly returns for AMZN, GOOG, and AAPL using the techniques we discussed in Chap. 2.

```
1   > data.amzn <- load.data("AMZN Yahoo.csv", "AMZN")
2   > data.amzn <- to.monthly(data.amzn)
3   > ret.amzn <- Delt(data.amzn[, 6])
4   >
5   > data.goog <- load.data("GOOG Yahoo.csv", "GOOG")
6   > data.goog <- to.monthly(data.goog)
7   > ret.goog <- Delt(data.goog[, 6])
8   >
9   > data.aapl <- load.data("AAPL Yahoo.csv", "AAPL")
10  > data.aapl <- to.monthly(data.aapl)
11  > ret.aapl <- Delt(data.aapl[, 6])
```

**Step 2: Calculate Portfolio Returns** We use cbind() to combine the return vectors for AMZN, GOOG, and AAPL. Then, we remove the first observation for December 2014 because that observation only has NAs and we only needed the prices for December 2014 in order to calculate the January 2015 returns. We then rename the columns to the tickers of the securities we are using.

```
1   > port <- cbind(ret.amzn, ret.goog, ret.aapl)
2   > head.tail(port)
3            Delt.1.arithmetic Delt.1.arithmetic.1 Delt.1.arithmetic.2
```

© The Author(s), under exclusive license to Springer Nature Switzerland AG 2021                455
C.S. Ang, *Analyzing Financial Data and Implementing Financial Models Using R*,
Springer Texts in Business and Economics,
https://doi.org/10.1007/978-3-030-64155-9

```
 4  Dec 2014                    NA                    NA                    NA
 5  Jan 2015         0.14235538            0.01542555            0.06142418
 6  Feb 2015         0.07229291            0.04467552            0.10077668
 7            Delt.1.arithmetic Delt.1.arithmetic.1 Delt.1.arithmetic.2
 8  Oct 2019        0.023474719            0.03372435            0.11068444
 9  Nov 2019        0.013587301            0.03559211            0.07755414
10  Dec 2019       -0.006436077            0.02957953            0.06368565
11  > port <- port[-1, ]
12  > names(port) <- c("AMZN", "GOOG", "AAPL")
13  > head.tail(port)
14                  AMZN         GOOG         AAPL
15  Jan 2015  0.14235538  0.01542555  0.06142418
16  Feb 2015  0.07229291  0.04467552  0.10077668
17  Mar 2015 -0.02120159 -0.01862464 -0.03137180
18                  AMZN         GOOG         AAPL
19  Oct 2019  0.023474719 0.03372435 0.11068444
20  Nov 2019  0.013587301 0.03559211 0.07755414
21  Dec 2019 -0.006436077 0.02957953 0.06368565
```

**Step 3: Construct Portfolio Returns**  We now create the hypothetical portfolio's returns using `rowMeans()`.

```
 1  > hypo.port <- data.frame(rowMeans(port))
 2  > names(hypo.port) <- "Port.Ret"
 3  > head.tail(hypo.port)
 4         Port.Ret
 5  1   0.07306837
 6  2   0.07258170
 7  3  -0.02373268
 8         Port.Ret
 9  58 0.05596117
10  59 0.04224452
11  60 0.02894303
```

**Step 4: Save Hypothetical Portfolio Returns to a CSV File**  We create the dataset *csv.port* by combining the date vector, which we obtain by using `index()` on the *port* dataset, and *hypo.port* dataset we created above. We then rename the first column in *csv.port* as Date. Finally, we use `write.csv()` to save *csv.port* in our working directory as a CSV file labeled Hypothetical Portfolio (monthly).csv.

```
 1  > csv.port <- cbind(data.frame(index(port)), hypo.port)
 2  > names(csv.port)[1]<-paste("Date")
 3  > head.tail(csv.port)
 4         Date    Port.Ret
 5  1 Jan 2015  0.07306837
 6  2 Feb 2015  0.07258170
 7  3 Mar 2015 -0.02373268
 8         Date    Port.Ret
 9  58 Oct 2019 0.05596117
10  59 Nov 2019 0.04224452
11  60 Dec 2019 0.02894303
12  >
13  > write.csv(csv.port,"Hypothetical Portfolio (Monthly).csv")
```

# Appendix D
# Constructing a Hypothetical Portfolio (Daily Returns)

In this appendix, we construct the daily returns of a hypothetical value-weighted portfolio for use in some of our analysis. For this analysis, we leverage off the value-weighted portfolio we created in Chap. 3.

**Step 1: Check if the *portfolios* Dataset is in the R Environment**  We created *portfolios*, which contains prices for an equal-weighted and value-weighted portfolio from January 2, 2019 to December 31, 2019.

```
1   > head.tail(portfolios)
2                   EW         VW
3   2019-01-02 1011.9233 1011.897
4   2019-01-03  960.4724  960.081
5   2019-01-04 1007.5588 1007.090
6                   EW         VW
7   2019-12-27 1462.572 1467.658
8   2019-12-30 1454.841 1460.877
9   2019-12-31 1459.378 1465.801
```

**Step 2: Create Initial Investment Value**  The portfolio values above do not start with the $1000 investment. The simplest way to create that would be to extract one row from *portfolios*, move the date back one day, and hard code the $1000 starting value.

```
1    > invest.date <- portfolios[1, 2]
2    > invest.date
3                    VW
4    2019-01-02 1011.897
5    >
6    > index(invest.date) <- index(invest.date) - 1
7    > invest.date
8                    VW
9    2019-01-01 1011.897
10   >
11   > invest.date[1, ] <- 1000
```

C.S. Ang, *Analyzing Financial Data and Implementing Financial Models Using R*,
Springer Texts in Business and Economics,
https://doi.org/10.1007/978-3-030-64155-9

```
12  > invest.date
13                VW
14  2019-01-01 1000
```

**Step 3: Combine Both Datasets**  Using rbind(), we stack the initial investment object to the value-weighted portfolio prices (Column 2) in the *portfolio* dataset.

```
1   > combine <- rbind(invest.date, portfolios[, 2])
2   > head.tail(combine)
3                  VW
4   2019-01-01 1000.000
5   2019-01-02 1011.897
6   2019-01-03  960.081
7                  VW
8   2019-12-27 1467.658
9   2019-12-30 1460.877
10  2019-12-31 1465.801
```

**Step 4: Calculate Portfolio Returns**  We use Delt() to calculate the returns of the hypothetical portfolio. We then remove the initial investment date in Line 12 and rename the portfolio return Port.Ret in Line 13.

```
1   > hypo.port <- Delt(combine)
2   > head.tail(hypo.port)
3                 Delt.1.arithmetic
4   2019-01-01                   NA
5   2019-01-02           0.01189697
6   2019-01-03          -0.05120679
7                 Delt.1.arithmetic
8   2019-12-27          -0.001856224
9   2019-12-30          -0.004619646
10  2019-12-31           0.003370195
11  >
12  > hypo.port <- hypo.port[-1, ]
13  > names(hypo.port) <- "Port.Ret"
14  > head.tail(hypo.port)
15                  Port.Ret
16  2019-01-02  0.01189697
17  2019-01-03 -0.05120679
18  2019-01-04  0.04896317
19                  Port.Ret
20  2019-12-27 -0.001856224
21  2019-12-30 -0.004619646
22  2019-12-31  0.003370195
```

**Step 5: Convert to Data Frame**  Before exporting to a CSV file, we need to move the date from the row labels into its own column. We do that in Line 1. Notice that *Date* has an uppercase first letter. We do this because when we combine the *Date* with *hypo.port* in Line 6, and the column label takes on the name of the object. We then rename the row labels as observation numbers in Line 7 using seq().

```
1   > Date <- as.Date(index(hypo.port))
2   > head.tail(Date)
3   [1] "2019-01-02" "2019-01-03" "2019-01-04"
4   [1] "2019-12-27" "2019-12-30" "2019-12-31"
5   >
6   > hypo.port <- data.frame(Date, hypo.port)
7   > rownames(hypo.port) <- seq(1, nrow(hypo.port), 1)
8   > head.tail(hypo.port)
9           Date    Port.Ret
10  1 2019-01-02  0.01189697
11  2 2019-01-03 -0.05120679
12  3 2019-01-04  0.04896317
13          Date      Port.Ret
14  250 2019-12-27 -0.001856224
15  251 2019-12-30 -0.004619646
16  252 2019-12-31  0.003370195
```

**Step 6: Save Portfolio Returns to CSV File**  We save *hypo.port* to a file **Hypo-thetical Portfolio (Daily).csv**, which will be saved in our R working directory.

```
1   > write.csv(hypo.port, "Hypothetical Portfolio (Daily).csv")
```

# Index

CPSIA information can be obtained
at www.ICGtesting.com
Printed in the USA
LVHW080448140921
697767LV00002B/145